The Microbiology of the Atmosphere

PLANT SCIENCE MONOGRAPHS

Advisory Board

Published so far

A PLANT SCIENCE MONOGRAPH

General Editor: Professor Nicholas Polunin,
M.S. (YALE), M.A., D.PHIL., D.SC. (OXON.)

The Microbiology of the Atmosphere
2ND EDITION

P. H. GREGORY PH.D., D.SC. (LONDON), D.I.C., F.R.S.

Formerly Professor of Botany, Imperial College of Science
and Technology, University of London. Lately Head of Plant
Pathology Department, Rothamsted Experimental Station,
Harpenden, England.

LEONARD HILL   An Intertext Publisher

Published by
Leonard Hill Books
a division of
International Textbook Company Limited
24 Market Square, Aylesbury, Bucks.

First published 1961
This edition 1973

ISBN 0 249 44110 1

Printed in Great Britain by
Clarke, Doble & Brendon Ltd., Plymouth

DEDICATED

to

my wife

MARGARET FEARN GREGORY

Contents

Preface to the Second Edition

This edition continues the endeavour to understand microbial events in the atmosphere, considered as a natural phenomenon. I thank readers who have pointed out errors in the first edition, or made suggestions for improvement, either in correspondence or conversation, including of course my colleagues at Rothamsted.

I am grateful to Mrs Maureen E. Lacey (née Bunce) who painted new and improved versions of her colour illustrations (Plates 6 & 7).

From the wealth of new material published since the first edition only representative papers have been cited. Chapters are added on rain-splash dispersal, and on inhaled microbes in relation to respiratory infection and allergy. A chapter has been added (still far too short) on survival in the atmosphere. For the second half of the chapter on infection gradients I have substituted material taken from my article in the *Annual Review of Phytopathology*, 1968, by kind permission of the Editors for Annual Reviews, Inc.

Besides endorsing the acknowledgements listed in the first preface, I also thank J. H. A. Dunwoody for work with the Rothamsted computer; R. H. Turner for scanning electron micrographs; D. R. Henden and D. H. Lapwood for help given in field work. As ever I am grateful to my wife for constant help and for critically reading the typescript.

PHILIP H. GREGORY

11 Topstreet Way
Harpenden, England
September, 1972

Preface to the First Edition

Aerobiology is usually understood to be the study of passively airborne micro-organisms – of their identity, behaviour, movements, and survival. One characteristic, which it shares with many other population studies in biology, is that the ultimate relevant unit consists of the individual cell or small group of cells. Analysis at the molecular or sub-atomic level is irrelevant to our present purpose. Like geography, aerobiology is an agglutinative study, drawing information from many kinds of scientific research. Although it already has its patron saint, Pierre Miquel, and its martyr, Fred C. Meier, aerobiology is best regarded as an activity whose material will in due course be incorporated into the main body of biological science – without, I hope, any necessity for splinter societies, journals, and international conferences.

This book amplifies and extends a course of Intercollegiate Lectures given to botanical students in the University of London in 1956. The theme, which has occupied me for over fifteen years, is as follows. Transport through the atmosphere is the main dispersal route for such organic particles as the spores of many micro-organisms. How do the properties of the atmosphere, and the properties of these particles themselves, affect their dispersal? How do the particles get into the air? How far, and in what numbers, are they dispersed? By what processes do they become grounded, so that they can continue growth? What is in the air, and how can we measure it? What are the practical consequences of this process for the micro-organisms themselves, and for man, other animals, vegetation, and crops?

Although there are one or two other books on airborne microbes, this is the first to treat the subject as a world-wide phenomenon. It is, perhaps, inevitable that it should be attempted by a mycologist. Few other biologists find their material so dominated by the atmosphere, and no other micro-organisms have so thoroughly exploited the possibilities of aerial dispersal as the fungi. One of the fascinations of the subject is the impact of facets of its knowledge on such apparently diverse topics as artificial rain-making, allergy, smoke screens, effluent of nuclear power-stations, crop protection, icing of aircraft, air hygiene, and many other topics. This book treats of the development and principles of aerobiology rather than applications; yet the stimulus to nearly all aerobiological work comes from applied science.

In this book the term 'microbe' is used freely when a general word is wanted; but, like the word 'spore', it has admittedly been stretched beyond its normal meaning. Airborne pollen of flowering plants must be included and is safely covered by the term 'spore' (botanically: 'micro-spore'); but are pollen grains and mushroom spores microbes? There is no other commonly accepted word that covers quite what is meant by the word 'spore' as used here: 'propagule', 'disseminule', 'biota', 'diaspore'? We have isolated part of the continuum for study but find we are not well-equipped verbally for the task of dealing with it. The microbial population of the atmosphere is referred to here as the 'air spora', using 'spora' as a word analogous to 'flora' and 'fauna'.

Botanical nomenclature has presented some difficulties: authorities have not been given for specific names, and the names used by other authors have usually been quoted as given in the original papers – without necessarily attempting to guess what was meant, or following the nomenclature fashionable in 1960. I have converted other workers' numerical data to the metric system, and temperature to the Celsius scale, to aid comparison, and have moreover assessed spore concentrations on the uniform basis of number per cubic metre.

Frequently, in making general statements, I have omitted safeguarding, but tedious, escape clauses: this has been done to spare the reader who will understand that biological generalizations abound in exceptions and complexities.

Interpretations in this book are mostly my own responsibility, but I am grateful for help received from many people during its preparation. In particular I offer my thanks to the following: G. Samuel and W. Buddin for introducing me to dispersal problems in the field; F. C. Bawden for encouragement in the study of aerobiology and for reading this book in manuscript; E. C. Large for advice on planning the book; D. A. Boalch (and many other librarians) for continual help with the literature; members of the British Mycological Society for named specimens of fungi, and H. L. K. Whitehouse for mosses; A. Horne, V. Stansfield, and F. D. Cowland for photography; R. Adams, G. C. Ainsworth, J. R. D. Francis, E. J. Guthrie, Elizabeth D. Hamilton, J. M. Hirst, C. T. Ingold, C. G. Johnson, F. T. Last, Kate Maunsell, T. Sreeramulu, and O. J. Stedman for discussion and help with aerobiological problems and applications; Audrey Baker, Beatrice E. Allard, and Marie T. Seabrook for clerical assistance;

and Maureen E. Bunce for experimental help, revision of the manuscript, and preparation of many of the illustrations – especially the paintings for Plates 1, 5, 6, and 7. I also wish to thank authors, editors, and publishers for permission to copy illustrations which are acknowledged in the text. Finally, for the calculations involved in Figures 24 to 27, and for those chapters needing the help of a mathematician, I have been fortunate in having the constant advice and willing help of my wife, Margaret F. Gregory, to whom I am most deeply grateful.

PHILIP H. GREGORY

Rothamsted Experimental Station
Harpenden, Herts., England.
September, 1960.

List of Figures in Text

List of Plates

I

Historical Introduction

The air we breathe, like our food and drink, varies in quality from time to time and from place to place. This fact was recognized many centuries before industrialized man assumed the right to pollute the atmosphere with poisonous chemicals and radioactive isotopes.

In Britain we hold that 'when the wind is in the East 'tis good for neither man nor beast'. Some places are noted for invigorating air, and some for relaxing air; but it is not yet clear whether these properties are associated merely with differences in temperature, humidity, and movement of a gaseous mixture consisting mainly of 78% nitrogen, 21% oxygen, and 0·03% carbon dioxide with traces of the inert gases, or whether some other factor or factors are involved.

SPECULATIONS ON THE ORIGIN OF DISEASE

Classical writers believed that the wind sometimes brought sickness to man, animals and crops. Hippocrates, the father of medical science, held that men were attacked by epidemic fevers when they inhaled air infected with 'such pollutions as are hostile to the human race'. A rival, though perhaps not entirely incompatible, view held that epidemics were the result of supernatural agencies, and were to be warded off or cured by taking appropriate action.

Lucretius in about 55 B.C. held quite modern views. He observed the scintillation of motes on a sunbeam in a darkened room and concluded that their movement must result from bombardment by innumerable, invisible, moving atoms in the air. This brilliant intuition enabled him to account for many interesting phenomena, including the origin of pestilences. We now know that bodies which transmit human diseases through the air are larger than those which Lucretius thought of as atoms – the mosquitoes carrying malaria, for instance, or the droplets which spread the common cold and influenza viruses indoors. But in his concept of baleful particles

1

carried in clouds by the wind, settling on the wheat or inhaled from the polluted atmosphere, Lucretius touched on some of the main problems existing in plant pathology and allergy today.

EARLY MICROSCOPISTS AND THE DISCOVERY OF SPORES

After Lucretius, more than 1500 years passed before men even began to be aware that the air teems with microscopic living organisms. The discovery had to wait almost until the invention of the microscope.

For a long time after Aristotle and Theophrastus, the lower plants lacking obvious seeds were believed to be generated spontaneously in decaying animal or vegetable matter. The same view was held of the origin of many of the lower animals. However, the minute 'seeds' or spores of several kinds of plants were observed in the mass long before the invention of the microscope allowed them to be identified and observed individually. What was more natural than to suppose that these minute particles were wafted about by the winds?

The discovery of reproduction of ferns is attributed to Valerius Cordus (b. 1515, d. 1564), and spores of the fungi seem to have been observed soon after this by a Neapolitan botanist, J. B. Porta, although the rusty-coloured spore deposits under bracket-fungi on beech trees must always have been familiar to the countryman.

It was P. A. Micheli (b. 1679, d. 1737), botanist to the public gardens at Florence, who first illustrated the 'seeds' of many fungi, including mushrooms, cup-fungi, truffles, moulds, and slime-moulds. Further, by sowing spores on fresh-cut pieces of melon, quince and pear, and reproducing the parent mould for several generations, he showed that the spores of some common moulds were, indeed, 'seeds' of the fungi. He noted, however, that some of his control slices also became contaminated, and he concluded that the spores of moulds are distributed through the air (see Buller, 1915).

The handmade lenses of Anton van Leeuwenhoek rendered visible the world of minute organisms whose existence had only been guessed at before, and whose significance in nature had scarcely even been imagined. He could just see bacteria, and in his letters to the Royal Society in 1680 he described some yeasts, infusoria, and a mould. From his experiments he came to doubt the current belief in spontaneous generation; it seemed more plausible to him to suppose that his 'animalcules can be carried over by the wind, along with the bits of dust floating in the air' (Dobell, 1932). The controversy over spontaneous generation was to last for a couple of centuries; but in the second half of the eighteenth century, ideas were developed by Nehemiah Grew and E. F. Geoffrey on the function of pollen of flowering plants. J. G. Koelreuter, in 1766, was perhaps the first to recognize the importance of wind pollination for some plants and of insect pollination for others. C. K. Sprengel in 1793 developed these views and concluded that flowers lacking a corolla are usually pollinated in a mechanical fashion by wind. Such flowers have to produce large quantities of light and easily-transported pollen, much of which misses its target or is washed out of the air by rain. Thomas A. Knight in 1799 reported that wind could transport pollen to great distances.

By the beginning of the nineteenth century, therefore, it was recognized that pollen of many, but by no means all, species of flowering plants, and the microscopic spores of ferns, mosses, and fungi – as well as protista [protozoa] – were commonly liberated into the air and transported by the wind. The potential sources of the air spora had been discovered and identified in the main before the year 1800, but their role remained obscure.

CONTROVERSY ON SPONTANEOUS GENERATION*

Leeuwenhoek doubted the belief, dating from Aristotle, that flies, mites and moulds were generated spontaneously by decaying animal and vegetable matter: animalcules carried by air provided an alternative explanation. J. T. Needham (b. 1713, d. 1781) had claimed that minute organisms would appear in heated infusions; but L. Spallanzani (b. 1729, d. 1799) showed, by a series of experiments, that when organic materials were subjected to sufficient heat treatment (with various precautions against contamination) they would neither putrify nor breed animalcules unless exposed to air. From this Spallanzani concluded that the microbes were present in the air admitted experimentally to his sterilized vessels. A rearguard action was fought to explain away these results. J. Priestley (b. 1733, d. 1804) and L. J. Gay-Lussac (b. 1778, d. 1850) claimed that heating the vessels drove out the air and that it was shortage of oxygen, not lack of 'seeds', which prevented heat-sterilized materials from generating a microbial population.

Meanwhile, Appert (1810) put heat sterilization on a commercial basis by applying it to food preservation; but the controversy lingered on, even into the present century, although the experiments and polemics of Louis Pasteur were decisive. Pasteur showed that food could be conserved in the presence of oxygen and that preservation depends on the destruction by heat of something contained in the air. In 1859 F. A. Pouchet, of Rouen, had raised the objection that a very minute quantity of air sufficed to allow the development of numerous microbes in heated infusions, and that the air would have to be a 'thick soup' of microbial germs.

In reply, Pasteur (1861) sterilized a series of evacuated flasks containing nutrient medium. So long as the flasks remained unopened they all remained

* See also Bulloch (1938) and Oparin (1957).

sterile; but, even when they were opened and air was admitted, he found that one or two of each batch would remain sterile on incubation. Pasteur replied to Pouchet, denying that only a minute quantity of air needs to gain access for a microbial population to develop and for putrefaction to take place. On the contrary, the cause of the putrefaction was discontinuous, and a sample of 250 cm³ of air might *or might not* contain germs.

Pasteur then showed, by opening batches of about forty such flasks in various sites, that the quantity of airborne germs differed in different places. In the open air in Paris he obtained bacteria, yeasts and moulds; but some flasks remained sterile. In cellars of the Observatoire, where the temperature was constant and the air still and dust free, many more flasks remained sterile.

On 5 November 1860, Pasteur deposited at the office of the Academy no fewer than 73 quarter-litre flasks, some of which he had opened to the air in batches of twenty at various heights above sea level ranging from the foothills of the Jura to high up on Mont Blanc, as follows:

Altitude (m)	Locality where air sampled	Number of flasks	
		contaminated	sterile
—	Country air, far from dwelling houses on the first plateau of the Jura	8	12
850	Jura mountains	5	15
2000	Montanvert, near Mer de Glace on Mont Blanc	1	19

The cause of this supposed 'spontaneous generation' was therefore not only discontinuous, but moreover, its concentration decreased with height above sea level.

F. A. Pouchet had admitted that among the dust particles of vegetable origin there were some spores of cryptogams, but he held that these were too few to account for the phenomenon of putrefaction.

Pasteur decided that he would abandon Pouchet's method, which relied on examining naturally occurring deposits of dust on the surface of objects, in favour of a new method of studying particles collected from actual suspension in air. Pouchet had drawn invalid conclusions from surface deposits because, according to Pasteur, the light air-movements which constantly play over surface deposits would pick up and remove the extremely minute and light spores of microbes more readily than any coarser particles. (It now appears, however, that the small numbers of the lighter bodies in surface deposits are due to the extreme slowness with which they are deposited, rather than to their preferential removal after deposition.)

Pasteur's apparatus for extracting the suspended dust in the air for microscopic examination was quite simple (Figure 1). A tube of 0·5 cm diameter was extruded into the open air through a hole drilled in a window frame several metres above the ground. The rear part of the tube was packed with a plug of gun-cotton to catch particles. Air was drawn through the apparatus by means of a filter-pump, and the volume of air was measured by displacement of water. Tests were made on air drawn from beside the Rue d'Ulm, and from the gardens of the École Normale in Paris. During aspiration, solid particles were trapped on the fibres of the

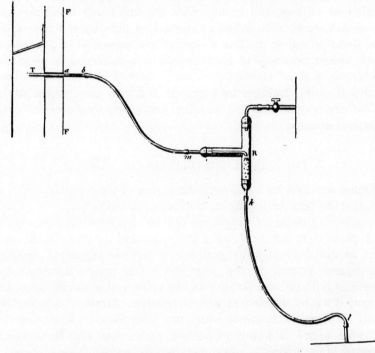

Figure 1. Pasteur's gun-cotton filter for airborne microbes. a = gun-cotton plug, 1 cm long, held in position by spiral platinum wire (b). FF = window frame drilled to allow passage of T = tube to exterior for sampling outdoor air. R(m.k.l.) = aspirator.

gun-cotton plug. After use, the gun-cotton was dissolved in an alcohol–ether mixture, the particles were allowed to settle, the liquid decanted and the deposit mounted for microscopic examination.

Pasteur, as usual, had little interest in the specific identity of his organisms; he was no taxonomist. The particles exactly resembled the 'germs' of lower organisms. They differed in volume and structure so much among themselves that they clearly belonged to very many species or even groups, including bacteria, moulds and yeasts. Their numbers contradicted the general conclusion that the smallest bubble of air admitted to a heat-sterilized medium is sufficient to give rise to all the species of infusoria and cryptogams normal in an infusion. This view was shown to be highly exaggerated, and Pasteur indicated clearly that it is sometimes possible to bring a considerable volume of ordinary air into contact with an infusion before living organisms develop.

Pasteur had demonstrated visually the existence of an air spora, he had pointed out that it should be measured while in suspension and not after deposition on surfaces, and he had made the first rough visual measurements of its concentration in the atmosphere of the City of Paris: a few metres above ground in the Rue d'Ulm, after a succession of fine days in summer, several thousands of micro-organisms were carried in suspension per cubic metre of air. He then abandoned the method – remarking, however, that it could doubtless be improved and used more extensively to study the effects of seasons and localities, and especially during outbreaks of infectious diseases.

THE GERM THEORY OF DISEASE

We must now look back and trace the growth of the microbial theory of disease that had been developing for more than a century.

The minute growths of fungus noticed for centuries on mildewed or 'rusted' plants were believed to be a consequence of the diseases; the dusty powder on rusted wheat was regarded as a curiously congealed exudation of the diseased plant itself. But might this not be putting the cart before the horse? Could the rust possibly be the cause of the disease instead of an effect? Perhaps the first to give reasonably affirmative evidence was Fontana (1767), who examined wheat rust with his microscope and described what he saw as a grove of parasitic plants nourishing themselves at the expense of the grain.

As further crop diseases were studied it became clear that, in some, infection is acquired by planting in contaminated soil, while others are carried on seed, and still others are spread in the wind by airborne fungus spores (see Large, 1940).

The discovery that microbes can cause disease in man and animals came somewhat later, and the first animal pathogens to be recognized as such were again fungi – no doubt because they were easier to find than bacteria. In 1835, Agostini Bassi showed conclusively, by inoculation experiments, that a specific mould is the cause of the 'muscardine' disease of silkworms which was then threatening the silk industry of Piedmont. Next, historically, came the recognition of the fungi causing favus, ringworm, and 'thrush' in man, as a result of the work of David Gruby and Charles Robin.

Pasteur had demonstrated that microbes are normally abundant in the air. Many of them can cause fermentation or putrefaction when introduced into sterile organic substrates; and it was natural to speculate that others might be the causes of epidemics of some of the so-called 'zymotic' diseases whose etiology was then unknown. Medical workers soon began a systematic search among airborne microbes for the unknown causes of infectious diseases.

The search was long and on the whole unfruitful because most epidemic diseases that attacked man were gradually traced to sources other than out-door air. However, in the course of the search, most of the important characteristics of the air spora were discovered – and then forgotten. The search occupied the last thirty years of the nineteenth century and coincided with the golden age of bacteriology. Listing the dates of some contemporary salient advances in bacteriology will help to give the background to this phase of aerobiology (*see* Bulloch, 1938).

PASTEUR, L. Microscopical and cultural demonstration of the existence of an
    air spora, and the fermentation of urea by a *Micrococcus*   1861–62
KOCH, L. Introduction of pure-culture methods, and demonstration of spore
    production in bacteria. Discovery of cause of anthrax   1876
    Statement of Koch's postulates   1878
    Introduction of gelatine to solidify media   1881
HANSEN, G. H. A. Discovery of cause of leprosy   1874
NEISSER, A. Discovery of the *Gonococcus*   1879
KOCH, L. Discovery of the tubercule bacillus   1882
    Discovery of the cholera *Vibrio*   1883
LOEFFLER, F. Discovery of bacillus of swine erysipelas   1885
NICOLAIER, A. Discovery of the tetanus bacillus   1885
KITSATO, S. ⎫
YERSIN, A. ⎬ Discovery of the plague bacillus   1894
IVANOVSKI, D. ⎫
BEIJERINCK, M. W. ⎬ Discovery of filterable viruses in plants   { 1892 / 1898 }

THE HYGIENISTS AND THEIR INVESTIGATION OF AIR

While the causes of infectious diseases of man and animals were being unravelled in laboratories and clinics, a series of field investigations into the air spora was in progress to find out whether fluctuations in number and types of microbes present in the atmosphere were connected with out-breaks of such diseases as cholera, typhoid and malaria.

Salisbury (1866) investigated the air spora in connexion with malaria in the Ohio and Mississippi Valleys, by exposing sheets of glass above marshy places during the night and examining them microscopically. He observed small, oblong, *Palmella*-like cells singly or in groups on the upper side of the glass sheets, but never in the droplets which formed on the under side. He believed that these cells were produced from a grey mould growing on the surface of prairie soil, and were in fact the mould's spores which were liberated at night and rose some 10 to 30 metres in the air, none being present during the daytime. Their liberation could be prevented by covering the ground with a layer of quicklime or straw.

Some form of the 'aeroconiscope', invented by Maddox (1870, 1871), was favoured by many investigators in this period. The model used by Cunningham (1873) consisted of a conical funnel, with the mouth directed into the wind by a vane, ending in a nozzle behind which was placed a sticky microscope cover glass. Dust particles, driven into the cone by the wind, were impacted on the cover glass (Figure 2). Cunningham's studies were made in two Calcutta gaols where cholera and other fevers were rife, and where medical statistics were available. He sampled for 24-hour periods, and illustrations of representative catches of airborne organisms, mainly fungus spores and pollens, were published in a series of splendid colour plates. He found no correlation between these micro-organisms and the incidence of fevers in gaols. Moist weather diminished inorganic dusts, but it appeared to increase the total number of fungus spores.

The most intensive sustained analysis of bacteria and moulds in the atmosphere was made in Paris during the last quarter of the nineteenth century. Largely through the influence of the chemist J. B. A. Dumas the Observatoire Montsouris was launched as a State institution in 1871 to make records needed for meteorology and agriculture. The Observatoire was housed in a palace in the Parc Montsouris, about 5 km south of the centre of Paris. One of its tasks was to be the microscopic and cultural study of the inorganic dust in the air, including both Mucedineae (moulds) and bacteria.

Observations were started in 1875 by M. Schoenauer. He was replaced after a year or two by Pierre Miquel (b. 1850, d. 1922), the distinguished bacteriologist, who continued in charge of the work which he termed 'Micrography' for thirty-four years. During the course of the survey, various methods were tested and discarded or improved; but all aimed at estimating the number of particles of various types contained in a measured volume of air. Moulds were at first estimated microscopically in a 24-hour deposit, obtained by impinging the air to be sampled on a glycerined glass slide which was placed horizontally 2 to 3 mm above a downward-facing orifice. The diameter of the orifice was from 0·5 to 0·75 mm. Suction of 20 litres per hour was maintained by a water-operated pump (Miquel, 1879). Miquel found that this apparatus yielded about 100 times more particles, for a given period of exposure, than the aeroconiscopes designed by Maddox and Cunningham, though for qualitative work away from the laboratory he still used a wind-operated trap of the Maddox type.

Bacteria, especially bacterial spores, could not be satisfactorily counted microscopically and Miquel was forced to estimate them by cultural

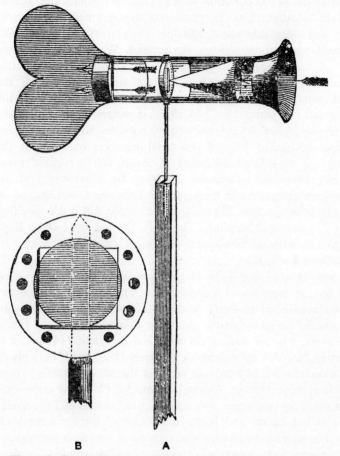

Figure 2. Cunningham's aeroconiscope. A = side view of apparatus (partly in section); B = face view of sticky surface behind
apex of cone (on larger scale).

methods. At first he drew known volumes of air through liquid media
(sterile beef extract, etc), partitioning the liquid either before or after
exposure into 50 or 100 vessels, and adjusting the volume of air sampled
so as to leave from a quarter to a half of the vessels sterile – in order to
get a reliable estimate of the number of bacterial particles in the volume of
air sampled. The numbers of microbes in the air varied greatly in the same
place at different times, and this variation was studied in relation to season,
weather, district, and altitude. Miquel was the first to make a long-term
survey of the microbial content of the atmosphere by volumetric methods.

Outdoors in the Parc Montsouris, Miquel estimated that the mould spores averaged about 30 000 per cubic metre in summer, sometimes increasing to 200 000 in rainy weather. In prolonged dry weather they decreased in number, and were only about 1000 per cubic metre in winter, with very few indeed when snow was on the ground. While rain was falling the number of mould spores usually decreased considerably, but afterwards their numbers recovered quickly – in fact, much more quickly than did those of particles of inorganic dust. Resting stages ('eggs') on infusoria were estimated at about 1 or 2 in 10 cubic metres of air. Pollen grains in June made up 5% of the airborne organic particles, while starch grains near habitations accounted for 1%. Bacterial numbers outdoors in the Parc Montsouris were at first estimated at about 100 per cubic metre; but improved culture media increased this figure by a factor of 7 to 10 times. The numbers of bacteria in the centre of Paris were, perhaps, 10 times as high again as in the Parc Montsouris, with more still inside dwellings, and still more in crowded hospitals. The work showed signs of settling into a steady routine with the publication of Miquel's *Les Organismes Vivant de l'Atmosphère*, Paris, 1883.

However, in 1883 and 1884 Miquel was stung into a burst of renewed activity by the intrusion of a rival centre for the study of hygiene which had been established in Berlin under W. Hesse, who used the new solid media which Miquel abhorred. With the collaboration of de Freudenrich in field work, Miquel studied the microbial population of the air at high altitudes in the Alps by volumetric methods (1884, p. 524); with the help of a sea captain, M. Moreau, the air over the sea was studied on voyages to Rio de Janeiro, Odessa, Alexandria, and La Plata; the micro-organisms brought down in rain water were caught, precipitated and counted; hourly changes in the mould and bacterial content of the air were studied on improved volumetric traps with sticky slides, or on paper impregnated with nutrient media and moved by clockwork. At Montsouris, fungus spores showed a diurnal periodicity with two maxima at about 8 and 20 hours, regardless of wind velocity. When he pressed the study of changes in spore content of the air with passage of time still further, Miquel found that the hourly reading was merely a smoothing of still shorter-term variations.

Trapping airborne bacteria at Montsouris on a moving paper disc impregnated with nutrient medium, Miquel (1885) observed a regular diurnal periodicity, with two maxima at approximately 07.00 and 19.00 hours and averaging about 750 per cubic metre, and with two minima at approximately 2 and 14 hours averaging about 150 per cubic metre. This periodicity was not related to wind direction, and was not altered by moderate falls of rain. In the centre of Paris the bacterial content also showed two daily maxima and two daily minima, but there the minima were about equal to the maxima at Montsouris, and the times of the maxima were closely related to activities in the City such as sweeping the street and the passage of horse-drawn traffic.

Miquel appears to have been overwhelmed by the richness of the information on the mould spore flora provided by his apparatus, for he promptly abandoned it, merely remarking that 'the micrographer who has the leisure could make some nice [curieuse] studies of this subject'. It was, however, not abandoned before important facts about the mould spora had been discovered by this excellent method.

Interest in the mould spora waned when it became clear that the devastating epidemic diseases prevalent from time to time in cities were not fungal in origin but were due to bacteria, and attention then became urgently focused on drinking water as the source of many of the current epidemic fevers abounding in Paris and other great cities. The laboratory at Montsouris then became the centre for the bacterial analysis of samples of drinking water sent from wells in Paris and other parts of France. When he retired in 1910 Miquel was made an officer of the Légion d'Honneur.

Meanwhile, in Germany, the work of W. Hesse (b. 1846, d. 1911) had proceeded along similar lines. Hesse's apparatus for air sampling consisted of a narrow horizontal tube, 70 cm long and 3·5 cm wide, containing a layer of Koch's nutrient gelatine. A known volume of air was aspirated slowly through the tube, and micro-organisms settled and grew on the medium. Most colonies developed near the entrance to the tube, and Hesse assumed that by the time the slow stream of air had reached the end of its 70 cm course all microbes would have been precipitated by gravity. Hesse found that moulds penetrated much further into his tubes than did the bacteria, and he made the important deduction that mould germs as found in the atmosphere are on average lighter than the bacterial germs. This led him to conclude that, whereas fungus spores were usually present in the air as single particles, the aerial bacteria mostly occur in the air either as large aggregates or attached to relatively large carrier particles of dust, soil or debris (Hesse, 1884, 1888). He also observed that most colonies consisted of a single species – bacteria usually in small colonies of pure culture, and fungi as isolated spores – and deduced that the airborne germs are not in the form of aggregates of different species.

Hesse's method was also used in London by Frankland (1886, 1887) and by Frankland & Hart (1887) on the roof of what is now known as the Old Huxley Building of the Imperial College of Science and Technology, and elsewhere. Simultaneous comparisons were made between the number of microbes per 10 litres (as indicated by colonies growing on Hesse's tubes of peptone gelatine) and the number deposited on horizontal dishes of the same medium, expressed as the number deposited per unit area per minute. Tests were made both outdoors and inside crowded or empty buildings.

Frankland noted that the number of colonies was greater when the mouth of the tube faced the wind rather than in other directions, so he standardized his method by always turning it at an angle of 135° to the wind. A control tube facing the wind but not aspirated was always used, and sometimes it had a substantial number of colonies. Frankland seems to have been the first to realize that aerodynamic effects are of major importance in techniques for trapping the air spora (*see* Chapter IX).

These methods for studying the air spora were continued into the present century, notably by Saito (1904, 1908, 1922) in Japan, by Buller & Lowe (1911) in the Canadian Prairies, and by Sartory & Langlais (1912) in Paris.

THE ALLERGISTS

The idea that men, other animals, and plants could become infected by microbes which set up pathological changes had been made acceptable by the analogy of sterile organic infusions that became seeded with putrefying microbes. The idea became widely accepted during the latter half of the nineteenth century and, once the causes of the common epidemic diseases had been established, advances in hygiene and therapy began to transform the social scene. Yet there remained some diseases for which no pathogenic or parasitic invader could be found. Some of these, such as pellagra and beri-beri, have now been traced to a variety of nutritional deficiencies. Another group, the so-called allergies, were at first difficult to grasp because a peculiar condition of the patient was a complicating factor. Allergic diseases, unlike those caused by invasion of the body by a pathogenic micro-organism, are due to a changed condition of an individual patient who has become sensitive and reacts adversely to substances, often in minute amounts, which normal individuals can tolerate easily. The substance or allergen can be taken into the body, for example in food, or by contact through the skin, or by inhalation from the air.

Hay fever was one of these puzzles. Long before Pasteur's epoch, hay fever had been attributed to inhalation of pollen; but it remained for Charles H. Blackley (1873), a Manchester physician, to prove by inhalation experiments on himself and others that this guess was correct, and to demonstrate by trapping methods that pollen was at times present in the air in large quantities. Blackley first tried Pasteur's gun-cotton filters and obtained some pollens, but too few to satisfy him. Finally he used four sticky horizontal microscope slides exposed under a roof supported by a square central post. The slides were placed at 'breathing level' (about 135 cm above ground), and he caught a maximum of 880 grains/cm^2/24 hours on 28 June 1866. In 1867 his maximum was only 106, and 1869 he placed his slides vertically in a vane shelter and gave no numerical data. He found that rain reduced the number of pollen grains caught to about 5% of the number caught in dry weather. He explored the air above ground level up to 430 metres by means of kites, and found that vertical slides facing the wind caught nearly 20 times as much pollen at the higher altitudes as at breathing level.

Blackley showed by means of his sticky slides that the air contains enough pollen during the grass-flowering season for large quantities to be deposited on exposed surfaces. He also gave himself an attack of bronchial catarrh by inhaling *Penicillium* and *Chaetomium* spores – an experiment which he said was too unpleasant to repeat.

According to Durham (1942), after Blackley's pioneer work no progress was made with these studies until the period 1910–1916, when fresh interest was aroused by the discovery that injections of pollen extracts can be used to desensitize patients who are allergic to pollen. Inhaled fungus spores were recognized as allergens following the work of Cadham (1924) and Feinberg (1935) in North America. When the study of airborne allergens was taken up again in the present century, it was unfortunate that the technique chosen should have been the so-called 'gravity-slide' adopted by Blackley – a method which Pasteur had abandoned in 1861 and which Miquel had roundly condemned as 'the simplest and most defective method' of collecting airborne particles.

By the early years of this century it became possible to assess the value of the ancient belief that the wind brings disease. Many diseases of crops, but very few diseases of man, have proved to be caused by minute particles carried on the wind. The particles are not some sort of invisible atoms as Lucretius thought; indeed, among the motes in the sunbeam, he may himself have been watching some of the baleful fungus spores and pollens which cause crop diseases and respiratory allergy.

Meanwhile evidence was accumulating that these particles might be carried by wind to distances vastly greater than had been imagined by the ancients. In dust deposited after transport for hundreds of kilometres by sirocco and trade winds, Ehrenberg (1849, 1872*a*, 1872*b*) found large quantities of protista [protozoa] and plant spores, and gradually he became convinced that viable microbes could survive transport through the atmosphere. When the *Beagle* was near the Cape Verde Islands, Darwin (1846) found the atmosphere hazy with dust from North Africa. In samples of this dust Ehrenberg found sixty-seven kinds of organisms, including freshwater infusoria and cryptogamic spores (Plate 1), and Darwin at once grasped the importance of the phenomenon in the geographical distribution of organisms.

AEROBIOLOGY

Interest in airborne microbes quickened in the twentieth century when heavier-than-air machines made it possible to explore the lower few kilometres of the atmospheric ocean. Fred C. Meier (*b*. 1893, *d*. 1938) of the United States Department of Agriculture introduced the word 'Aerobiology' to describe a far-reaching project of research on microbial life in the upper air. Unfortunately, when only preliminary abstracts of his work had been published (cf. p. 183) he was lost on a flight over the Pacific Ocean (Haskell & Barss, 1939). Eventually, after this devastating setback, the American Association for the Advancement of Science published a Symposium on extramural and intramural aerobiology (Moulton, 1942) and the new discipline was firmly launched.

II

Spores: Their Properties and Sedimentation in Still Air

Some microbes become airborne while in an actively metabolizing phase. More commonly they occur as *spores*, which are hardier, metabolically less active and often better adapted to aerial dispersal. This phase is exemplified by endospores of bacteria, spores of Actinomycetes, fungi, ferns, mosses, pollen (microspores) of flowering plants, and by protozoan cysts.

Functionally spores are of two distinct types. Resting spores (memnospores) serving to extend the life of the organism through time, and dispersal spores (xenospores) adapted for dissemination over the biosphere (Gregory, 1966). Here we are concerned only with dispersal spores, and only with their properties relevant to suspension in the atmosphere.

PROPERTIES OF SPORES

Structure. A spore may be a single cell, or consist of a few cells. In fungi the 'cells' may be incompletely separated by perforate septa. An outer cell wall, which fine-structure studies show is multilayered, surrounds cytoplasm with nuclei or nuclear material, often with reserve materials such as glycogen or lipids. Unlike vegetative cells, vacuoles of aqueous fluid are normally absent from spores, and the contents tend to have a lower water content than the rest of the organism. Some fungal spores, as pointed out by Ingold (1956), contain gas bubbles when in dry air, and probably this is not uncommon.

Spore wall. Normally consisting of resistant material (chitin, cellulose, sporopollenin), the wall becomes greatly thickened in some species, e.g. *Helminthosporium* (Plate 6, Figure 51). The surface may be hydrophobic or hydrophilic. Some spores have a sticky surface; many are hygroscopic.

Spore colour. Spores of many species are colourless and transparent (hyaline); others are coloured, yellows, reds, browns and purples predominat-

15

ing. Pigment may be in outer layers of the wall, or in the inner wall only as in *Ganoderma* (Plate 7, Figure 34), or in the cytoplasm. Colour is a useful aid in identification.

Surface texture. Some spores have smooth walls, even under the electron microscope. Others are rough, spiny or variously textured (Plate 2).

Aggregation. Normally spores are present in the air dispersed as single units, but, depending on the liberation mechanism, in some species aggregates of several spores forming one 'dispersal unit' are common, as for example in *Cladosporium* (Plate 6, Figure 53). The 'drop excretion mechanism' of basidiospores appears to be the most efficient mechanism for launching single spore units (Chapter IV).

Density. The specific gravities of spores have mostly been found to be greater than water, commonly 1·1 to 1·2. However, lower densities are recorded, and this is probably attributable to buoyancy adaptations, such as internal gas bubbles, bladders of conifer pollen (Plate 8, Figure 8). or the empty cell fragments of the spores of *Coccidioides immitis.*

Shape. Spore shape varies widely according to species. The smallest spores. such as those of bacteria and Actinomycetes, tend towards a spherical general form. Larger spores of fungi also tend to be sub-spherical, or somewhat cylindrical. Some are extremely elongated and thread-like with a length many times their breadth (e.g. *Claviceps purpurea*, Plate 6, Figure 11). Some are multiradiate (e.g. *Tetraploa*, Plate 6, Figure 54). Typical shapes are illustrated in Plates 6 and 7, all at a magnification of 1000 diameters.

Contrary to the views of many systematists, colour, size, shape and texture of fungal spores must be looked upon as probably functional adaptations resulting from interplay of tensions in growth (Savile, 1954), modified in evolution by requirements of the liberation process, of flotation in air, and ultimately of deposition and survival.

Size. Any significant persistence in suspension in the atmosphere is only possible for organisms at or below the limit of human vision, say 50 μm (*see* Goetz, 1965). The scale of sizes in Table 1 shows, on a logarithmic scale, where spore sizes that concern us fit into the gamut of terrestrial magnitudes (*and see* Figure 3).

Surface charge. As typical of colloids, spores when moist are surrounded by an electrical double layer, the sign varying with species and conditions (Douglas *et al.*, 1959; Fisher & Richmond, 1969). This charge determines

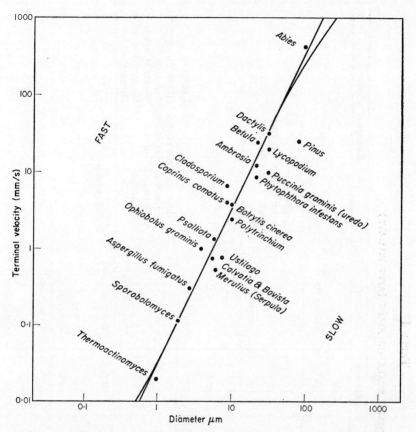

Figure 3. Observed terminal velocities of fall of spores and pollen (cm/s) related to *diameter* (μm). The straight line represents expected terminal velocity of smooth sphere (density 1·00) from Stokes's law. (The curved deviant at top right is Davies's correction, and that at bottom left is Cunningham's correction.)

direction of migration of spores, in *aqueous* suspension, in an electro-phoresis apparatus.

Additionally, spores may (or may not) carry an acquired charge, arising by some means or other during liberation, or perhaps picked up by capture of atmospheric ions. Basidiospores are found to be charged, when tested immediately after liberation from an agaric fruit-body (Buller, 1909). In the bracket fungus, *Ganoderma applanatum*, spores are mostly, if not all, positively charged (Gregory, 1957); those of *Merulius (Serpula) lacrymans* normally carry a negative charge, of the order of a few electrons per spore (Swinbank *et al.*, 1964).

TABLE I

LOGARITHMIC SCALE OF SIZES (APPROXIMATE)

Radiations	Inorganic objects	Metres	Sub-unit	Biological objects
		10^4		
	Troposphere (Thickness)	10^3	kilometre	
Radio waves		10^2		Larger trees
		10^1		Larger mushroom rings
		10^0	metre	Larger vertebrates Man
		10^{-1}		
U.h.f.	Raindrops	10^{-2}		
		10^{-3}	millimetre	(Aerobiological range)
Infrared (far)	Drizzle (mist)	10^{-4}		Pollens and fern spores (Rafts for bacteria and viruses)
	Cloud droplets	10^{-5}		Fungus and moss spores Sneeze droplets
Infrared (near)	Aitken condensation nuclei	10^{-6}	micrometre	Actinomycete spores and bacteria Human skin scales
Visible light		10^{-7}	(micron: μm)	Droplet nuclei
Ultraviolet		10^{-8}		
Far ultraviolet		10^{-9}	nanometre	Virus particles
X-rays (soft)		10^{-10}	(Ångstrom)	
X-rays (hard)		10^{-11}		
Gamma rays		10^{-11}		
Cosmic rays		10^{-12}	picometre	

Size, density, surface roughness, and possibly electrostatic charge, combine to control the speed at which a spore falls in still air – its terminal velocity of sedimentation.

SEDIMENTATION IN STILL AIR

All the particles with which we are concerned are heavier than air. In still air they sink with a constant and characteristic 'terminal velocity'.

Stillness as a quality of air is only relative. In the laboratory we can make the air as still as possible by eliminating draughts and convection currents, only to find an intense underlying activity revealed by the scintillation of motes in a beam of light. The motes are small enough to be jerked irregularly by the impact of gas molecules; but they are too large to be transported bodily by molecular diffusion, and most of the phenomena of colloidal suspensions are irrelevant to the air spora. We shall meet some analogies with the diffusion of a gas, however, in studying the diffusion of a cloud of spores in the atmosphere.

In this study we shall usually ignore the underlying molecular activity of the medium, and consider a patch of air as 'still' if it is not being transported bodily at more than a certain speed. Outdoors this speed might be 10 cm/s; in a room it might be 1 cm/s; and under carefully controlled conditions in special apparatus, a higher standard might be expected. For the present we must leave the definition vague, and simply regard air as 'still' when, in a particular context, the effects of wind, turbulence, and molecular activity are negligible. Knowledge of the properties of small particles in still air throws light on the behaviour of spores in moving air outdoors.

FACTORS DETERMINING VELOCITY OF FALL

One effect of its molecular activity is that the air is viscous, i.e. it resists the movement of solid particles. A small particle liberated into the air from a resting position tends to fall with an acceleration due to gravity; however, the resistance of the air increases faster than the speed of fall, and a state of balance is soon reached in which the particle stops accelerating and continues to fall through the air at a constant *terminal velocity*.

The terminal velocity of smooth spheres with diameters between 1 μm* and 100 μm is satisfactorily predicted by Stokes's law (for smaller particles Cunningham's correction becomes applicable, and for larger particles C. N. Davies's (1952) correction; non-spherical particles have to be treated

* μm = 10^{-6} metres (1/1000 millimetre).

experimentally at present). Stokes's law can conveniently be given in the form:

$$v_s = \frac{2}{9} \frac{\sigma - \rho}{\mu} \cdot gr^2$$

where at ordinary surface temperature and pressure:

v_s = terminal velocity (velocity of sedimentation) in cm/s

σ = density of sphere in g/cm³ (water = 1·00)

ρ = density of medium (air = 1·27×10⁻³ g/cm³)

g = acceleration of gravity (981 cm/s²)

μ = viscosity of medium (air at 18° C = 1·8×10⁻⁴ g/cm/s) (*N.B.* This must be distinguished from the symbol μm, the micron.)

r = *radius* of sphere in cm (*N.B.* It is important to use radius and not diameter.)

For a water droplet falling in air, $v_s = 1·2×10^{-2}r^2$ cm/s, when the radius is expressed in microns (μm). A fog droplet of 10 μm radius (i.e. 20 μm diameter) has a calculated terminal velocity of 1·2 cm/s.

The diameters of particles constituting the air spora vary from approximately 1 μm to 100 μm, and their place in the size spectrum of atmospheric bodies is indicated in Figure 3. Some are approximately spherical, others differ widely from spheres, some spores are filamentous, perhaps one hundred times as long as wide (*see* Appendix 1, p. 299, Plates 6–8).

The pollens and spores with which we are concerned belong to the size range where Stokes's law is valid, but they are seldom anything like smooth spheres. Stokes's law has given values of at least the right order, however, for spores whose terminal velocities have been measured experimentally (Figure 3). At first sight the pollen grains of some species of conifers appear to fall unexpectedly slowly, but these grains have conspicuous air sacs which greatly reduce the density of the individual particle.

Although the densities of spores of few species have yet been measured, there are reasons for expecting spores to be much less dense than mineral particles and indeed to resemble water droplets in density. The few determinations which have been made, relative to water = 1, are given in Table II.

The properties of spores are not invariable, but may alter with external conditions – sometimes enough to have a marked effect on their terminal velocity. For instance, the spores of the toadstool *Amanitopsis (Amanita) vaginata* were recorded by Buller (1922) as falling at 0·5 cm/s when observed immediately below the gill after liberation, but they became desiccated on continuing to fall through dry air and soon slowed down to one-third of their initial speed. Durham (1943) gave laboratory determinations of densities of pollens, and also some probable outdoor values at ambient humidity, shown in parentheses in Table II.

TABLE II

OBSERVED DENSITIES OF POLLENS AND SPORES (WATER $= 1 \cdot 00$)

Species	Density	Author(s)
(Angiospermae)		
Acnida tamariscina	1·0	(1)
Alnus glutinosa	0·75	(2)
Alnus glutinosa	0·97	(1)
Ambrosia elatior	0·63 (0·55)	(1)
Ambrosia bidentata	0·56 (0·50)	(1)
Betula verrucosa	0·81	(2)
Broussonetia maculata	1·1	(10)
Corylus avellana	1·00	(2)
Dactylis glomerata	0·98	(2)
Fagus sylvatica	0·71	(2)
Fraxinus americana	0·90	(1)
Iva xanthifolia	0·79	(1)
Juglans nigra	0·93	(1)
Phleum pratense	1·00 (0·90)	(1)
Quercus imbricaria	1·04	(1)
Salsola pestifer	1·0 (0·90)	(1)
Typha angustifolia	0·75	(2)
Typha latifolia	1·61	(2)
Xanthium commune	0·52 (0·45)	(1)
Zea mays	1·10 (1·00)	(1)
(Gymnospermae)		
Juniperus communis	0·40	(2)
Picea excelsa	0·55	(2)
Pinus sylvestris	0·39	(2)
Pinus montana	0·50	(2)
Taxus baccata	0·58	(2)
(Pteridophyta)		
Lycopodium sp.	1·175	(3)
(Bryophyta)		
Polytrichum sp.	1·53	(3)
(Fungi, etc)		
Alternaria oleracea	1·1	(8)
Amanitopsis vaginata	1·02	(4)
Coprinus plicatilis	1·21	(5)
Erysiphe polygoni (conidia)	1·094	(6)
Glomerella cingulata	1·1	(8)
Lycoperdon sp.	1·44	(3)
Lycoperdon sp.	0·73	(10)
Monilia fructicola	1·1	(8)
Neurospora sitophila (conidia)	1·1	(8)
Peronospora destructor	1·34	(6)
Psalliota campestris	1·20	(5)
Puccinia graminis tritici	0·807–0·862(?)	(7)
Sphaerotheca macularis	1·10–1·11	(5)
Stemphylium sarcinaeforme	1·1	(8)
Uromyces phaseoli	1·36	(6,5)
Ustilago nuda	0·56	(10)
Bacillus cereus (dormant spores)	1·28	(9)

Author references (1) Durham, 1943; (2) Pohl, 1937; (3) Zeleny & McKeehan, 1910; (4) Buller, 1909; (5) Jhooty & McKeen, 1965; (6) Yarwood, 1952; (7) Weinhold, 1955; (8) Miller, McCallan & Weed, 1953; (9) Black & Gerhardt, 1962; (10) Starr & Mason, 1966.

Four methods have been used for measuring terminal velocity (v_s), and observed values have been collected in Table III.

TABLE III

OBSERVED TERMINAL VELOCITIES OF POLLENS AND SPORES

Species	v_s cm/s	Author(s)
(Flowering plants)		
Abies pectinata	38·7	(11)
Alnus viridis	1·7	(10)
Ambrosia trifida	1·56	(13)
Ambrosia artemisii	1·56	(13)
Betula alba	2·4	(10)
Carpinus betulus	2·2–6·8	(9, 10, 11)
Corylus avellana	2·5	(10)
Dactylis glomerata	3·1	(10)
Fagus sylvatica	5·5	(10)
Larix decidua	9·9–22·0	(9, 10)
Larix polonica	12·3	(11)
Picea excelsa	8·7	(10)
Pinus cembra	4·5	(11)
Pinus sylvestris	2·5	(10)
Quercus robur	2·9	(10)
Salix caprea	2·16	(11)
Secale cereale	6·0–8·8	(9)
Tilia cordata	3·24	(11)
Tilia platyphylla	3·2	(10)
Ulmus glabra	3·24	(11)
(Pteridophyta)		
Lycopodium sp.	1·76–2·14	(2, 5)
(Bryophyta)		
Polytrichum sp.	0·23	(2)
(Fungi)		
Alternaria sp.	0·3	(6)
Alternaria (tenuis type)	0·4–0·55	(15)
Amanita rubescens	0·15	(1)
Amanitopsis vaginata	0·29–0·61	(1)
Aspergillus fumigatus	0·03	(16)
Bolbitius vitellinus	0·7	(15)
Boletus badius	0·11	(1)
Boletus felleus	0·12	(1)
Botrytis cinerea	0·22–0·45	(15)
Bovista plumbea	0·24	(5)
Bovista plumbea	0·07	(15)
Calvatia gigantea	0·075	(15)
Coccidioides immitis	0·003	(12)
Cladosporium spp.	0·07	(15)
Coprinus comatus	0·4	(1, 15)
Coprinus plicatilis	0·43	(1)
Cronartium ribicola	2·03	(3)
Erysiphe graminis	1·2	(7)
Galera tenera	0·21	(1)
Helminthosporium sativum	2·0–2·78	(5, 6)
Hemileia vastatrix	0·6–1·2	(17)
Lophophodermium pinastri	0·43	(14)

TABLE III—*continued*

Species	v_s cm/s	Author(s)
Lycoperdon pyriforme	0·05	(5)
Lycoperdon sp.	0·047	(2)
Marasmius oreades	0·13	(1)
Merulius lacrymans	0·05	(15)
Monilia sitophila	0·16	(5)
Ophiobolus graminis	0·01	(15)
Paxillus involutus	0·11	(1)
Phytophthora infestans	0·85	(15)
Pluteus cervinus	0·067	(1)
Polyporus squamosus	0·10	(1)
Psalliota campestris	0·13	(1)
Psathyrella candolleanum	0·2	(15)
Puccinia coronata avenae II	1·0	(4)
Puccinia graminis secalis II	1·06	(4)
Puccinia graminis secalis I	1·02	(4)
Puccinia graminis tritici II	0·94–1·25	(4, 8)
Puccinia graminis tritici I	1·06	(4)
Puccinia triticina II	1·26	(4, 5)
Rhytisma acerinum (straight)	0·13	(15)
Russula emetica	0·16	(1)
Scleroderma aurantium	0·1	(15)
Sporobolomyces sp.	0·012	(15)
Streptomycetes from hay	0·02	(15)
Tilletia tritici	1·41	(5)
Ustilago tritici	0·07	(5)
Ustilago zeae	0·3	(6)
Venturia inaequalis	0·2	(15)

Author references (1) Buller, 1909; (2) Zeleny & McKeehan, 1910; (3) McCubbin, 1918; (4) Ukkelberg, 1933; (5) Stepanov, 1935; (6) J. J. Christensen, 1942; (7) Yarwood & Hazen, 1942; (8) Weinhold, 1955; (9) Bodmer, 1922; (10) Knoll ex Rempe, 1937; (11) Dyakowska ex Erdtman, 1943; (12) Dimmick, 1965; (13) Raynor & Ogden, 1965; (14) Lanier *et al.*, 1969; (15) P. H. Gregory & D. R. Henden, *unpublished*; (16) J. Lacey, *unpublished*; (17) Bowden *et. al.*, 1971.

Method 1. Timing by direct observation. The simplest method is to time the fall over a short, measured distance in a small chamber of still air by direct observation with a horizontal microscope. It was used in the pioneer work of Buller (1909), and by Yarwood & Hazen (1942). So far the method has been used only for small, slowly-falling spores, because large ones travel too fast to be timed by direct observation. The method could no doubt be extended to fast-moving spores by photographing with a flash of known duration.

Method 2. Timing in fall tower. Most use has been made of the technique of releasing spores or pollen at the top of a column of still air in a vertical cylinder and finding the time they take to arrive at the bottom. This is the method used by Zeleny & McKeehan (1910), McCubbin (1918), Ukkelberg (1933), Stepanov (1935), and Weinhold (1955).

McCubbin and Ukkelberg report results of similar type. The numbers of wheat-rust spores reaching the bottom of the cylinder in successive intervals of time showed a negative skew distribution. Ukkelberg was able to show that part of this skewness was due to the presence of clumps of spores that fell faster than single units. It is also clear that, with both urodospores and aecidiospores of rust fungi, a large number of single spores fall very slowly. Measurements are needed to test whether, within one species, the single spores arriving first at the bottom are larger than those arriving at the end of the test. Another possibility is that small eddies may have hastened the fall of some spores and retarded others. A more serious defect of the method is that a vertical circulation of air by convection in the cylinder might bias the result by introducing a systematic acceleration or retardation of fall. This drawback could be overcome by establishing a small temperature difference between the top and bottom of the cylinder, so that the stratified air would be stabilized as in a 'temperature inversion' (p. 36). A thermostat could produce artefacts from convection currents set up by rhythmic temperature changes. Buller (1909) emphasized the difficulty of reducing air to anything like stillness, even in closed beakers.

Method 3. Inertial separation. Using inertial devices described in Chapter IX, which fractionate particles according to a property related to terminal velocity, estimates of v_s may quickly be obtained. Such devices are the cascade impactor (May, 1945) and the Andersen sampler (Andersen, 1958).

Method 4. Stirred settling. Successive measurements of concentration of a spore cloud, gently stirred (e.g. by convection) have been used by Dimmick, Hatch & Ng (1958) to measure terminal velocities. The equation for die-away of concentration in a closed space, as a result of sedimentation allows terminal velocity of the particle to be calculated (*see* Chapter XIII; and Green & Lane, 1957). Some values obtained by this method (Gregory & Henden, unpublished) are included in Table III.

In air, spores gain or lose water rapidly and the effect of spore hydration on terminal velocity, noted earlier by Buller, is evidently complex. Weinhold (1955) reported that with uredospores of *Puccinia graminis tritici*, changes in volume and weight occurred within 3 minutes of transfer to air of different temperature and humidity. He claimed that, contrary to expectation, spores stored at 5% relative humidity fell at 1·25 cm/s, in spite of being smaller and less dense than spores stored at 80% relative humidity which fell at 1·1 cm/s. Increasing the humidity of the air through which the spores fell increased the terminal velocity, which was: 1·03, 1·22, 1·23, and 1·54 cm/s at relative humidities of 24, 45, 52, and 80% respectively. With increasing temperature, terminal velocity decreased from 1·06 cm/s at 23·4° C to 0·94 cm/s at 39·9° C.

Bulk sedimentation of a spore cloud. Stokes's law applies to *isolated* bodies falling in air. With a cloud of falling particles above a certain limiting concentration the air may be dragged down too, and then the mixture of air plus spores will fall as a current at a speed exceeding the terminal velocity of separate particles (Fuchs, 1964). Stepanov (1935) reported on the velocities of ascending air currents required to keep a spore mass in suspension and was apparently measuring bulk sedimentation. Chamberlain (1967) calculates that a spherical volume of air of 1 metre diameter, containing only 10 mg of suspended particles, could acquire a bulk sedimentation velocity of 20 mm/s. Chamberlain suggests that this phenomenon may account for the unexpectedly fast velocity of deposition of *Lycopodium* spores at distances close to the point of release, found in experiments by Gregory *et al.* (1961; *and see* p. 112). The bulk sedimentation effect must be guarded against in attempts to measure terminal velocity by timing descent down a fall-tower, and it may account for the anomalous fall speed of coffee rust *(Hemileia vastatrix)* uredospores reported by Nutman *et al.* (1960); as discussed by Bowden *et al.* (1971).

We still have few observations on the rate of fall of greatly elongated spores found in such fungal genera as *Ophiobolus, Epichloë, Geoglossum,* and *Cordyceps,* whose shape makes Stokes's law inapplicable. Falck (1927) calculated terminal velocities for a number of species with approximately ellipsoidal spores on the assumption (derived from experiments with wax models in syrup) that the expected velocity, $v_e = v_s \sqrt[3]{(a/b)}$, where v_s is the fall velocity of a spherical particle of the same volume, and a and b are the *half*-axes of the ellipsoid. McCubbin (1944a) also stressed our lack of observations on asymmetrical spores, and provisionally suggested a method of calculating terminal velocity on the assumption that surface drag accounts for most of the retardation. He showed that observed terminal velocities of most spherical and ovoid spores fitted the approximate formula,

$$v_s = \frac{\text{length} \times \text{width}}{40}$$

where velocity is in millimetres per second, and spore dimensions are in microns (μm). Fusiform spores were treated as consisting of an intercalated cylinder of length l μm, between two axial cones, each of axial length x μm. (v_s being again in mm/s.) (*See* Ingold, 1965, p. 14.) The problem is re-examined by Lanier & Zeller (1968) and Lanier *et al.* (1969), who give tables calculating terminal velocities of cylinders and ellipsoids over a wide range of radii.

During fall in still air, an asymmetrical particle will assume a characteristic orientation. Hydrodynamical theory requires that the orientation assumed will be that in which the resistance of the air to the motion of the particle is greatest. This phenomenon can be observed with the naked eye if minute airborne particles are watched in a beam of light in a darkened room.

We know little as yet about spore orientation in air. Buller (1909) observed that some slightly elongated spores tend to fall with their long axis horizontal, as is to be expected for dynamical reasons. Sometimes factors other than shape seem to influence the orientation of an asymmetric spore. When Yarwood & Hazen (1942) watched the smooth conidia of *Erysiphe graminis*, measuring 32×20 μm, during fall in vertical glass tubes 7 mm in diameter, they saw that half the spores fell with the long axis horizontal and the other half with it vertical. This might indicate an uneven distribution of materials of different density in the cell contents; but, more likely, the vertical position was due to drag at the glass wall boundary, because if the tube is made even narrower, all the spores fall vertically. I have seen the filamentous ascospores of *Cordyceps gracilis* similarly oriented whilst being carried up by convection currents beside a vertical glass surface. Further, while watching the tailed spores of the puffball, *Bovista plumbea*, falling in a small chamber on the stage of a horizontal microscope, the tail was seen to trail behind the spherical spore. In Chapter VII it will be indicated that spores tend to be deposited with characteristic orientation on a surface.

NON-VERTICAL SEDIMENTATION: GLIDING MOTION

In still air spherical particles should fall vertically. This would not necessarily be true of some of the highly irregular shapes encountered among fungus spores. The deviation of a trajectory from the vertical is particularly strongly pronounced for needle-shaped particles (Fuchs, 1964, p. 35).

Whereas symmetrically shaped spores will fall vertically in still air under gravity (in a characteristic orientation if non-spherical), a spore which is asymmetrical either in its shape or its centre of inertia may spin on its axis or move sideways (Kunkel, 1948; Fuchs, 1964). On a large scale the phenomenon can be studied in steerable parachutes (Brown, 1951). Thus the direction in which a non-spherical particle sediments is not necessarily vertical, and the possibility that some curiously shaped or appendaged spores can drift laterally into crevices and under overhanging surfaces needs exploring.

To summarize: Stokes's law holds for small smooth spheres. Few pollens or spores are spheres, but many are microscopically smooth. Others, when highly magnified, are seen to possess warts, spines or other projections, or even to be pitted. Roughnesses would be expected to increase friction during movement through air and to retard fall, but we have no experimental evidence of this.

Viewed over the whole range of spore and pollen sizes say from 4 to 100

μm diameter, and of terminal velocities of from 0·05 to 10 cm/s, it is clear that Stokes's law gives a good idea of terminal velocity in still air, but that asymmetry and surface roughness may play a part as yet unmeasured.

EFFECTS OF SEDIMENTATION

The effects of spore fall in still air can be observed indoors, particularly if a room is left closed and unoccupied – a fact noted early in the study of air hygiene by workers using Hesse's horizontal tube method of air sampling, or some modification of it (*see* Chapter I, p. 11). Although all these investigations suffer from the defect of being based on highly selective culture media, all workers agree that wind or crowds stir up microbes, and that these soon settle inside buildings where the air is left undisturbed.

At the Royal Institution in London, England, Tyndall (1881) made a close study of microbes in the air in relation both to the question of spontaneous generation and to the antiseptic surgery which was being developed by Lister at that period. Tyndall showed that the air of a darkened room scattered a powerful beam of light. Gas molecules did not appreciably scatter light. Scattered light always arose from suspended *particles*, some of them too fine to be described as motes. By passing a beam of light through windows in the side walls of a glass-fronted box, he showed that, after a day or two, the air became optically empty, the particles having settled on the floor, walls and even the roof of the box. At the same time Tyndall found that the air, previously full of microbes, had become sterile. The ability to generate life was associated with the presence of light-scattering particles, and the air of small spaces could be sterilized by mere sedimentation. Tyndall had the curious idea that microbes remained associated in the air in clouds, much as fish are associated in shoals, and he explained that some of Pasteur's flasks (p. 4) must have been opened within clouds, while others were opened between clouds of floating microbes. We now think of micro-organisms as tending to be distributed in the air at random (Horne, 1935), but, under certain conditions, it may be that Tyndall was right.

Outdoors the effects of terminal velocity are usually masked by the speed and turbulence of the wind. However, conditions are sometimes tranquil enough for its effects to be detected. One example was found by Rempe (1937) of Göttingen, who made a series of aeroplane flights both by day and by night to study the distribution of tree pollen over German forests. By trapping on sticky cylinders, he obtained evidence that pollen grains of different sizes and terminal velocities differed in their relative abundance with altitude, even by day (Table IV).

c

TABLE IV

POLLEN DISTRIBUTION AT DIFFERENT ALTITUDES
(After Rempe, 1937; day flight A 6)

Pollen	Approximate diameter (μm)	Per cent of total pollen at height 10–40 metres	2000 metres
Betula (birch)	22	29·0	73·3
Carpinus (hornbeam)	37	55·0	10·0
Fagus (beech)	38	11·5	3·3
Others	—	4·5	13·3

By night it sometimes happened that pollen grains were partially sorted out according to size, even within a single species, as shown by the mean diameters of birch pollen on the night flight A 10 (Table V). The size range recorded by Rempe varied from 23 μm to 27·5 μm, so it is evident that even at night the sorting effect was not great – a difference in altitude of 1000 metres was associated with a decrease of only 4·5 μm (or 1/9th of the mean diameter), and with an estimated terminal velocity differing between 1·6 and 2·3 cm/s.

TABLE V

SIZE OF *Betula* POLLEN AT DIFFERENT ALTITUDES
(After Rempe, 1937; night flight A 10)

Altitude (metres)	Mean diameter (μm)	Estimated terminal velocity (cm/s)
1000	23·0	1·6
800	24·5	1·8
500	26·7	2·1
200	27·5	2·3
10–40	27·2	2·2

Quite insignificant convection currents may be enough to counteract the terminal velocity of small spores. Falck (1904) believed that the fruit-bodies of the larger fungi generate sufficient heat to induce convection currents which would carry their spores upwards. The temperature of an insulated mass of *Polyporus squamosus* increased nearly 10° C in 10 hours, and he regarded parasitism of fruit-bodies by maggots as a heat-generating adaptation favouring dispersal. Buller (1909) justly criticized this view, but field experiments are needed to determine whether the pilei of agarics modify air flow by their own heat, by absorbing solar radiation, or by their shape generating stationary eddies in an air stream. The colours of agarics are usually considered to be functionless, but the presence of dark colours among species inhabiting burnt ground suggests that this character may have been selected during evolution. It is not impossible that both colours and shapes of agaric fruit-bodies are partly adaptive.

III

The Atmosphere as an Environment

Aerobiology is a synthesis: just as the geographer draws upon astronomy and geology, so the aerobiologist draws on many sources. To understand the environment of the air spora we must go to meteorology. A fuller account of the relevant features of the atmosphere can be found in works on dynamical meteorology, e.g. Meteorological Office, London (1960), Geiger (1965), Sutton (1953), and the book by the United States Atomic Energy Commission (1968) with many excellent diagrams.

THE ATMOSPHERE AND ITS LAYERS

The atmosphere is usually recognized as layered. Some of its main features are illustrated in Figure 4, in which altitude is represented on a logarithmic instead of the usual linear scale in order to allow the various layers to be shown together on one page, and to illustrate vividly how the properties of the atmosphere change most sharply near the ground.

Barometric pressure, density of the air, and (as a rule) temperature, decrease with increasing height above the Earth's surface. These changes are all quantitatively important in aviation, and calculations are based on a table of an agreed 'International Standard Atmosphere'. Changes in temperature, humidity, density, and viscosity of the air with changes in altitude will have complex effects on suspended microbial spores and pollens, but are not likely to affect their terminal velocity greatly.

In Figure 4, the three vertical panels of the diagram represent conditions in contrasting weather types. The central panel represents a dull, windy day, with a cloud layer shielding the ground from direct sunlight (conditions on a cloudy night are not very different). The right-hand panel represents a sunny day, and the left-hand panel a still, cloudless night. The thickness of each individual layer varies according to conditions; the boundaries between them vary in definiteness; sometimes transitions are imperceptible, but there is sometimes even a visible interface between layers. The layers

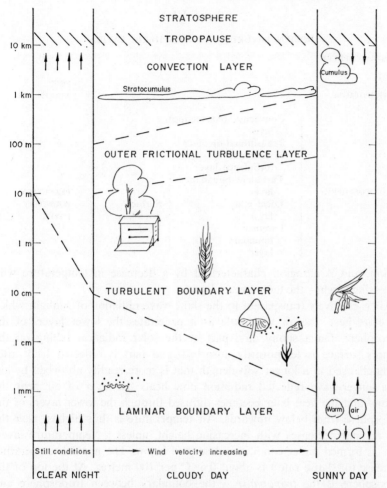

Figure 4. Diagrammatic representation of layers of the atmosphere (with logarithmic vertical scale).

are variously named in the literature and this may be confusing unless the following approximate equivalents are borne in mind. (*See* p. 32, Table VI.)

It is convenient to describe these layers in the reverse order, from ground level upwards, beginning in the troposphere with the laminar boundary layer.

THE TROPOSPHERE

The troposphere is the name given collectively to the lower layers of the atmosphere extending from the ground up to a height of approximately

TABLE VI

NOMENCLATURE OF ATMOSPHERIC LAYERS

STRATOSPHERE				
TROPOPAUSE				FREE ATMOSPHERE
TROPOSPHERE	Convective (convection) layer			
	Transitional or outer frictional turbulence layer			
	Turbulent boundary layer	Surface boundary layer		PLANETARY BOUNDARY LAYER
	Local eddy layer			
	Laminar boundary layer			

10 km, and is a region characterized by a decrease in temperature with increasing height – the temperature lapse.

Air is relatively transparent to the short-wave radiation of sunlight which therefore heats the air very little as it penetrates the lower layers of the atmosphere. On a sunny day, part of the solar radiation falling on the Earth's surface is temporarily absorbed, and part is reflected back after being changed to a longer wavelength that is more readily absorbed by air. This longer-wave reflected radiation now heats the layer of air near the ground and the heat later becomes diffused through the lower layers of the atmosphere from below upwards. Air temperature is thus highest near the ground and decreases with increasing height, unless a 'temperature inversion' is formed under conditions described below. The normal temperature decrease (or 'lapse rate') is about $0.6°$ C per 100 metres. At the top of the troposphere is the *tropopause* – the boundary between troposphere and stratosphere.

The troposphere comprises the following five layers.

LAMINAR BOUNDARY LAYER

In contact with the surface of the earth and all projecting bodies is a microscopically thin layer of air held firmly by molecular forces. Except for molecular diffusion this layer is still and windless.

Above this windless film the atmosphere is usually in motion, set going either by pressure differences of distant origin, or by convection currents produced by local heating. The lowest layer of moving air, next to the still layer, is known as the 'laminar boundary layer' (or sometimes the 'laminar layer'). This again is a thin layer, of the order of a millimetre thick, in which there is no turbulence and the air flows in streamlines parallel with the nearest surface; heat, gases, and water vapour can move across the streamlines by molecular diffusion. Wind speed is negligible at the still surface film, and in the laminary boundary layer wind speed increases *linearly with height* (momentum being transmitted through the layer by molecular diffusion only). Particles, droplets, or spores getting into the laminar layer will sink through it, following trajectories determined by wind speed and gravity, and will come to rest at the Earth's surface.

A laminar layer also exists at the interface around any solid body, and much of the foregoing description applies equally to the air layer at the surface of a leaf or stem.

The thickness of the laminary boundary layer varies with the wind speed and with the roughness of the adjacent surface. In a high wind it may be thinned down to a fraction of a millimetre, and turbulent air from the next higher layer may then reach down nearly to the Earth's surface. In very calm weather the laminar layer may thicken considerably.

In contrast with the relatively equable air at a metre or two above the surface, the eco-climate of the laminar boundary layer is violently change-able (Monteith, 1960). Unless protected by a layer of vegetation, small organisms at ground level may be subject to extreme heat from the sun's rays by day, followed by a rapid drop in temperature as heat is lost by radiation to a clear sky at night.

The laminar boundary acts as a dust trap. Particles which have sunk through it and come to rest in the still or slowly moving air at the surface are out of reach of eddies – until some unusual condition arises which thins the laminar layer enough for eddies to penetrate down and sweep away dust particles. High winds may do this; or local heating of the sur-face, perhaps on a micro-scale, may set up 'dust-devils' – smaller or larger whirlwinds raising dust into the air.

LOCAL EDDY LAYER

For biological purposes we need to add the 'local eddy layer'. Even in streamlined air, local stationary eddies may exist behind small roughnesses; and, as will be shown on p. 44, air flow over a cup-shaped depression may set up a rotation pattern sufficient to throw dust up from a bowl. This layer is probably important in nature, where ideally smooth surfaces are rare. A special type of boundary at the top of a plant layer or crop has been called the 'outer active surface', or in forests, the 'crown layer'.

TURBULENT BOUNDARY LAYER

In this layer, where flux of momentum decreases linearly with height, solid obstacles, arising at the surface in the laminar boundary layer, project into the wind and cause eddies which may break away from the surface and travel down wind. A surface is aerodynamically *smooth* in conditions when the laminar layer is thick enough to submerge projections from the surface; but if the irregularities project through the laminar layer, the surface is considered *rough*. As the thickness of the laminar layer depends both on the wind speed and the stability of the atmosphere, it is clear that a particular surface such as a grass sward or a hairy leaf may be aerodynamically smooth under one set of conditions and rough under another. Each surface has a characteristic roughness parameter. Air flow over calm water may be smooth; but, except at extremely low wind speeds, flow over land is normally rough and disturbed by surface irregularities which cause turbulence.

Eddies of two types may occur; local, or stationary eddies which may arise on both the windward and leeward sides of a bluff obstacle, and eddies which break away and travel with the wind in the obstacle's wake. The forward velocity of a turbulent wind is thus the net result of a complex movement; the wind has vertical and lateral components as well as the forward horizontal movement. Further, turbulence in the vertical and horizontal directions may differ in intensity and it is then said to be non-isotropic.

Occurrence of mechanical or frictional turbulence depends on the wind speed being fast enough, and the object large enough, to cause eddying. Whether or not flow will be turbulent under a given set of conditions can be calculated by the method of Osbert Reynolds, who found that flow is turbulent when the Reynold's number, defined as Re = (length × wind velocity) ÷ kinematic viscosity, exceeds about 2000. Here 'length' is taken as a characteristic dimension of the object; and kinematic viscosity for air under average surface conditions equals $0·14$ cm^2/s. Thus for a leafy bush 100 cm high in a wind of 100 cm/s, we have:

$$Re = \frac{100 \text{ cm} \times 100 \text{ cm/s}}{0·14 \text{ cm}^2/\text{s}} = 71\,000,$$

so flow would be expected to be turbulent.

In the turbulent boundary layer, properties such as temperature, concentration of water vapour, and wind velocity, change much less rapidly with increasing height than in the boundary layer beneath. Eddies mix the different parts of the layer much more rapidly than do the slow processes of molecular diffusion. Particles can also be carried by eddies upwards and laterally in a manner impossible in the laminar layer. In the turbulent boundary layer the wind velocity, temperature, and water vapour show a change which is *linear with the logarithm of the height*. In this layer diurnal changes of temperature are less pronounced than in the laminar boundary layer underneath, and diurnal changes decrease still further with increasing height until, at the top of the next layer, they have almost disappeared.

An increase in wind speed increases the thickness of the turbulent boundary layer both downwards, by thinning the layer, and upwards by pushing into the transitional layer as turbulence increases. The turbulent boundary layer is thinnest on clear calm nights and thickest on hot sunny days, when it may reach to a height of 150 metres.

The turbulent boundary layer is the part of the atmosphere most familiar to us. While our feet are planted in the violently fluctuating climate at ground level, our heads, and the weather-recording instruments of the conventional Stevenson's screen, inhabit the relatively equable turbulent layer.

TRANSITIONAL OR OUTER FRICTIONAL TURBULENCE LAYER

Here frictional turbulence, generated in the layer below, still dominates vertical diffusion, but it dies out gradually until, at the top of the layer, both turbulence and diurnal changes disappear. Both layers may be dusty, and the top of the transitional zone is sometimes visible as a distinct dust horizon at 500–1000 metres, marking the upper limit to which spores and other particles are raised by frictional turbulence (though much greater heights may be attained by convection).

In dynamical meteorology this zone is defined as the region where the wind structure is determined partly by surface friction, and partly by the Earth's rotation.

CONVECTIVE LAYER

This layer extends from about 1 km above ground to the top of the troposphere at about 10 km. As in all layers constituting the troposphere, the temperature continues to decrease with height to the top of the convective layer, though diurnal temperature variation is almost absent. Frictional turbulence does not reach here, but, as already indicated, particles from the Earth's surface can be carried into this layer by large-scale convection currents when the ground is heated by sunshine.

The height above the ground attained by a mass of heated air before it loses buoyancy, depends on the temperature gradient and water-vapour content of the air at the time, as explained in the standard works on meteorology. Ascent may be halted if there is a temperature inversion layer in the upper atmosphere. Under conditions of thermal instability, 'bubbles' of heated air may arise intermittently from areas where the ground or vegetation is being heated by the sun. The bubbles may rise to the convective layer, carrying spores and other particles as well as water vapour to the level at which cumulus clouds are formed, and at times reaching to the base of the stratosphere (Mason, 1957).

NIGHT RADIATION AND TEMPERATURE INVERSION

At night, wind speeds tend to diminish; the laminar boundary layer then becomes thicker than by day and the turbulent boundary layer may become thinner, being reduced to perhaps only 10 to 15 metres in thickness.

These changes may be carried still further if the sky is cloudless, thus allowing radiation from the ground to escape into space. Loss of heat by radiation cools the ground and this in turn cools the air lying nearest to the ground. Thus, instead of temperature decreasing with increasing height, a 'temperature inversion' is set up: over cold air near the ground lies air at a higher temperature – up to a certain height, the top of the inversion, above which the usual lapse rate is again encountered.

As the air is coldest and densest at the bottom of the inversion, gravity tends to prevent it from ascending and mixing with warmer air above. The air in the inversion becomes stratified according to temperature and remains very stable, in contrast to the instability that is apt to develop when the ground is heated. Such a layer of cold, heavy air may flow slowly downhill as a nearly laminar katabatic wind, filling hollows with cold air and aiding the formation of frost pockets (Geiger, 1965).

In a temperature inversion, spores and dust particles tend to settle out, leaving the air relatively clean although the air above the inversion may continue to carry a normal spore-load.

ROLE OF CONVECTION

When the surface of the ground is heated by sunshine the lowest layer of air may be heated in turn. If a large temperature lapse-rate is established, the atmosphere becomes unstable, because the less dense ground layer of air tends to rise and carry its load of microbes upwards, being replaced by cooler air from above. The pattern of this overturn is not yet clear. A regular 'cellular' pattern of ascending and descending air was formerly envisaged, but now the ascending air is pictured as taking the form of 'chimneys' or of 'bubbles'. Cone (1962) describes the fully developed bubble or 'thermal' as an ascending vortex ring of warm air around which a current of cooler air is entrained as a shell. This outer cool shell descends on the outside of the vortex ring and ascends on the inside, giving a maximum lift at the centre of the ring, a position utilized by large soaring birds.

Glider pilots are familiar with the properties of warm ascending currents of air or thermals, as described by Yates (1953). In still air a glider sinks at about 90 cm/s (about 20 times the terminal velocity of a pine pollen grain). On dull days thermals do not develop. They reach their maximum upward velocity of 3 m/s, or up to 25 m/s in cloud, at midday in summer. Yates indicates that, depending on the type of soil, on wind strength, and on sun height, a sizeable thermal is released from an area of 1·25 km² every 5 to 15 minutes in summer. At a height of 300 metres, thermals may be 300 metres in diameter, though they are probably often smaller when lower down, whereas at still greater heights a diameter nearer 2 km was reported by Ludlam & Scorer (1953). Their vertical movement may cease at a temperature inversion, or may continue to from 3000 to 15 000 metres. The temperature in a thermal appears to average 1–2° C higher than the surrounding air through which it is ascending.

The theory of thermals is still a matter of controversy, but there seems no doubt that air rises from some surfaces more readily than from others. Green vegetation and wet soils may be relatively cool, while a ripe cereal crop, buildings, roads or bare rock, may heat up rapidly in the sun and become the source of warm rising air. Thermals can also arise at a cold front, and glider pilots regard hilly country as the best source of thermals.

The pattern by which cool air sinks to replace the thermal is also a matter of debate. Downward draughts reported in the neighbourhood of thermal upcurrents appear to be comparatively feeble. Probably the downward movement occurs over a much wider area than the thermal, in the form of a slow sinking over a comparatively broad area of the atmosphere. The sinking speed may be comparable with that of a fungus spore (Hirst, 1959), but the local rising velocity may commonly be 100 or more times this velocity. Some species of birds soar in large thermals, just as dragonflies do in smaller ones near the ground. Other birds haunt thermals to prey on insects carried upwards (Scorer, 1954). Clearly thermals can act energetically in raising microbes to great heights.

THE STRATOSPHERE

In this region, which extends upwards from the tropopause to the limit of the atmosphere, the temperature lapse-rate, characteristic of the troposphere, is zero or may even be reversed. The height of the tropopause varies with season, latitude and other factors. The bottom of the stratosphere may be found at an altitude of about 10 km, though under special conditions it may become lower temporarily even reaching to ground level (Danielsen, 1964).

The dust of the stratosphere is believed to be mainly meteoric in origin and to have entered the Earth's atmosphere from space. It is generally believed that terrestrial dust, including organic spores, is almost confined to the troposphere – except for occasional incursions in air currents dragged up into the stratosphere by volcanic eruptions (or hydrogen bombs). However, studies of atmospheric circulation, discussed by Machta (1959) and Goldsmith & Brown (1961), may point to an exchange of air between troposphere and stratosphere – with rising air over the equator and descending air in middle latitudes.

CIRCULATION OF THE ATMOSPHERE

Under the influence of pressure differences resulting from solar heating, and friction between wind and the rotating Earth, a general pattern of atmospheric circulation is set up. The surface winds shown in atlases are the ground-level part of a three-dimensional system that is still being explored (Ludlam, 1967). The pattern still being worked out shows a complex circulation with air sometimes moving comparatively slowly and sometimes at high velocity in jet streams. Over the equator air generally tends to ascend and flow discontinuously towards the poles, which are themselves regions of generally subsiding air (Palmén, 1951). Across the equator there is relatively little interchange of air between northern and southern hemispheres.

AIR MASSES

The fact that air may have the same temperature and humidity over a wide geographical area has given rise to the concept of the discrete air mass, with properties different from adjacent air masses and separated from them by 'fronts'. When an air mass remains stationary for some time, it acquires a temperature and humidity dependent on the surface on which it rests. These characteristics will be retained for some time after the air mass has moved on into a new region. Air masses are classified, therefore, according to their origin, and we may have, for example: polar maritime, tropical maritime, polar continental, and tropical continental types, as well as air masses of intermediate origin (Belasco, 1952). The different air masses interest us, not only because they bring different kinds of weather, but also because they might conceivably bring an air spora characteristic of their place of origin.

IV

Spore Liberation

Up to now we have considered only the physical properties of spores and of their environment. Spores, however, are parts of living organisms whose evolution has been extensively moulded by the environment. The air spora comes mainly from species which are highly adapted towards using wind energy for their dispersal. The physical properties of the atmosphere make dispersal possible, but also set problems to organisms using it. Adaptations for wind transport have been evolved in many widely-separated taxonomic groups.

The process of wind dispersal of spores has three principal stages. (1) *Spore liberation*. The processes by which pollen grains and spores 'take-off' into the air from the structure where they have been formed are described here and in Chapter V. (2) *Dispersion*. Chapter VI describes the transport of spores by gentle air currents or strong winds, and the diffusion of an airborne spore cloud. (3) *Deposition*. Chapters VII and VIII will deal with the processes by which spores leave the air and land again on a surface – a necessary prelude to the germination of a pollen grain or a mould spore on its substratum.

The pollen or spore output of some species is notoriously large. For instance Pohl (1937) estimated, for the dominant species encountered by pollen analysts, the pollen production per stamen, flower, inflorescence, and branch, revealing an annual production averaging many millions per square metre of ground covered (*see also* Erdtman, 1943). According to J. J. Christensen (1942), a field of wheat moderately affected by *Puccinia graminis* would produce at least 25 million uredospores per square metre. Buller (1909) estimated that one giant puff-ball *(Calvatia (Lycoperdon) gigantea)* produced 7×10^{12} spores, and that the yearly output of a single fruit-body of the bracket fungus, *Ganoderma applanatum*, might be $5 \cdot 46 \times 10^{12}$ spores (Buller, 1922, p. 137). The spore output of mosses and ferns is also enormous.

PROBLEMS OF 'TAKE-OFF'

As described in Chapter III, the surface of the ground or plant is covered by a thin layer of still air and by the laminary boundary layer of slowly moving air; a spore will fall through this composite zone under the influence of gravity. To tap the energy of moving air for dispersal, a spore must overcome the adhesive forces that tend to keep it in contact with neighbouring spores or with its substratum. It must cross the still- and laminar-air layers at the interface between the ground or other surface and the atmosphere, in order to enter the freely moving air of the turbulent boundary layer, where it stands a chance of being carried into higher layers of the troposphere.

Many species that are distributed as spores have not solved this problem, but instead have become adapted for dispersal by some other agency such as water, insects or other animals. There are more insect-pollinated (ento-mophilous) *species* of flowering plants than wind-pollinated (anemophilous) species, though in the temperate regions at least there are more wind-pollinated *individuals* because of the preponderance of grasses and anemo-philous trees. We may wonder how important in practice is the occasional dispersal of a spore by some agency other than that to which it is adapted. However, it is a fundamental principle that the better a species is adapted to dispersal by one agency, the poorer are its chances of dispersal by another agency, unless, like many fungi, it produces spores of several distinct types that are specialized for different dispersal mechanisms.

If we wish to control the dispersal process, a precise knowledge of the mechanisms involved is preferable to the vague idea that the spores will get there somehow anyway! Success in colonization or fertilization depends on logistics – on getting enough material to the right place at the right time.

Energy is required to detach spores from their source. It may be an active process through which, by some explosive or hygroscopic mechanism, spores are discharged by energy operating through the parent structure. Or it may be passive, by the energy of an external agent, usually wind, or the kinetic energy of falling raindrops. Seasonal development of the parent structure and maturation of the spores commonly determine what organisms are in the air at a particular time, but other factors modify this pattern. The working of the various discharge mechanisms is more or less affected by external conditions, and the result is that the output of spores of a particular species varies greatly from time to time. Conversely, all the individuals of one species in an area may behave in unison, so that the composition of the air spora differs vastly on different occasions.

TAKE-OFF MECHANISMS IN CRYPTOGAMS ETC

Spore- and pollen-liberation mechanisms have formed the subject of classical researches in biology for over a century. The wealth of information in the scattered literature on land plants is reviewed by Ingold (1965, 1971), and knowledge about bacteria by Wells (1955). In the present connection we are concerned with those aspects of the mechanism which determine when, and under what conditions, spores get into the air.

VIRUSES

The viruses seem little adapted to independent air dispersal. With several viruses, however, there is now circumstantial evidence of spread by airborne infection, all implying the existence of a 'take-off' mechanism, though the efficiency of such a mechanism still remains in doubt. Some viruses infecting the animal respiratory tract are forced into the air on droplets during coughing and sneezing. Circumstantial evidence suggests that fowl pest virus may become airborne on dust particles rising by convection from intensive poultry houses (C. V. Smith, 1964; *and see* Langmuir, 1961, and Chapter XV). Most bacterial and plant viruses, if they occur in the air at all, only get there on 'rafts' of debris or in water droplets. Soil and plant debris probably rafted tobacco necrosis virus which was reported as airborne by K. M. Smith (1937). Schisler, Sinden & Sigel (1963) find that a virus disease of mushrooms can be transmitted by infected mushroom spores which, being self-launching, must provide the virus with a highly effective raft.

Some of the so-called 'polyhedral' viruses infecting insects are exceptional. Reports of outbreaks among insect pests of forest trees in eastern Europe speak of copious yellow deposits of the polyhedral bodies, shed by parasitized insects, which coat the surfaces of vehicles travelling through the forests. Study of air dispersal might explain some of the anomalies in the behaviour of insect viruses. To prevent contamination by an airborne infective particle of the dimensions of a virus may well require quite unusual experimental precautions.

BACTERIA

Moving air does not normally detach bacterial cells or spores from the surface of a colony, at least when it is slimy, and in the absence of an active discharge mechanism natural processes capable of producing an aerosol of single bacterial cells are unknown. Bacterial endospores (which biologically appear to be 'memnospores') are not more easily airborne than are vegetative bacterial cells (Lamanna, 1952). Mechanical disturbance of dust, clothing, surgical dressings, etc, however, carries into the air contaminated particles of substratum acting as 'rafts' and bearing clumps of bacteria (Bourdillon & Colebrook, 1946). Rafts of soil or dust particles are raised by wind, by 'dust-devils' when the ground is heated by solar radiation, and by human and animal activity, such as cultivation of bare ground. Rain-splash, breakers, and sea-spray continuously throw minute, potentially bacteria-laden droplets into the atmosphere (Chapter V). Droplets expelled by coughing and sneezing are important indoors (*and see* p. 211), yet processes which put bacteria into the air are still imperfectly understood (Lidwell, 1967; Williams, 1967). This is also true of the yeasts whose frequent abundance in the air remains unexplained, except for the Sporobolomycetaceae which show the basidiospore discharge mechanism (pp. 48–49).

ACTINOMYCETES

The mycelial organization of the Actinomycetes allows the Streptomycetaceae to develop aerial hyphae bearing dry powdery spores – an example of the elevated sporophore device, common in more elaborate organisms, for raising the spore-producing organ above the substratum towards the moving layers of the atmosphere. Biologically, spores of the Streptomycetaceae are clearly xenospores, unlike the endospores of typical bacteria which are memnospores.

Launching processes in the Actinomycetes seem not to have been studied in detail. Hesseltine (1960) suggests that dispersal of Streptomycetes is associated with the presence of water. In a study of liberation of Streptomycetes from fallow soil, Lloyd (1969) recorded from nil to 129 propagules per cubic metre over undisturbed soil. Numbers were increased by mechanical disturbance. Rain impact put some single spores into the air, but the majority appear to have been attached to soil particles. By contrast, the organisms causing Farmer's lung disease are shed freely into the air when infested fodder is shaken into the air in the dry condition (Gregory & Lacey, 1963; Lacey & Lacey, 1964).

MYXOMYCETES (MYCETOZOA, MYXOGASTRALES)

The slime-moulds are a group thoroughly adapted for wind dispersal. Some, such as *Reticularia*, merely expose a dry, powdery spore mass on a cushion raised above the substratum. Others, such as *Stemonitis* and *Trichia*, expose small, dry spore masses on stalks at most a few millimetres high. The spores are set free by twisting movements of hygroscopic elaters as the humidity of the air alters (Ingold, 1939), or in a few species, spores may be removed by eddies from shallow wind-cups. A few species are splash-dispersed (Dixon, 1963). An active discharge mechanism was discovered in *Schizoplasmodium cavostelioides* by Olive & Stoianovitch (1966).

FUNGI

Adaptations facilitating air dispersal show more diversity in the fungi than in any other group, except, perhaps, the flowering plants with their wealth of adaptations for seed dispersal. These adaptations range from the often passive but quite effective processes in the Fungi Imperfecti, to the spectacular ballistic feats of the ascus gun. The various mechanisms have been summarized by Dobbs (1942a, 1942b) and Ingold (1939, 1953, 1960, 1965, and 1971), and they formed one of the main topics of the classical work of Buller (1909–1950).

In contrast, spores of many other species of fungi rarely get into the air but are carried by insects, on seeds, or in soil. Mere dispersal by insects may be relatively unimportant; but, where the insect activity inoculates the substratum or host, it is a mechanism homing on its target in a manner comparable in efficiency with insect pollination of flowering plants.

PASSIVE MECHANISMS

Passive liberation, by the action of external energy, depends on 'spore presentation' (Hirst, 1959).

(i) *Shedding of spores under gravity.* Stepanov (1935) concluded that spores of some *Cunninghamella* species, and of some Fungi Imperfecti, including *Botrytis cinerea, Monilia sitophila, Helminthosporium sativum,* and the macroconidia of *Fusarium,* could be shed under gravity. However, as he also showed that minor air currents could release spores of some of these fungi, the effect remains uncertain.

(ii) *Shedding in convection currents.* Stepanov (1935) placed open petri-dish cultures at the bottom of glass cylinders 10 to 12 cm high in which convection currents were induced by differential heating. Sticky slides or a surface of inverted sterile medium at the top of the cylinder trapped spores which might have become detached and carried aloft by convection. With temperature differences of the order of 10° C, conidia of *Monilia sitophila* and *Botrytis cinerea* were freely transferred upwards, but *Colletotrichum lini* were not. Smaller temperature differences, such as resulted from the slight heat produced by a mould culture or an electric lamp shining on the floor, were ineffective.

(iii) *Blowing away* ('deflation'). This commonly occurs with dry-spored fungi including moulds, smuts, and the uredospores of rusts. The spores are often presented to wind on an elevated sporophore, any stem or leaf patho-gen usually being adequately raised on its host tissue. Quantitative studies so far are insufficient to lead to a theory of 'deflation', but there is good evidence that the higher the wind speed the more spores are carried away, a process that can easily be watched in the open air by observing a fresh fructification of the slime-mould, *Reticularia*.

Stepanov (1935) was apparently the first to use a small wind-tunnel to blow spores at controlled wind speeds. Using either cultures or plants infected with pathogenic fungi, he found that the minimum wind speed required to remove spores differed according to the organism being tested: for *Botrytis cinerea* it was 0·36–0·50 m/s; for *Monilia sitophila, Ustilago* spp., uredospores *Puccinia triticina (recondita)* and *Helminthosporium sativum* it was 0·51–0·75 m/s; for aecidiospores of *Puccinia coronifera (coronata)* and *P. pringsheimiana*, 0·76–2·0 m/s; for *Cunninghamella* sp., 1·5–1·75 m/s. On the other hand, *Phytophthora infestans* and *Fusarium culmorum* spores were not removed at any speed tested up to 3·37 m/s. More spores were removed in turbulent than in streamlined wind.

Zoberi (1961) studied blowing away of spores from tube cultures of *Thamnidium, Phymatotrichum, Trichothecium, Piptocephalis, Trichoderma, Mucor* and *Mycogone*. In general spore liberation occurred more freely the higher the wind speed and the drier the air. With prolonged blowing the rate of liberation decreased with time. Spores could not be detached from *Mucor rammanianus* at any speed tested.

Substantially similar results were obtained by Gregory & Lacey (1963) when mouldy hay, associated with cases of Farmer's lung, was shaken in a small wind-tunnel and the numbers of mould and Actinomycete spores liberated were estimated after increasing intervals of time. When tested by blowing for an hour, two thirds of the total spores released were obtained in the first minute (although many more were removed by subsequent washing with water). Although the total number of spores blown away in a given time is roughly proportional to the wind speed (Figure 5) the liberation 'die-away curve' is similar in winds between 0·6 and 4·9 m/s. The fact that a sample of hay which had been blown for 30 minutes at 1·2 m/s and had liberated 50 million spores per gramme, in the course of a die-away to nearly zero, produced a second typical liberation curve for another 55 million spores when blown for another 30 minutes at 4·9 m/s is illustrated in Figure 6. This suggests that part of the mechanism of deflation is that increasing wind speed decreases the thickness of the boundary layer of air at the leaf surface, exposing more deeply immersed spores to the pruning action of eddies penetrating the laminar layer (*see* Carter *et al.*, 1970a, 1970b).

A special structure facilitating deflation is the 'wind-cup' described by Brodie & Gregory (1953). Flow of air over a cup-shaped structure produces a double eddy system which can effectively remove dry spores contained in

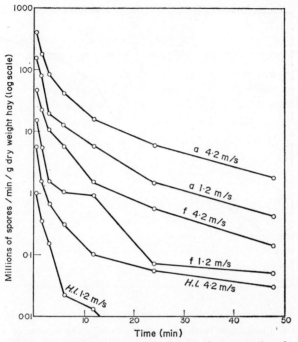

Figure 5. Number of spores released from sample of mouldy hay (associated with Farmer's lung disease), blown at wind speeds of 1·2 and 4·2 m/s, plotted against time, on a loglinear scale, and classified into: f = total moulds; a = total Actinomycetes (and bacteria); and H.l. = *Humicola lanuginosa*. (From Gregory & Lacey, 1963, reproduced by permission from *Transactions of the British Mycological Society*.)

the cup, as shown by wind-tunnel experiments with smoke and *Lycopodium* spores. Soredia were also removed from the podetia of a *Cladonia* at 1·5–2·0 m/s, and spores were removed from the cupulate sporangia of certain Myxomycetes at 0·5 m/s.

(iv) *Mist pick-up.* Dry, or even humid, wind fails to detach spores of some fungi which are nevertheless readily removed from their conidiophores by collision with minute droplets carried by mist-laden air. This method is known to function with two important crop pathogens, *Cercosporella herpotrichoides* (Glynne, 1953), and *Verticillium albo-atrum* (R. R. Davies, 1959), and it may play a part also in the dispersal of *Cladosporium*.

Raindrops can bring about the passive liberation of fungus spores by several distinct processes, including the bellows mechanism, 'rain tap and puff', and rain splash.

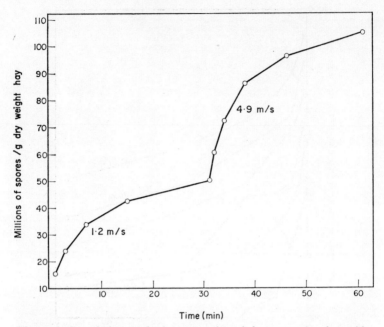

Figure 6. Cumulative total of spores released from sample of mouldy hay, blown first in wind of 1·2 m/s for 31 minutes, followed by wind of 4·9 m/s for a further 31 minutes. (From Gregory & Lacey, 1963.)

(v) *Bellows mechanism*. Certain Gasteromycetes, including the puff-balls such as *Lycoperdon perlatum* and *L. pyriforme*, and the earth-stars (*Geastrum* spp.), have a thin, flexible, waterproof wall covering the spore mass. Indenting this wall forces out a jet of air laden with spores. Contact with animals operates the bellows efficiently, but must be a relatively rare event in nature. Raindrops or run-off drops from trees also operates the bellows mechanism, and as one fruit-body would be hit many thousands of times in a season, rain is probably the most effective agent in the field (Gregory, 1949). In India W. H. Long & Ahmad (1947) found that the bellows of *Tylostoma* is operated by wind-blown sand grains in addition to raindrops. Rain-operated bellows also aid dispersal of some Myxomycetes, such as *Lycogala* and *Fuligo* (Dixon, 1963).

(vi) *Rain tap and puff*. Detachment of spores when leaves are vibrated by falling raindrops has long been suspected. The process has been studied by Hirst & Stedman (1963) in conjunction with an entirely new raindrop-actuated mechanism which they named 'puff'.

The puff phenomenon occurs when a raindrop strikes a *dry* surface.

By contrast, when a raindrop falls into a liquid or strikes a water film over a *wet* surface, we observe the phenomenon of 'rain-splash' which may disperse spores by an entirely different mechanism – the topic of Chapter V.

When a large drop, falling at terminal velocity, hits a dry, inelastic surface, the liquid in the drop spreads radially over the surface, at an initial speed of over 70 m/s. The speed decreases as the liquid spreads out over a radius of over 2 cm, the whole process being completed in about 2 milliseconds. The rapid movement of liquid must disturb air within the laminar boundary layer. Tests showed that the resulting puff transfers spores into the turbulent air, where they can be dispersed by eddy diffusion.

(vii) *Hygroscopic movements* of conidiophores, which may result in detachment of spores during violent twisting, occur in a number of Fungi Imperfecti and Phycomycetes. The effect depends on rapid changes in atmospheric humidity and is often most marked in the morning hours.

ACTIVE MECHANISMS

All active mechanisms for spore liberation depend on the fungus having a sufficient water supply. The more ephemeral fruit-bodies develop after rain and discharge spores for a short period only. Some more durable fungi can be dried but will discharge again when re-wetted; others again can draw on an extensive mycelium deep in the substratum and may be almost independent of the weather. Distances of ejection vary with different organisms and have been compiled by Spector (1956, p. 153).

(viii) *Squirt-gun mechanism.* This is found in many Ascomycetes in which the ascus, which contains the ascospores, typically swells at maturity and finally bursts at the tip, propelling the spores into the air to a distance varying from a fraction of a millimetre to several centimetres. The larger the projectile, the further it tends to be shot (Ingold, 1965).

Four clearly distinct types of liberation are recognized in the Ascomycetes by Ingold (1953) as follows:

'1. In the *Discomycete* type the spore-producing surface, consisting of asci intermixed with parallel paraphyses, is more or less exposed, most often as a lining to a shallow cup-shaped apothecium. The extensive exposed hymenium allows opportunities for "puffing" – the simultaneous bursting of numerous asci.

2. In the *Pyrenomycete* type the asci are contained in a small flask-shaped structure (perithecium) which opens to the outside by a minute ostiole. Before each ascus can discharge the spores, its tip must reach the ostiole, and the canal of the neck is usually so narrow that normally only one ascus can emerge at a time.

3.   In the *Erysiphales* type the fruit-body is a cleistocarp. This is rather like a perithecium but is completely closed; there is no ostiole. In this type the swelling asci must first burst the wall of the cleistocarp before they can emerge and discharge their spores.

4.   In the *Myriangium* type, though the hymenium is exposed in a structure like a small apothecium, the spherical asci are embedded in a plectenchymatous tissue and are free to discharge only when this gradually undergoes gelatinization.'

Some Ascomycetes which lack explosive asci may liberate spores in slime to be dispersed by rain-splash. Other species, again, may be either explosively or slime dispersed, according to the conditions obtaining.

(ix)   *Squirting mechanisms*, which propel spores violently into the air, occur among Phycomycetes in *Pilobolus*, *Basidiobolus*, and *Entomophthora muscae*, as well as in the imperfect genus *Nigrospora* (Webster, 1952).

(x)   *Rounding-off of turgid cells* acts as a discharge mechanism when the flattened double walls between two turgid cells suddenly separate. By this means spores of some Phycomycetes can be ejected up to a centimetre into the air. The same mechanism acts to eject aecidiospores when aecidia of rusts are moistened. Discharge of all these types is favoured by high humidities. By contrast, aecidiospores of the rust fungi are discharged under conditions unlike those normally favouring dispersal of uredospores.

(xi)   *Water-rupture*, similar to that well known in fern sporangia (p. 50), where a drying cell suddenly expands as a gas bubble appears, is a launching mechanism demonstrated for a number of Fungi Imperfecti, including *Alternaria* and *Helminthosporium turcicum* by Meredith (1963, 1965).

(xii)   *Basidiospore discharge*. This is a highly characteristic process which is found with the same essential features almost throughout the Basidiomycetes. The basidium is a cell producing one or more sterigmata, at the end of each of which one basidiospore is formed asymmetrically. Typically, when the spore is mature, a drop of liquid is excreted at the hilum end of the spore and almost immediately the spore moves off to a distance of a fraction of a millimetre.

In species which form their basidia over exposed surfaces, as in many of the lower Basidiomycetes, the spore after discharge has a chance of being picked up by an eddy.

The higher Basidiomycetes often show great elaboration of a stalked fruit-body with the basidia lining the vertical surfaces of folds, gills, pores or spines. Here, in cavities protected from wind and adverse conditions, the basidiospores are liberated into still air and fall under the influence of gravity into the moving air-current below the cap-shaped or bracket-shaped fruit-body. Spore discharge in the higher Basidiomycetes often goes on continuously throughout almost the entire life of the fruit-body – to all appearances little affected by wind or humidity. Experiments by Zoberi (1964, 1965), however, indicate that temperature and humidity play a part in controlling the rate of basidiospore discharge.

Just how the basidiospore leaves the sterigma remains a major puzzle of mycology; several explanations have been advanced, but none seems entirely satisfactory (*see* Ingold, 1971).

The same mechanism apparently occurs in the mirror-yeasts (Sporobolo-mycetaceae), which probably evolved from lower Basidiomycetes (unlike the Saccharomycetaceae, which are clearly Ascomycetes), but to avoid pre-judging the issue by calling the spores of the mirror-yeasts 'basidiospores', the term 'ballistospore' has been coined to include all spores showing the drop-excretion mechanism. (But 'ballistospore' is a biased word, as it prejudges the nature of the mechanism, and is inapt if the bubble-bursting hypothesis of Olive (1964) is confirmed: rocketry rather than ballistics would then be involved. On the other hand, projectiles from the ascus gun clearly merit the term!) A moist substratum is necessary for spore-discharge in the Sporobolomycetes.

LICHENS

The fungal components of lichens discharge ascospores from typical perithecia or apothecia, or basidiospores from basidia. Besides these, frag-ments of the thallus including both fungal and algal cells are blown about freely. Groups of algal cells surrounded by fungal hyphae, separating from the lichen thallus as 'soredia' are also blown away. Bailey (1966) used a small wind-tunnel to study 'deflation' of soredia and fragments from *Lecanora conizaeoides*, *Parmelia physodes*, *Pertusaria amara*, and *Cladonia impexa*. With winds of from 2·9–5·1 m/s, specimens which had been stored dry before use shed their soredia copiously as soon as blowing started, but the output died away as blowing continued. By contrast, wet lichen thalli shed few soredia, but reached a peak of liberation after a short period of drying.

Lichen soredia may possibly aid the dispersal of algae when they become grounded in a habitat moist enough for the algal component to dominate the fungal, and where the resulting colony will be an alga instead of a lichen. Again, some terrestrial and epiphytic algae may crumble and blow away.

ALGAE

Adaptations facilitating take-off into the air are not reported in the algae, though some of the simpler types of algal cells get into the air regularly

(Gregory *et al.*, 1955; Hamilton, 1957, 1959). Pettersson (1940) suggested that *Chlamydomonas nivalis* is carried away from its habitat on snow-fields and glaciers in melt water, and then becomes airborne by splash in mountain torrents. *Cephaleuros parasiticus*, a green alga causing 'red rust' on leaves of tea, cocoa, and many other tropical plants, has aerial filaments ending in sporangia which can become airborne in wind, and are also moved by rain-splash.

BRYOPHYTES

Spores of mosses and liverworts are formed in sporangia which are typically raised on stalks above the substratum, but the structure of the sporangium is quite different in the two groups. The moss sporangium is a firm 'box' opening at the top, whereas the liverwort sporangium breaks open completely, exposing the spores in a mass of stiff threads (the elaters).

In the simpler liverworts the spores may be blown away by the wind from the mass of elaters, or the elaters may twist hygroscopically, actively throwing the spores into the air. In most leafy liverworts, however, a spring mechanism, released by water-rupture in the drying elaters, throws the spores into the air (Ingold, 1939, 1965), while in *Frullania* the sporangium explodes by an efficient spiral spring mechanism which is also released on drying.

The mosses liberate spores from the stalked sporangium (capsule) by two principal methods. *Sphagnum* has an 'air-gun' mechanism (Ingold, 1965). An air space below the spore mass is compressed by transverse contraction of the drying sporangium wall, internal pressure increases and, finally, the top of the sporangium breaks, ejecting a spore cloud to a height of 15 or more centimetres.

Most of the other mosses have flask-shaped sporangia, which open gently at the top when mature. In some genera the mouth of the sporangium is surrounded by one or more rows of triangular teeth which move hygroscopically, closing the mouth at high humidities. To what extent spore liberation depends in nature on the shaking of the sporangium in the wind, and what role is played by hygroscopic movements of the teeth in actively throwing out spores, is still a matter of controversy; but evidently spore liberation is checked by high humidities and low wind speeds.

PTERIDOPHYTES

Spores of the Pteridophytes (ferns and their allies) are formed on the fronds within a closed sporangium, from which they are dispersed into the air by a 'sling' mechanism which depends on water-rupture under great tension as the maturing sporangial wall dries (*see* Ingold, 1939). Pettersson

(1940), in Finland, found that effective scattering of fern spores takes place outdoors only when the relative humidity of the air falls to 76% or even to 60%, according to the species.

POLLINATION OF PHANEROGAMS

Insects and wind are the chief agents for cross pollination of flowering plants. Other pollinating agents that are effective in a far smaller number of species include water currents and humming birds. There are probably ten times as many entomophilous (insect-pollinated) as anemophilous (wind-pollinated) species of flowering plants in the world as a whole. The characteristics of wind-borne pollen become clear when contrasted with insect-borne pollen (Table VII). There are many exceptions to the generalizations in this table and, in particular, some plants make the best of both methods. Both anemophilous and entomophilous plants often protect their pollen

TABLE VII

TYPICAL CHARACTERISTICS OF ANEMOPHILOUS AND ENTOMOPHILOUS PLANTS

	Wind-pollinated	*Insect-pollinated*
Flowers	Lack conspicuous and attractive petals, scent and nectar	Often with bright colours, scent; nectar attractive to insects.
Flower position	Projecting into air; hanging from bare branches before leaves open (catkins); on erect stalks (grasses, etc) or at ends of branches (conifers).	Tend to be exposed to view, but not exposing anthers to wind. Flowers usually maturing when plant in full growth and insects abundant.
Prevention of self-fertilization	Male and female organs often in separate flowers or inflorescences, or on separate plants. If flowers hermaphrodite, one sex commonly matures before the other, or if sexes are in separate inflorescences, the female is often above the male.	Flowers usually hermaphrodite, with structural or genetic barriers to selfing.
Pollen	Often shed into the air in vast quantities. Shape rounded, often nearly spherical or ellipsoidal. Size range narrower than for entomophilous pollen and seldom less than 15 μm. Surface typically smooth as seen under the microscope, non-sticky, easily separating into single grains in air.	Usually restricted pollen production with little shedding. Shape very variable. Size very variable, 3 to 250 μm, but often less than 15 μm. Surface typically rough, spiny or warted, often oily or sticky, tending to adhere in clumps.

from rain, and many store it within the flower for some time after shedding from the anthers. Anemophilous pollen is not generally shed into very calm or very damp air.

GYMNOSPERMS

Conifer pollen, instead of being formed in stalked anthers like that of Angiosperms, is produced in two or more pollen sacs on the lower side of the male cone-scales. The pollen grains are large and often bear two conspicuous air-filled bladders which decrease the density of the particle and so retard its fall under the influence of gravity.

In *Pinus*, cone scales of the erect male cone separate as they mature, and pollen shed from the paired sacs falls into small hollows on the upper surface of the cone scale below. From these hollows the pollen is blown away when the wind reaches sufficient velocity. Some other conifers have hygroscopic mechanisms protecting their pollen from rain and allowing its release only in dry weather. In *Taxus*, *Thuja*, *Cupressus* and *Juniperus*, the pollen is not winged. In *Juniperus* the expanded ends of the cone scales interlock closely in damp weather, separating again in dry air and allowing pollen to be blown out.

ANGIOSPERMS

Details of flowering-plant mechanisms are given by Marilaun (1895), Knuth (1906), Erdtman (1943, 1952, 1957), Wodehouse (1945) and others.

(i) *Grasses, rushes, sedges and their allies.* The Gramineae, Cyperaceae, Typhaceae, and Juncaceae are typically wind-pollinated. From the raised inflorescences of grasses, the anthers are extruded on long filaments to which they are so lightly attached that they vibrate in the slightest wind. Often, as in *Arrhenatherum*, the end of each pollen sac bends up (Figure 7), forming a spoon into which pollen is shed from a slit, and where it accumulates until blown away by the wind. Either very damp or very dry weather may delay both extrusion of the stamens and splitting of the anthers. Except for rye and maize, most cultivated cereals are self-fertilized and shed little pollen, but pasture grasses are free shedders.

In central Europe, Kerner von Marilaun (1895) found that different grasses flowered for brief periods of only 15 to 20 minutes daily, and at characteristic times of the day:

hours

04–05 *Poa, Koeleria, Avena elatior (Arrhenatherum)*
05–06 *Briza, Deschampsia caespitosa, Triticum, Hordeum*
06–07 *Secale, Dactylis, Andropogon, Brachypodium, (Bromus?), Festuca* spp., *Holcus* (1st anthesis)

hours

07–08	*Trisetum, Alopecurus, Phleum, Anthoxanthum*
08–09	Exotic types in Europe: *Panicum, Sorghum*
09–10	*Setaria italica, Gynerium (Cortaderia) argenteum*
11–12	*Agrostis* spp.
12–13	*Melica, Molinia, Nardus, Elymus, Sclerochloa,* some *Calamagrostis* spp.
14	A few *Bromus* spp.
15	A few *Avena* spp.
16	*Agropyrum*
17–18	*Deschampsia flexuosa*
19	*Holcus* (2nd anthesis)

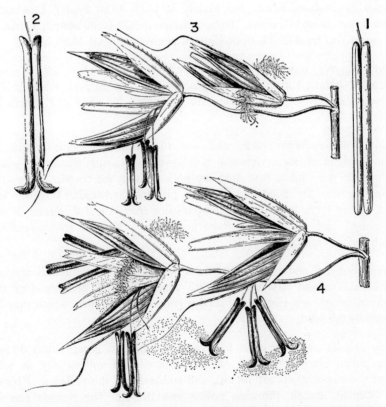

Figure 7. Anthesis of false oat-grass *(Arrhenatherum elatius)*: (1) closed anther; (2) open anther; (3) spikelets on a calm day; and (4) spikelets in a wind. (Reproduced from Marilaun's: *Natural History of Plants*, by permission of Messrs Blackie & Son Limited.)

This timetable does not necessarily apply elsewhere, and, in Nebraska, Jones & Newell (1946) found a less precise timing and showed that anthesis is determined by temperature. They distinguish cool-season grasses from warm-season grasses. Cool-season grasses include: *Festuca elatior* (anthesis at 13.30–15.00 hours); *Agropyrum* spp. (14.00–18.30 hours); *Bromus inermis* (14.30–19.00 hours); *Poa pratensis* (during the night); *Secale cereale* (02.30–11.30 hours, maximum at 06.00–08.30 hours). The warm-season group includes: *Bouteloua gracilis* (03.00–09.00 hours, maximum 04.30–05.30 hours during darkness); *Buchlöe dactyloides* (06.30–13.00 hours, maximum 07.00–08.30 hours); *Panicum virgatum* (10.00–12.00 hours, delayed in cool season); *Zea mays* (07.30–16.00 hours, maximum 08.30–11.00 hours).

Hyde & Williams (1945, p. 89), from the cooler climate of Wales, report both discrepancies and agreements with von Maurilaun's timetable: *Holcus lanatus* (04.00–06.00 hours, but mainly at 14.00–19.00 hours); *Cynosurus cristatus* (05.00–06.00 hours); *Arrhenatherum* (07.00–08.00 hours); *Trisetum flavescens* (before 08.00 hours); *Festuca pratensis* (08.00–14.00 hours).

(ii) *Aquatic monocotyledonous herbs* include a few other wind-pollinated plants, for example *Triglochin* and *Sparganium*, while in the genus *Potamogeton* some species are pollinated by wind and others by water.

(iii) *Entomophilous herbs and low shrubs* include some species in which a phase of insect visitation is followed by an opportunity for wind-pollination, the anthers first shedding pollen within the corolla; but as the flower matures, the elongating filaments protrude and scatter pollen in the wind. These types include the semi-parasites *Bartsia* and *Lathraea* (Rhinanthaceae), and the heaths *Calluna* and *Erica* – but not *Rhododendron*, which has very sticky pollen.

(iv) *Tropical and sub-tropical trees* include few anemophilous species, but *Casuarina* and *Myrothamnus* are wind-pollinated, and some of the palms, although entomophilous, shed a good deal of pollen, which may be carried by the wind.

(v) *Nettles and their allies* form an anemophilous group which do not store pollen after dehiscence of the anthers. The anthers dry as they mature, tissue tensions are set up, and suddenly, as the pollen sacs burst, the filaments uncoil, throwing pollen into the air. The process can be watched on a still, dry day when small puffs of pollen appear as the nettle flowers explode, but in damp air dehiscence of the anthers is inhibited. The mechanism occurs in *Urtica*, *Parietaria*, *Morus* and *Broussonetia*. A sifter mechanism similar to that of the grasses occurs in *Cannabis* and *Humulus*.

(vi) *Herbs with inflorescences elevated* above the general level of the foliage include a number of anemophilous types such as *Mercurialis*. In *Plantago* and *Globularia* the anthers, which are exposed in cups, close their slits in moist weather but shed their pollen in dry air. Upward-facing cups occur also in *Poterium* and *Sanguisorba*. Sifter mechanisms occur in some species of *Rumex* and *Thalictrum*. Other conspicuous pollen shedders occur in the Chenopodiaceae *(Beta, Salsola, Chenopodium)* and in the Amaranthaceae, and also in some groups within the Compositae – especially *Ambrosia* and *Artemisia*.

(vii) *Deciduous trees* of temperate regions form a biological group. Typically the male flowers are aggregated into pendulous catkins, usually appearing shortly before the leaves expand. In *Alnus, Betula, Castanea, Corylus, Fagus, Juglans, Populus (Salix* is both insect-pollinated and a wind-shedder), and *Quercus*, pollen is protected from rain after shedding while temporarily stored on the upper scales of the flower standing underneath – until it is blown away in a manner reminiscent of *Pinus*. *Platanus* closes its catkins by a hygroscopic mechanism, so that pollen is not merely protected from rain but can be blown away only in dry weather. *Hippophaë* pollen is shed into the base of the flower while it is still in bud. At maturity the perianth lobes remain united at the top but separate at the base, leaving slits through which pollen can be removed by the wind. In another group, including *Fraxinus, Buxus, Phyllyrea* and *Ulmus*, the anthers project as upward-facing cups from which pollen is removed by wind.

The result of interaction between conditions in the external environment and the liberation mechanisms is a great fluctuation in the abundance of spores or pollen of any species at different times and in different places (Chapter X). Further, internal rhythms in the parent organism may control maturation and liberation of fungus spores as discussed by Ingold (1971, pp. 214–38).

The take-off mechanisms briefly sketched in this chapter, with others (doubtless including some still undiscovered), are not mere curiosities of natural history. On the contrary, they are highly efficient processes that restrict spore liberation to limited meteorological conditions. Pollen and spores of mosses and ferns tend to be shed into dry winds. Ascomycetes and lower Basidiomycetes are more likely to discharge spores when the substratum is wet. Spore shedding in higher Basidiomycetes is less affected by air humidity and wind speed. Spores of some Fungi Imperfecti may depend on wind for removal, or on changes in humidity, or on rain for splash dispersal (Chapter V). Soil- and dust-borne bacteria and protista [protozoa] are probably borne aloft in high winds from heated or mechanically disturbed ground. The nature of the 'take-off' mechanism profoundly affects the occurrence of different kinds of spores or pollens in the atmosphere – with consequent significance for hay-fever patients, seed crops, plant diseases, evolution, and geographical distribution.

V

Liberation in Splash Droplets

Microbes are not normally removed from liquids or damp surfaces by wind (Nägeli, 1877, pp. 109, 111). But processes that break up the liquid may put microbes into the air in small droplets.

Rain-splash has been recognized for a long time as an agent in the dispersal of some of the saprophytic and parasitic slime-spored fungi. It is also effective in spreading some bacterial plant pathogens, but the way it functions is more obscure than the air dispersal of dry-spored fungus pathogens. Rain-splash functions outdoors.

Indoors the most important droplet-making processes are generated by man in sneezing, coughing and talking. Splashing liquids from laboratory and industrial processes may spread viral or bacterial pathogens (Rosebury, 1947; Darlow & Bale, 1959).

SPLASH DISPERSAL

Rain-splash may be doubly effective in dispersal. In still air an organism is spread locally around the point of impact in large droplets. In wind, droplets arising from rain-splash may evaporate to 'droplet-nuclei' and are then carried by air currents.

PIONEER EXPERIMENTS OF FAULWETTER AND STEPANOV

Faulwetter (1917a, b) studied the action of wind-blown rain in disseminating angular leaf spot of cotton, a bacterial disease caused by *Xanthomonas malvacearum*, and provided a plausible explanation of the way in which bacterial and fungus crop pathogens are spread, which had previously been obscure. Faulwetter inoculated a few plants in a cotton field, and from the direction and distance of subsequent spread of the disease, and from weather records, he concluded that the falling raindrops splash up water from a bacteria-laden surface film on a leaf. Wind then transports this ascending splashed water, possibly even in a chain action with bacteria carried away from the original lesion, deposited, and then splashed up again and carried further, and 'so on, until a dilution too great for further infection is obtained'. (Faulwetter, 1917a.)

Impressed, both by the definiteness of the results from his inoculations and by the absence of any other information on the splash of water droplets, Faulwetter (1917b) turned his attention to a study of the mechanics of splash in the laboratory. Single drops of water were dropped from burettes onto glass plates, or onto blotting paper which was either dry or wetted with eosin or dilute acetic acid. The scatter of droplets thrown up by the splash was detected by resulting stain on other glass plates or litmus blotters arranged around the point of impact.

The first tests were done in still air. Following impact of a single drop 0·02 cm³ (diameter = 3·4 mm) falling 4·9 metres on to a wet horizontal glass plate, 800 splash droplets were thrown off, the modal horizontal distance being between 10 and 13 cm (mean horizontal distance, 18 cm). The droplets differed in size, smaller droplets being more abundant than large ones near the point of impact. The maximum distance travelled in the plane of the source was 144 cm, or 185 cm if caught at a level 60 cm below the point of impact. Distance travelled by the splash droplets was affected by the size of drop, depth of film, elevation and inclination of the surface of impact, and velocity of wind.

When a drop splashed on a surface inclined at 45°, Faulwetter reported that droplets travelling along the median line 'ascend little, if at all, but travel outward and downward. Those going to the sides ascend at first and then fall, similarly to the splash from a level surface'.

When splash occurred in the moving air stream of a 46 cm electric fan (giving a maximum speed of 4·5 metres per second at a distance of 1·5 metres) the smaller droplets were carried further than the larger ones – reversing the situation found for splash in still air. Droplets were splashed as far as 5·5 metres horizontally. Faulwetter pointed out many factors which would operate under field conditions to modify the results of such laboratory tests, including collision between splash droplets and raindrops, the range of drop size in natural rain, the texture of leaves and the variable thickness of film held on the uneven leaf surface, the effects of the leaf yielding on its petiole after impact, movement of leaf in the wind, and the greater terminal velocities attained by larger drops falling freely out of doors.

Faulwetter was convinced that splash droplets are composed of water from the surface film only, not from the falling drop, and he assumed that bacteria in the film would be carried in the splash droplets (recent work shows that he was wrong in the first conclusion, but right in the second). He considered that his data would account quantitatively for the abundant infection by splash droplets across two to three rows of cotton plants observed in field experiments, and this conclusion has been thoroughly substantiated.

Further, he pointed out that local dissemination of the same type may also be expected, even in the absence of rain, when foliage is wetted to run-off as in dew or fog. Drops falling from leaves can be larger than rain-drops, because usually they do not fall far enough to reach their terminal velocity and break up. If falling from only 30 cm they are able to scatter splash droplets over an area 50 to 80 cm in diameter, even in the absence of wind.

Stepanov (1935) knew that certain fungi resist removal from their sub-stratum by dry winds, and he was also aware of Faulwetter's work on splash dispersal of *Xanthomonas malvacearum*. He therefore bombarded well-developed cultures of fungi on sterilized ears of wheat or on potato tubers (as appropriate) with falling drops of water to simulate the effect of rain. The rebounding splash droplets were caught on clean glass slides at 20 to 25 cm distance. Droplets from bombardment of *Fusarium moniliforme* contained an uncountable number of microconidia; *Fusarium culmorum* var. *lateius* with well-developed sporodochia gave many conidia, but *Phytophthora infestans* gave considerably fewer. Stepanov concluded that spores of *Colletotrichum lini*, conidia from the pionnotes and sporodochia of *Fusarium*, and many other sticky spores, could be detached from their substratum by falling raindrops and would be distributed over a wide area by strong winds.

Although most workers agree that the slime-spored fungi are seldom moved by dry wind but are easily removed by rain, there remain many records where typical slime-spored fungi have been removed by dry air (Burrill, 1907; Burrill & Barrett, 1909; Brandes, 1919). Possibly *Fusarium* behaves differently in different states; this economically important genus clearly warrants an extended study to explain this discrepancy.

THE MECHANISM OF RAIN-SPLASH

PROPERTIES OF RAIN

Raindrops differ in size, and at any given time a rain shower has a characteristic spectrum of drop-size distribution (Mason, 1957). Drops may evaporate while falling through drier air below the base of a cloud (Best, 1950, 1952); those smaller than 0·2 mm diameter seldom reach ground level, and Findeison (1939) regarded 0·1 mm as the lower limit for pre-cipitation to occur. In general raindrops reaching the ground vary between 0·2 and 5·0 mm in diameter (Hudson, 1963; Caton, 1966), small drops being much more numerous than large ones. Rain from low or shallow stratified

D

clouds which give rise to orographic or frontal rain, are often characterized by a wide size-range, with many small drops and few exceeding 1 mm in diameter. By contrast showers from convective clouds, as from cumulus in thundery weather, may have many drops larger than 2 mm in diameter, smaller drops not falling fast enough to get down through the upward convection current (Best, 1950). Cloud droplets are typically in the range 10 to 100 μm, and for practical purposes may be regarded as in suspension.

The velocity with which drops fall through the air under gravity varies greatly with size (Gunn & Kinzer, 1949; Laws, 1941). Terminal velocities of raindrops have been determined experimentally because, as they are not rigid spheres, they cannot be simply calculated from Stokes's law. During fall a drop becomes flattened on the advancing face, and if over one millimetre in diameter it may oscillate. Blanchard (1950) found that drops up to 8 mm in diameter may remain intact when falling through non-turbulent air, but drops more than 5 mm in diameter are very sensitive to microturbulence and soon break up; thus an upper limit is set to the size of natural raindrops.

EVENTS DURING A SPLASH

A raindrop splashes in the twinkling of an eye, and our knowledge of it comes from inference and high-speed photography. In a careful and well-documented study Worthington & Cole (1897) let water drops fall into a *deep* layer of liquid, and secured a remarkable series of photographs by accurately timed spark discharge (Plate 3). Edgerton & Killian (1939) published beautiful photographs taken by electronic flash but without quantitative data, showing the three principal cases of splash from a liquid drop: falling (1) on a dry plate; (2) on a thin film; and (3) into a deeper liquid layer.

These photographic studies show that the events of a splash are a sequence of considerable regularity, under the combined action of the momentum of the falling drop, friction between the drop and the liquid or solid surface on which it impinges, and surface tension. In the following description a distinction is maintained between the falling 'incident drop' and the resulting 'splash droplets'.

(1)   The simplest case of a drop falling a short distance on to a dry surface is illustrated by Edgerton & Killian (1939), by Kientzler *et al.* (1954) and by Levin & Hobbs (1971). When the drop touches the plate the liquid spreads out as a thin disc. As the liquid spreads farther from the point of impact the advancing edge of the disc rises as a crater-like wall which continues to increase both in circumference and depth, and its rim thickens by surface tension into a ring. With further widening of the crater and its annular ring, Plateau's surface tension effect causes the rim to thicken into

beads, which then break up into rays, and these in turn break up into droplets (some with intermediate secondary droplets) which are then shot into the air as more liquid flows up the rays from below.

(2) When a drop falls from a small height into deep liquid, it also throws up a crater which grows wider, higher and thicker, and jets of liquid are shot out from its rim. The interior of the crater is lined with the original liquid which formed the incident drop. The crater then begins to subside and a rebound takes place which raises a central pillar of liquid (the top of which may split off as a separate drop) with some of the liquid of the incident drop on the top of the pillar. As the central pillar subsides it becomes a 'cake' whose edge becomes the first well-marked ripple (Worthington, 1908).

(3) When a drop falls from a greater height (say 100 cm) the drop punches a sheer hole, the liquid of the drop streams up the cliff-like walls of the hole, and a protruding crater with radial jets is formed as before. But now a new phenomenon arises as the mouth of the crater begins to close over the cavity, forming a bubble, because the annular rim of the cup contracts more quickly than the rest of the wall and so tends to close over the mouth of the crater. The bubble may re-open and make way for the rebound column, or it may remain closed until punched open by the head of the column. The bursting of such bubbles in sea water has been studied by Mason (1957).

Cases 2 and 3 are particularly relevant to dispersal of bacteria and aquatic micro-organisms (Maguire, 1963).

(4) When a drop falls on a surface covered with a thin liquid film an intermediate situation arises, highly relevant to bacterial and fungal spore dispersal. There is no rebound column (Raleigh jet) or bubble formation, and it is evident that the surface film is pushed outwards by the descending drop; the resulting radially moving mass of liquid is then deflected upwards where surface tension moulds it into a crater which breaks up into jets and rays of droplets as before (Figure 8). The cup is fully formed and jets are beginning to emerge at 0·0007 second after the incident drop touches down, and the jets are at their maximum development at 0·0035 second.

EXPERIMENTAL STUDY OF SPLASH

SPLASH FROM A THIN HORIZONTAL FILM

Dispersion of spores of *Fusarium solani* by splash from a thin horizontal film was studied by Gregory, Guthrie & Bunce (1959). Splash droplets were trapped on horizontal microscope slides coated with a gelatine film, dyed with naphthol green B, as described by Liddell & Wootten (1957). On touching this film a small droplet spreads out leaving a clear spot about

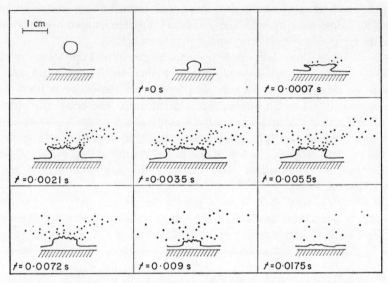

Figure 8. Splash from impact of water drop (5 mm diameter) falling with velocity of 440 cm/s on a thin film of water (drawn from stills from ultra-high-speed film made for the author by Mr E. D. Eyles at Kodak Research Laboratories, Harrow).

three times its original diameter. Microscopic examination shows the boundary of the spot clearly defined by its sharp edge, and from the diameter of the spot the diameter of the droplet can be read from a calibration curve based on spots from droplets of known diameter. The naphthol green method proved tricky to use and improved methods make use of photographic emulsions with their carefully standardized gelatine films (Koenig & Spyers-Duran, 1961). Simpler still is measurement of drop impression by phase contrast microscopy (May, 1961).)

In these tests on *Fusarium solani*, which were all done in the still air of the laboratory, water drops were allowed to fall from known heights onto films of a suspension of macroconidia spread on a horizontal glass surface. The total number of droplets thrown up from a single splash crater was affected by several factors (Table VIII). Both the total number of droplets produced and the number of droplets carrying spores increased as the diameter and velocity of the incident drop increased, and as the thickness of the film decreased, over the range tested.

The number of droplets recorded in one splash varied from about 100 to over 5000. The total volume of droplets ejected can be roughly estimated from the data, and appears to vary from one quarter to over twice the volume of the incident drop.

TABLE VIII

ESTIMATED TOTAL NUMBERS OF DROPLETS, AND NUMBERS CARRYING
MACROCONIDIA OF *Fusarium solani*, RESULTING FROM A SINGLE SPLASH
(Gregory, Guthrie & Bunce, 1959)

Height of fall (m)	Diameter of incident drop (mm)	Target film thickness					
		0·1 mm		0·5 mm		1·0 mm	
		Total droplets	Droplets with spores	Total droplets	Droplets with spores	Total droplets	Droplets with spores
2·9	2	309	82	172	26	175	24
2·9	3	828	348	563	109	487	82
2·9	4	1470	567	993	211	817	127
2·9	5	4110	1560	2470	829	1760	522
7·4	3	1180	559	1020	283	1030	470
7·4	4	4830	2010	3020	603	2040	391
7·4	5	5260	2133	3580	1690	2160	518

Droplets resulting from a single splash also varied over a wide range of
diameters, and the limits of the range appeared to be from 5 μm to 2400
μm. The smallest droplets were the most numerous, and the data on droplet diameter is most conveniently summarized by plotting the cumulative
percentage of the number of droplets up to the limit of each size group on
logarithmic–probability paper. The number median (50% size) can then be
estimated from a freehand curve through the points, and shows that over
most of the range of conditions tested, 50% of the droplets had diameters
less than some value between 60 and 80 μm. Few of the smallest droplets
carried spores, and the median diameter for spore-carrying droplets varied
between 140 μm and 510 μm in different experiments. This proportion,
however would depend on the concentration of spores in the target film.

Part of the kinetic energy of the incident drop is used in breaking up the
liquid of drop and film into small droplets. Part is used in projecting these
droplets into the air. Similarly the kinetic energy of falling rain may transport soil particles, and the role of rain in soil erosion was studied by Laws
(1939) and Ellison (1944).

In the laboratory, splash in still air projected droplets to horizontal
distances of up to 100 to 125 cm, but 50% of the droplets travelled only
8 to 21 cm in still air. The smallest droplets, with a relatively large drag,
were soon slowed by air resistance and fell close to the target, whereas the
larger droplets (including many carrying spores) travelled farther.

When the numbers of droplets falling on annuli one centimetre wide
around the point of impact with target film are compared it is clear that

the total number of droplets deposited declined steadily with increasing distance from the centre. But the numbers of droplets carrying *Fusarium solani* and *Gloeosporium album* conidia had maxima at about 10 to 30 cm from the centre of impact.

The vertical range of droplets ejected in a splash was about half the horizontal range – few reached a height greater than 40 cm. The greatest height was reached by droplets of intermediate diameters between 164 and 655 μm, and droplets larger or smaller than this had a more limited vertical range.

Laboratory tests under these simplified conditions point to the extreme potency of splash as a spore dispersal mechanism. Outdoors the simple picture must be complicated by other factors, and extensive research is needed. However, the conditions obtaining in the experiments described above may be quite relevant for describing spread in a closed space among vegetation, in close-spaced crops, in contaminating a seed crop with seed-borne pathogens, and in spread of infection throughout a bush or tree. The horizontal range of droplets is known to be much greater when splash occurs in wind (Weston & Taylor, 1948). Further, small droplets evaporate rapidly; for instance water droplets 90 μm in diameter have a life of only one second in air at 50% relative humidity (Henderson, 1952). Droplets borne aloft by wind must usually evaporate rapidly, and from that moment any spore or particle picked up in the process of splashing would travel just like a dry-airborne spore. However, any surface-active substances carried in the droplet might significantly retard evaporation, and hasten its fall.

Besides wind, two other factors must operate to modify the simplified picture of splash obtained in the laboratory. Depending on the intensity of the rain, falling drops will in turn collect droplets, and prevent their spore load (if any) from travelling farther on the wind. By contrast, as already suggested by Faulwetter, a spore deposited in an exposed place after travel in a splash droplet is still available for re-splash, though spores deposited in fissures or on the rough ground are probably not easily re-suspended. Field experimentation during natural rainstorms is needed to work out the importance of these factors. In wind the smaller droplets may well be blown through the trajectories of the larger droplets, and after evaporation become truly airborne.

SPLASH ON NATURAL TARGETS

In nature splash often occurs on a rough, sloping surface, or with other complexities not included in the tests described above. With drops falling on a twig, inclined at 45° and covered with conidial cushions of the coral spot fungus *(Nectria cinnabarina)*, droplets were detected at up to 50 cm distance in all directions. Each splash produced about 2600 droplets, *all* carrying the oval to oblong conidia, 2 to 7 μm of the *Nectria*. (Gregory, Guthrie & Bunce, 1959.)

Splash from diseased plant tissue was also studied by Weston & Taylor (1948) who showed that a single drop falling 4 metres on a leaf infected with *Botrytis* could contaminate an area of over 2·5 square metres. In one splash droplet 156 conidia were counted. In a shower of rain at least 32 square metres were contaminated in 45 minutes.

THE SPLASH-CUP MECHANISM

The splash-cup is a device, studied particularly by Brodie (1951, 1957), which is widespread among lower as well as higher plants, by which the energy of falling raindrops throws relatively large bodies to distances of a metre or two. Examples are the peridioles of the bird's nest fungi (Nidulariaceae), the gemmae of *Polyporus conchatus*, and droplets bearing spermatozoids of the Bryophytes. However, the particles scarcely become airborne as they follow a definite trajectory.

HOW SPLASH PICKS UP MICROBES

The mechanism by which splash droplets pick up microbes from thin films seems to be as follows.

On impaction with the target surface the incident drop flattens and sweeps the surface liquid and particles outwards. We commonly see a clear, spore-free circle, about 1 cm wide at the point of impact on a spore film after a splash. Surrounding this clear area there are often ten or more small radial elliptical zones, also free of spores, whose origin is unexplained. Worthington & Cole's photographs (e.g. Plate 3) indicate that liquid from the incident drop lines the inside of the crater, and it seems likely that in its upward and outward movement this inner wall entrains some of the surface film, with its suspended particles, to form the outer wall of the crater. The subsequent break-up of this double wall under surface tension would account for the intimate mixture of dyes found in droplets resulting from a red ink drop splashing on a green ink film; and also for the presence of spores in splash droplets, no matter whether they come from the incident drop or from the surface film Gregory *et al.*, 1959.

Whether or not a splash droplet contains spores presumably depends on the concentration of the target film, and on the proportion of droplet volume contributed by liquid from the incident drop and from the target film respectively. It is possible that the shape of some elongated or curved spores may be related to their pick-up by droplets, e.g. the so-called 'A' and 'B' conidia of *Phomopsis*, and the sickle-shaped conidia of *Fusarium*. Further study is needed to show whether shape and other properties of the particle affect its chance of being carried in a droplet of approximately similar dimensions.

DRIP-SPLASH

Raindrops falling at terminal velocity are not the only drops capable of dispersing slime spores. Faulwetter (1917a) realized that drops from leaves of plants which fall only short distances can be larger than raindrops because they do not reach the limiting terminal velocity at which drops break up if they are more than 5 mm in diameter. With large drops falling only 31 cm on a thin film of water, he scattered droplets over an area 50 to 75 cm in diameter: in Faulwetter's view these droplets could contain spores or bacteria from the surface film, but not organisms picked up from the leaf or stem by the water before dripping. But had he tested falling drops of spore suspension he would have observed that the resulting splash droplets also carry spores (e.g. from mycelia growing on the leaf) as shown by Gregory, Guthrie & Bunce (1959) for *Fusarium solani*.

Splash by drip from vegetation probably resembles the simplified laboratory experiments more closely than rain-splash. Because of the slower wind speed encountered among vegetation the drop tends to fall vertically instead of slanting as in wind-blown rain. Further, drops falling from vegetation tend to be large and uniform in size, whereas in rain the fall of one incident drop large enough to cause a splash is normally accompanied by many small scrubbing drops which tend to remove splash droplets from the air.

Trees drip when wetted to run-off by intercepted rain, drizzle or fog droplets, or by accumulating dew (Kittredge, 1948). A closed canopy may intercept perhaps 90% of the total precipitation. A dry canopy will retain all the precipitation when rain starts to fall, but when the foliage becomes saturated, surplus water will drip to the ground from leaves and twigs. A proportion of the intercepted rain, varying from 1 to 16% will reach the ground by 'stem flow' instead of drip, but the effect of closed canopy is to convert most of the rain from a very mixed size spectrum (all falling at terminal velocity and with small, inefficiently-splashing drops preponderating) to a shower of larger, splash-active drops falling at less than terminal velocity.

Fog-drip from leaves and twigs in the absence of rain is increased with height of the shrub or tree, and with the foliage area. In some coastal and mountain forests the amount of fog-drip received at ground level may be two or three times the total amount of precipitation in the open – a further example of the canopy converting fog into splash-active incident drops. The large drops that preponderate in rain-drip and fog-drip may have influenced the evolution of slime-spored pathogens common on trees and shrubs. It may also account for the firm texture of the upper part of the peridium of *Lycoperdon pyriforme* of woodlands in contrast with the softer *L. perlatum* which is more abundant in fields.

PICK-UP BY MIST

A further possibility to be considered outdoors is the ability of mist droplets carried by wind across a surface, to sweep up conidia from elevated conidiophores. Glynne (1953) could not detach conidia of *Cercosporella herpotrichoides* from the substratum by puffs of dry air, but, when she puffed fine spray from an atomizer, many conidia were carried away in the droplets. Similarly, R. R. Davies (1959) could not detach conidia of *Verticillium albo-atrum* by currents of dry air, but found that they were removed easily from the upright conidiophores by fine mist-like droplets from an atomizer (*and see* Carter *et al.*, 1970b).

Whereas slime spores produced on flat surfaces can be dispersed by splash, those on erect heads (the 'stalked spore-drop' of Ingold, 1961) may be insect dispersed or swept off by mist droplets. The essential distinction between splash and mist dispersal is that splash results from impact and disruption of a large drop that has fallen through air under gravity, whereas mist consists of droplets borne along more or less in suspension in moving air.

SIGNIFICANCE OF DROPLET SPREAD OUTDOORS

Splash droplets seem the most effective agents for launching microbes formed in sticky layers, for instance many fungi, including some Fungi Imperfecti, and also some Ascomycetes whose ascospores may be discharged dry into air in some conditions and exuded in slime under other conditions. Splash has the interesting property of effecting vertical dissemination of pathogens (in the trajectories of large droplets up and down a bush or tree) probably much better than the conical diffusion of small wind-blown particles. Splash is also effective with some plant pathogenic bacteria, bacteria rafted on soil particles, and aquatic micro-organisms.

Rain-splash must be the commonest droplet-producing event out of doors. With an average rainfall of 100 cm per annum, each square metre of ground would receive annually, something of the order of a thousand million raindrops large enough to produce a splash.

Splash from cascading rivers and rivulets probably makes a significant contribution to the air spora where the falling water has a high microbial content. Splash of snow melt-water is probably essential for the natural circulation of *Haematococcus (Chlamydomonas) nivalis* which produces the widespread phenomenon of red snow in polar and alpine regions (Pettersson, 1940).

EXAMPLES OF RAIN-SPLASHED ORGANISMS

Organism	*Author(s) cited*
Pseudomonas phaseolicola	Walker & Patel, 1964
Xanthomonas malvacearum	Faulwetter, 1917a; Brown, 1942, etc.
Streptomycetes	Hesseltine, 1960
Lichens (soredia)	Bailey, 1966
Phytophthora infestans	Stepanov, 1935
Phytophthora palmivora	Dade, 1927 (drip); Thorold, 1955 (splash)
Chaetomium globosum	Dixon, 1961 (whole perithecia)
Eutypa armeniacae	Carter, 1965 (secondary dispersal of ascospores)
Guignardia laricina	Yokota, 1963
Mycosphaerella musicola (cercospora state)	Calpouzos, 1955; Meredith, 1962a
Giberella zeae (conidial)	Nisikado, Inouye & Okamoto, 1955
Nectria galligena (conidial)	Lortie & Kuntz, 1963
Botrytis cinerea	Weston & Taylor, 1948; Jarvis, 1962
Cercosporella herpotrichoides	Glynne, 1953
Colletotrichum spp.	Stepanov, 1935; Mathur, 1951
Fusarium spp.	Stepanov, 1935; Bywater, 1959; Siddiqi, 1961
Gloeosporium laeticolor	Katajima, 1951
Pithomyces chartarum	Crawley, Campbell & Drew Smith, 1962
Phyllosticta spp.	Rolfs, 1917; Bilgrami, 1963
Rhabdospora (Septocyta) ramealis	Koellreuter, 1951; Oort, 1952
Aleurodiscus	Brodie, 1957
Hemileia vastatrix	Nutman, Roberts & Bock, 1960
Nidulariaceae (peridioles)	Brodie, 1957

Industrial processes generating droplets have been associated with outbreaks of virus diseases. Factory methods for processing and rendering carcasses have been implicated in producing aerosols spreading infection by psittacosis and Q-fever pathogens in man (Langmuir, 1961). The practice of spraying farm effluent into the air over grassland appears to be a highly efficient way of polluting the atmosphere with microbes and viruses (cf. Chapter XVII; and Gregory, 1971).

SIGNIFICANCE OF DROPLET SPREAD INDOORS

Take-off in droplets is of special interest indoors for the part it plays in spreading certain diseases in man. Wedum & Kruse (1969) emphasize the dangers inherent in many common laboratory procedures such as pipetting, centrifuging and opening screw-cap bottles – all of which may contaminate indoor air with microbe-laden droplets. Darlow & Bale (1959) found that splash from flushing of water closets can produce aerosols in the size range small enough to be inhaled into the lower respiratory tract. This constitutes a potential mode of transmission of viruses and bacteria present in excreta.

SNEEZING AS A DROPLET LAUNCHING MECHANISM

Sneezing, coughing and talking are now known to be processes generating aerosols. Early work, based on inefficient sedimentation sampling methods, grossly underestimated the numbers of small droplets produced by a sneeze. Improved methods, especially the Wells air-centrifuge and high-speed photography, revived interest in the sneeze as a means of producing potentially infective aerosols.

A sneeze is characterized by a sudden inspiration of air, followed by a vigorous expiration, accompanied usually by nearly closing the mouth and teeth. The more nearly the mouth and teeth are closed, the greater the velocity of the escaping air through the remaining orifices, and the more effective the atomization of the fluids in the mouth (Jennison, 1942).

As noted earlier, droplets are not removed by winds of normal speed flowing over a liquid surface, but at high speeds this no longer holds. The minimum droplet size produced from water atomized at 100 metres per second is 10 μm, and sneezing and coughing may produce such velocities. Mucus and saliva are more viscous than water and will produce larger droplets under the same conditions. High-speed photographs show that a cloud of droplets is ejected at velocities up to 40 metres per second to a distance of 50 to 100 centimetres from the mouth. A few large droplets fall to the ground, but most of them normally evaporate in about a quarter of a second and remain suspended as 'droplet nuclei'. A 500 μm diameter water droplet takes about half a minute to evaporate, but one of 100 μm takes only about one second. A single sneeze may produce a million droplets. The smallest observed were about 10 μm in diameter, and perhaps 80% were less than 100 μm and therefore destined to become droplet-nuclei in the 2 to 10 μm range – little larger in fact than any contained microbial particle. Most of the spray originates from the front of the mouth, and comparatively few droplets come from the nose (Jennison, 1942; Duguid, 1946; Wells, 1955).

The number of droplets produced by a cough is two orders of magnitude fewer than a sneeze. The mouth is often opened when coughing, and more pharyngeal secretion is atomized than in sneezing.

Talking produces a greater proportion of large droplets, most of which are ejected while enunciating the consonants, especially, f, p, t and s.

Lidwell (1967) shows numbers of droplets and resulting droplet-nuclei of different sizes from a single sneeze (10^6 droplets), two coughs or speaking loudly 4000 words (10^4 droplets). He also shows how the proportion of droplets carrying one or more infective organisms varies according to the concentration of the microbial suspension in the mouth from which the droplets are emitted. With more dilute suspensions few of the smaller droplets carry microbes, but as the concentration is increased, the greater the chance of small droplets carrying inoculum.

We can speculate that the ability to make the patient sneeze or cough has been evolved by certain viruses and bacteria into just as effective a launching mechanism as some of the complicated structural take-off devices described in Chapter IV. As with dry-liberation, the processes of drop formation will control the occurrence of another set of organisms in the air, again with profound effects on the health of man, animals and crops.

VI

Horizontal Diffusion

We have now described the particles composing the air spora and the relevant properties of the atmosphere. What happens to the particles after they have been launched into the atmosphere? Common sense tells us that they become dispersed – in the sense that their concentration per unit volume of air decreases with increasing distance from the point of liberation.

Tyndall (1881) believed that airborne microbes float through the atmosphere in miniature clouds. He explained Pasteur's demonstration of non-continuity in the spontaneous generation controversy by postulating that Pasteur had sometimes opened his flask in the midst of a bacterial cloud and obtained life, and sometimes in the interspace between two clouds and obtained no life. In hospital practice, opening a wound during the passage of a bacterial cloud would have an effect very different from opening it in an interspace between clouds.

It was not necessary to draw this conclusion, however, as Pasteur's results could be explained equally well if microbes were randomly distributed in the air. Evidence for random distribution was obtained by Horne (1935), who applied Fisher's χ^2 test to catches on 1000 or more petri dishes of sterile media which had been exposed in Kentish orchards by N. W. Nitimargi. The observed frequencies of total bacteria or total moulds, or of any genera or species tested separately, did not depart significantly from the Poisson distribution. Horne concluded that micro-organisms are distributed at random in the air, and that, for making valid comparisons between populations of airborne microbes at different places and times, analysis of variance could legitimately be applied to plate counts.

We no longer think of airborne spores as travelling in invisible flotillas, keeping station, mission-bent. But in another sense, and another scale, spore clouds can clearly be recognized, both near their source (Chapter X) and even at considerable altitudes over the sea (p. 17). It is even possible that within a cloud the individual particles may be kept apart by repulsion of like surface charges.

71

DISPERSION OF THE SPORE CLOUD

It is still convenient to speak of clouds of spores; not, indeed, keeping together in the manner of locust swarms, but tending to become dispersed while passively suspended in the atmosphere. Sampling a region small enough in relation to the size of the cloud may then reveal a random distribution of particles.

Dispersion of the spore cloud can be deduced from early observations on the distribution of rust on rye by Windt (1806), who observed that rust was severe near barberry bushes which are now known to be the alternate hosts for the fungus: 'the effects are striking and desolating at a distance of ten to twelve paces; I have also perceived them visibly at 50, 100, 150 paces and a final attack at above 1000 paces'. Similarly, dispersion of the pollen cloud made it possible for Blackley (1873) to advise his hay-fever patients to keep away from grass fields during the flowering season of the grasses. Attempts to formulate the process of spore dispersion through the atmosphere have been based on geometrical, empirical, or meteorological considerations.

The geometrical approach is suggested by analogy with the laws of radiation. Nägeli (1877) stated that the amount of dust which comes on an air current from one place falls off with the inverse square of the distance, whereas E. Fischer & Gäumann (1929) stated that, with linear increase of the distance, the chance of infection by rust spores decreases in cubic progression. Kursanov (1933) stated that, in the absence of wind, the number of fungus spores would fall off inversely as the cube of the distance from the source. The ideas that underlie the geometrical approach are simple. Spores travel away from the point of liberation: at greater distances the volume of air which they can occupy increases as the *cube* of the distance, or, alternatively, the surface of the ground on which they could fall increases as the square of the distance. A third possibility would be a simple inverse relationship with distance, as the areas of successive equidistant annuli around a point increase in arithmetical progression.

The geometrical method is unsatisfactory because, although in a general way it illustrates the features of dispersion, it is not clear why spores should travel in the manner predicted. The particles interesting to us here are passively borne and do not behave like radiations, because the air which carries them is not in the process of being continuously generated at some point in the atmosphere; consequently some totally different concept is needed.

The approach by empirical curve-fitting has been based on field records of dispersal gradients, such as the scatter of seeds or seedlings on the ground, contamination of seed crops by foreign pollen, or the incidence of plant diseases. Using such data, a curve is fitted to the observed points, either graphically or by the statistical method of least squares, and an attempt is made to find an empirical formula to fit the curve. These methods will be referred to in more detail in Chapter XVI, after the subject of spore deposition has been discussed. In general the empirical method has the advantage that an equation can usually be obtained, containing at most three parameters, which gives a good fit to any one set of field data. On the other hand it is difficult to compare results obtained by different workers. Their parameters are calculated from the data and correspond to no obvious natural phenomena; consequently it is difficult to use empirical formulae to predict a dispersal pattern under conditions differing from the original one.

In the long run a more ambitious approach seems essential, with the aim of developing formulae whose parameters correspond to factors of the environment, and which takes into account the total number of microbes liberated (if known), allow for variations in weather, and use a standard unit of distance.

DIFFUSION AS A RESULT OF ATMOSPHERIC TURBULENCE

Watching the drift of smoke from a bonfire or factory will convince the observer that wind, instead of having a steady streaming motion, is characteristically turbulent as described in Chapter III. According to Brunt (1934), large numbers of small-scale eddies, whose periods are of the order of 1 second, are usually present in the turbulent boundary layer, and at least two-thirds of the eddying energy is associated with eddies of less than 5 seconds. The action of these very numerous eddies of varying size on the very numerous spores produced from plant sources makes some regularity in the dispersal pattern possible.

The study of eddy diffusion has proved difficult, but it provides the most promising approach to the elucidation of dispersal. Before describing the methods in detail, a few general notions – familiar to physicists, but mostly unfamiliar to biologists – must be introduced.

We are attempting to discover laws governing spore diffusion in the atmosphere. In nature this is often a complex process, as there are obstacles preventing the free flow of air. We therefore use a device familiar to physicists: making a simplified model in the hope that, if we can understand the process of diffusion under simple conditions, we shall be able to attack the more complex situations found in nature. The assumptions needed for the simplified model are as follows.

(i) *The field*. Diffusion is assumed to be taking place in three dimen-sions in the atmosphere over a plane surface which is of indefinite extent, free from topographical irregularities – not necessarily 'smooth', but, if aerodynamically rough, then uniformly so.

(ii) *Co-ordinates*. To describe movement over the plane surface we need a system of co-ordinates. Their origin, *'O'*, is conveniently taken to be the point of liberation of the pollen or spores. The *'x'* axis is horizontal and positive in the downwind direction, and the *'y'* axis is also horizontal but at right angles to the direction of the wind. Lengths above and below the origin are measured on the vertical *'z'* axis.

(iii) *Sources*. Particles are liberated into the air from a source. The simplest form of source is a 'point source', and this may either liberate a number *'Q'* spores or other particles at a single instant (an 'instantaneous point source'), or it may be a 'continuous point source' emitting Q spores per second.

Instead of a point source we may have a 'line source'. For simplicity we assume that the line is horizontal and is emitting Q spores per centimetre of its (effectively) infinite length. The line source in turn may be either instantaneous or continuous. Furthermore, we may have, a 'strip source', or an 'area source' emitting Q spores per square centimetre, or a 'volume source'. Real sources in the field that correspond approximately to these ideal sources would be a single plant (point), a hedge (line or strip), a ground crop (area), and an orchard or forest stand (volume). The dimen-sions of the source must be treated as relative to their distance: thus a field would be regarded as effectively a point source when considered from distances many times its own width.

All these sources may be instantaneous or continuous. The cloud from an instantaneous source is a puff or spherical cloud, whereas the conical cloud arising from a continuous point source is familiar in the smoke plume from a chimney. A continuous point source can be viewed as made up of a succession of overlapping instantaneous emissions.

(iv) *Standard deviation*. Suppose that a 'puff' of spores has been liberated at an instant from a point source into a wind, and has become subject to the action of atmospheric eddies which move individual spores apart at random. After a short time the particles composing the cloud will show a *scatter* around their origin (Figure 9). At any instant such a cloud has two characteristics which we could compute if we had all the data: (1) the mean position of the particles, i.e. the centre of the cloud, which can be expressed as a point on the system of x, y and z ordinates; and (2) the standard deviation, σ, of the particles from their mean position

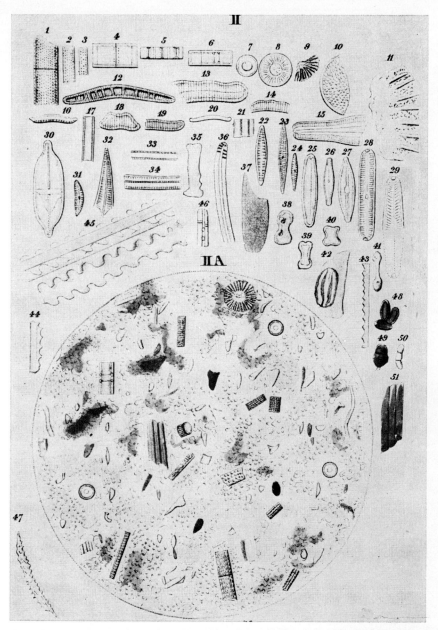

Plate 1. Ehrenberg's illustration of sample of dust collected by Charles Darwin on the *Beagle* near the Cape Verde Islands, January 1833.

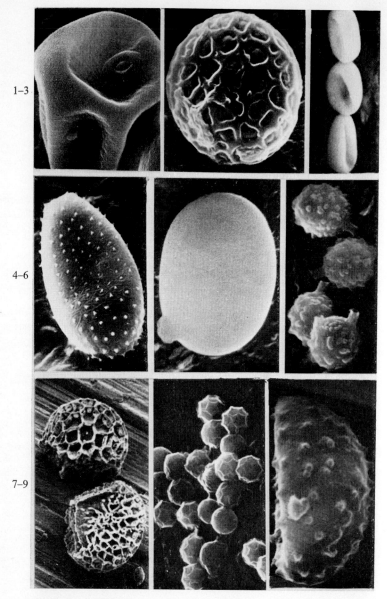

Plate 2. Surface texture of spores (Scanning electron micrographs by R. H. Turner, Rothamsted Experimental Station, Harpenden).

1 *Lagurus ovatus* grass pollen, ×1328
2 *Tilletia caries* bunt spore, ×1993
3 *Cryptostroma corticale* conidium, ×1993
4 *Puccinia graminis* uredospore, ×1632
5 *Agaricus bisporus* basidiospore (after normal discharge), ×8167
6 *Calvatia gigantea* basidiospore, ×2968
7 *Lycopodium* sp. spore, ×790
8 *Thermoactinomyces sacchari* (J. Lacey, culture A.978: type), ×5633
9 *Ustilago avenae*, smut spore, ×7060

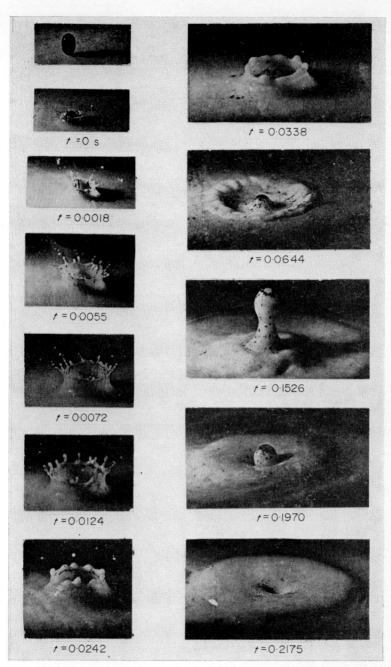

Plate 3. Photographs by Worthington & Cole (1897) showing splash of a water drop weighing 0·2 g (coated with lamp-black) falling 40 cm into a mixture of milk and water. Mag. $\times \frac{5}{8}$

Plate 4. The multi-stage liquid impinger (May, 1966). Three models: A, 50 litres/min; B, 20 litres/min; C, 10 litres/min. *(Reproduced by permission: Crown Copyright Reserved.)*

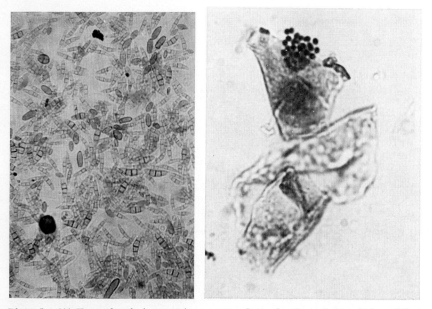

Plate 5. (A) Example of air-spora in summer after rain. Part of deposit from Hirst spore trap in a wheat crop at Rothamsted Experimental Station, August 1965, 17.30 hr GMT. (Mag. ×179) (Gregory & Henden); (B) Photomicrograph of epidermal scales in a slit-sampler deposit, collected from room air, showing bacterial colony developing after 3 hour incubation on nutrient agar. *(Photo by courtesy of R. R. Davies.)*

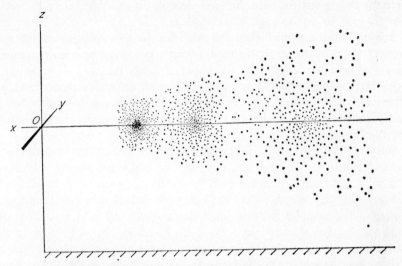

Figure 9. Diffusion of spore cloud during horizontal travel in wind. $O =$ origin of co-ordinates at source point of liberation; x, y, $z =$ down wind, cross wind and vertical axes respectively. Growth of cloud is measured by increase in standard deviation after the centre of the cloud has travelled to three positions down wind.

When the cloud has travelled further down wind it will have a new mean position, and during the time the cloud has been travelling it will have been further diluted by eddies, its particles will have got further apart, and consequently their standard deviation will have become larger.

The next problem is to find a relation between the standard deviation (σ) and the distance travelled. How does σ grow as x grows? Since the First World War this problem has excited the interest of many workers attempting to predict the concentration of gas clouds, smoke screens, smoke trails and crop pathogens.

The subject was pioneered by the Austrian Meteorologist, Wilhelm Schmidt (1918, 1925), who put forward a theory similar to those being developed almost simultaneously in Britain by G. I. Taylor and L. F. Richardson. Schmidt supposed that, with a given state of turbulence of the air, diffusion of particles proceeds like the diffusion of heat in a solid, but with an atmospheric turbulence coefficient A/ρ replacing the coefficient of thermal conductivity. He showed that for these conditions $\sigma^2 = 2(A/\rho)t$, where $t =$ time. His work is now mainly of historical interest, but we should note one interesting feature: according to Schmidt the standard deviation squared is proportional to the *time* during which diffusion has been taking place, so that on his theory the standard deviation will not be constant at a given distance, but will depend on the time taken to reach that distance, i.e. on the speed of the wind.

Schmidt also assumed that the particles in the diffusing cloud are brought to ground level by their fall under gravity, and he used measured values of terminal velocity to fix dispersal limits for various organisms.

As Dingle (1957) points out, a difficulty with Schmidt's theory, and of others derived from it, is that the eddy diffusion coefficient, instead of being nearly constant as in molecular diffusion processes, varies through eight orders of magnitude.

Sutton (1932) recognized that diffusion in the atmosphere differs from molecular diffusion of heat in a solid in one important respect. Diffusion in a solid is constant (depending on the mean free path of the molecules) however long the diffusion has been going on. Diffusion in the atmosphere is much more complex because atmospheric eddies are of a vast range of sizes, varying from a centimetre or so up to eddies that we recognize as fluctuations in wind direction, and even cyclones and anticyclones. Sutton realized that the size of eddy effective at a given moment in *diluting* a cloud is one of the same order as the size of the cloud itself at that moment. Thus a 1-cm eddy would not effectively dilute a cloud 1 metre in diameter, and a 1000-metre eddy would merely carry a 1-metre cloud around bodily, without diluting it. The eddy that dilutes a 1-metre cloud is itself of the order of 1 metre. This led Sutton to an equation for the standard deviation which is fundamentally different from that of Schmidt: $\sigma^2 = \frac{1}{2}C^2(ut)^m$, where $t =$ time; $u =$ wind speed; C is a new coefficient of diffusion with dimensions $(L)^{1/8}$; and m is a number varying between 1·24 in extremely stable, non-turbulent wind, and 2·00 under conditions of extreme turbulence. The value for normal overcast conditions with a steady wind is $m = 1·75$.

Because wind speed multiplied by time equals distance we can write Sutton's formula as: $\sigma^2 = \frac{1}{2}C^2x^m$. This suggestion that σ^2 is a function of the distance, x, is not unreasonable, because the surface roughnesses which generate frictional turbulence are spread out along the distance travelled by the cloud. It is moreover a tempting theory for studying dispersal in the field as a past event, because we do not need to know the wind speed operative when dispersal took place.

Values of C decrease with height because conditions at greater heights in the free atmosphere are less favourable for the formation of eddies. Values for m appear to increase with longer sampling periods, and Sutton suggests that m itself is a function of time. In making continuous observations over a long period on the density of a cloud (and this is exactly what a surface exposed to a passing cloud does do) Sutton suggests that the random element may become smoothed out, so that, after a sufficiently long period, $m = 2{\cdot}00$. These possibilities should be borne in mind when the density formulae described below are applied to some biological data where the sampling period is very long.

In practice it is found that near ground level, diffusion takes place faster on the x and y axes than on the vertical z axis. Turbulence is then said to be 'non-isotropic', and C has to be represented by its components: C_x, C_y and C_z. Haugen, Barad & Antanaitis (1961) found that values for m differ also in the horizontal and vertical directions. The number m is an indicator of the degree of turbulence of the air, and is, as a first approximation, independent of the mean wind velocity. It is primarily affected only by those factors which tend to damp out or enhance turbulence, such as the vertical temperature gradient and the roughness of the ground. For conditions of spore dispersal tests it seems appropriate to assume values of: $C_y = 0{\cdot}5{-}1{\cdot}0$ (metres)$^{1/8}$, $C_z = 0{\cdot}1{-}0{\cdot}2$ (metres)$^{1/8}$, and $m = 1{\cdot}75{-}2{\cdot}00$.

Expressions for the concentration of particles in a cloud emitted from various types of source were deduced by Sutton (1932), and are analogous to heat-conduction equations, as follows:

(i) *An instantaneous point source*, such as a puff of Q grammes of smoke, or a number Q of spores emitted at an instant of time. Here the concentration in the cloud is given by

$$\chi = \frac{Q}{\pi^{\frac{3}{2}}C^3 x^{\frac{3}{2}m}}\exp.\left\{-\frac{r^2}{C^2 x^m}\right\},$$

where r is the distance from the centre of the puff or cloud.

(ii) *A continuous point source*, such as a factory chimney emitting Q particles per second. Here, to obtain an integral that can be handled conveniently, the assumption is made that the spread of the cloud laterally and vertically is small compared with its spread down wind. When emission has continued long enough for the distribution to reach a steady state, the concentration is given approximately by

$$\chi = \frac{Q}{\pi C^2 u x^m}\exp.\left\{-\frac{y^2+z^2}{C^2 x^m}\right\}.$$

The cross-wind concentration shows a 'normal' distribution of particles. On the axis of the cloud ($y = z = 0$) the concentration is given by the simpler expression

$$\chi = \frac{Q}{\pi C^2 u x^m},$$

and because, according to the theory, m cannot exceed $2{\cdot}0$, the fall-off in concentration on the axis of a point-source cloud cannot be more rapid than the inverse square, no matter how turbulent the wind may be.

(iii) *A continuous line source at right-angles to the mean direction of the wind*, emitting Q particles per second per centimetre, and assuming the line to be of infinite length

$$\chi = \frac{Q}{\sqrt{(\pi)}Cux^{\frac{1}{2}m}}\exp.\left\{-\frac{z^2}{C^2x^m}\right\}.$$

Values obtained by Sutton suggest that, as a rough and ready rule, a finite line source behaves as a line of infinite length for distances of travel of the cloud up to 4 times the actual length of the line. For points on the *xOy* plane

$$\chi = \frac{Q}{\sqrt{(\pi)}Cux^{\frac{1}{2}m}}.$$

Sutton's statistical method does not exhaust the possible approaches to the problem of atmospheric diffusion, and attempts to find a still more useful model continue (*see* H. L. Green & Lane, 1957). From the theory of T. von Kármán, Calder (1952) developed an equation which is said to give better predictions than Sutton's theory up to distances of 100 metres, but not for greater distances.

FIELD EXPERIMENTS ON DIFFUSION OF SPORE CLOUDS

Several experiments have been reported that give data from which it is possible to test the applicability of eddy diffusion theories to diffusion of a microbial cloud in a horizontal direction.

Stepanov (1935) used artificial sources of spores that were liberated at a point in the open air. He trapped the spores on glass slides coated with glycerine jelly, placed on the ground at various distances from the source and in various directions relative to the wind. At the end of the experiment cover-glasses were placed on the slides, and the number of single spores per unit area was counted (spore clusters were disregarded).

In Experiment 1 (28 July 1933), on a lawn near the Middle Neva River, Elagin Island, Leningrad, approximately $1·2\times10^9$ spores of *Tilletia caries* were disseminated into the air through gauze, at a height of about 80 to 120 cm above the ground. According to anemometer readings the wind varied from 0·5 to 4·0 metres per second, but sometimes fell to a complete calm; its direction was also variable. Two glass slides were placed at each trapping position, the numbers of spores trapped being shown in Table IXA.

Experiment 2 (5 September 1933) was made at the same place as the previous one. This time a mixture of spores of *Tilletia caries* and *Bovista plumbea* was disseminated through a small sieve at a height of about 150

TABLE IXA

RESULTS OF DISPERSAL OF SPORES OF *Tilletia caries*
Experiment 1 (Stepanov, 1935)

Number of spores per cover-glass 18×18 mm (average of 2)

Angle of slide to wind	At 5 metres from place of dispersal of spores	At 10 metres from place of dispersal of spores	At 15 metres from place of dispersal of spores	At 20 metres from place of dispersal of spores
$-20°$	204	23	4	0
$-10°$	435	45	19	8
$+30°$	964	212	207	49
$+45°$	1198	587	87	142
$+55°$	659	123	77	15
$+65°$	341	24	26	7
$+75°$	365	5	26	53
$+85°$	20	10	9	14

cm. The scattering of the spores occupied 15 minutes, after which 30 to 35 minutes were (perhaps unnecessarily) allowed to elapse for the deposition of the spores. During this period the wind mostly varied from 2·3 to 3·0 m/s, but was sometimes calm. As shown in Table IXB, three slides were placed at each trapping position. Approximately $1·8 \times 10^9$ spores of *Tilletia* were used but those of *Bovista* were unfortunately not estimated.

Experiment 3 (16 April 1935) in the same location was made with a mixture of spores of *Tilletia tritici*, *Bovista plumbea*, and *Lycopodium clavatum*, scattered through a fine sieve at a height of 0·8 to 1·2 metres

TABLE IXB

DISPERSAL OF MIXED SPORES OF *Tilletia caries* AND *Bovista plumbea*
Experiment 2 (Stepanov, 1935)

Number of spores per cover-glass 18×18 mm (average of 3)

Angle	Tilletia				Bovista			
	5 m	10 m	20 m	40 m	5 m	10 m	20 m	40 m
$-45°$	3·0	0·3	0·7	0·0	0·3	0·0	0·0	0·0
$-30°$	128·0	2·3	0·3	0·0	7·0	0·3	0·0	0·0
$-15°$	43·3	54·7	4·7	0·3	7·0	4·0	0·0	0·0
$0°$	206·0	204·0	5·3	8·3	17·0	0·0	1·7	0·0
$+15°$	623·0	115·3	31·3	1·3	46·0	16·0	0·7	0·0
$+30°$	877·7	216·7	49·0	7·0	81·3	20·3	6·3	0·0
$+45°$	911·7	89·7	207·0	9·3	70·0	10·7	7·3	2·7
$+60°$	245·7	48·0	3·0	2·3	27·7	17·3	0·7	0·0

over a period of 17 minutes. Wind velocity varied from 2·02 to 2·21 m/s, but was sometimes strong and sometimes calm. At each of four points, at distances of 10, 20, 30 and 40 metres down wind of the point of spore liberation, were placed 50 slides coated with glycerine jelly. At the end of the experiment the total number of spores deposited on an area of approximately 162 cm² was counted at each of the four distances as shown in Table X (Stepanov, 1962).

TABLE X

DISPERSAL OF A MIXTURE OF SPORES OF *Tilletia tritici*, *Bovista plumbea* AND *Lycopodium clavatum*

Experiment 3 (Stepanov, 1962)

Number of spores per 162 cm²

Species	Number of spores liberated	Distance from source (m)			
		10	20	30	40
Tilletia tritici	175×10^7	14 642	5865	2007	936
Bovista plumbea	137×10^7	8712	4587	1850	697
Lycopodium clavatum	12×10^7	1486	779	311	155

Stepanov's results led him to an empirical law of spore dispersal which was expressed as: $y = C + a/sx$, where $y =$ the distance at which the spores were trapped, $x =$ the number of spores deposited per unit area of trap surface, $s =$ area of trap surface, and C and a are parameters dependent on the conditions of the experiment. The number of spores deposited is thus regarded as varying inversely as the first power of the distance from an origin of co-ordinates that is not coincident with the source.

It will be shown later that Stepanov's formula, which is the first fruit of the experimental approach to the problem, needs modification if it is to describe spore dispersal over a wide range of conditions (Gregory, 1945). First it will be necessary to re-examine the data from Stepanov's Experiments 1 and 2 in the light of Sutton's theory of eddy diffusion.

Stepanov's observational data enable us to test whether the standard deviation, σ, of the spores from their mean position agrees with Sutton's form: $\sigma^2 = \frac{1}{2}C^2 x^m$, or with the older theories of diffusion where $\sigma^2 = 2Kt$. The data also allow us to estimate the parameters m and C, which can then be compared with values obtained by meteorologists for similar conditions. Examination of Tables IXA and IXB shows that the spores at any one distance do not lie in a smooth normal frequency distribution, but are significantly clumped. This is probably because the duration of the dispersal operation was insufficient to smooth out the action of a few larger scale eddies.

The standard deviations of spores lying at each distance from the source have been calculated for Table XI, where for convenience the deviations from the mean position at each distance were measured along the arc with the point source as centre. The standard deviation at each distance was calculated from the usual formula, $\sigma = \sqrt{[(x-\bar{x})^2/(n-1)]}$. This is not strictly legitimate, because the trapped spores are a systematic instead of a random sample of the population and should be regarded as estimates of the ordinate of a normal frequency curve. However, the formula clearly gives a useful approximation – which would have been better if the traps had extended farther laterally, and if data for some of the intermediate radii had not been missing.

TABLE XI

CALCULATION OF PARAMETERS FOR SUTTON'S DIFFUSION EQUATION FROM STEPANOV'S DATA (1935)

Experiment Number			Distance from source (m)				
			5	10	15	20	40
1	Tilletia	σ	2·25	2·97	5·03	6·64	—
		log	0·3522	0·4728	0·7016	0·8222	—
2	Tilletia	σ	1·81	3·62	—	4·87	14·79
		log	0·2577	0·5587	—	0·6875	1·1699
	Bovista	σ	1·790	3·673	—	5·343	—
		log	0·2529	0·5651	—	0·7277	—
	log distance (m)		0·6990	1·0000	1·1761	1·3010	1·6021

Equation for regression line:
$a = 0\cdot5971$ (S.D. $0\cdot02$), $b = 0\cdot8812$ (S.D. $0\cdot072$), $y = 0\cdot5971 + 0\cdot8812\,(x-\bar{x})$.
Whence $C = 0\cdot637$ (metre)$^{\frac{1}{8}}$, and $m = 1\cdot76$

Both experiments were done in the same place, and with comparable wind velocities, and when the values for log σ are plotted against log x the points are found to lie reasonably close to a straight line. The slope of this line is incompatible with the older diffusion theories, but corresponds with Sutton's formula for σ, where with Stepanov's data, $C = 0\cdot64$ (metre)$^{1/8}$, and $m = 1\cdot76$. Sutton's work was apparently unknown to Stepanov when these experiments were done, and so the data could not at the time have been analysed in terms of eddy diffusion. However the agreement between experiment and theory provides evidence that spore dispersal in air is mainly controlled by eddy diffusion of the type postulated by Sutton. The values for C and m obtained from Stepanov's experiments agree well with those found in spore dispersal tests by other workers.

E. E. Wilson & Baker (1946b) liberated *Lycopodium* spores at 2·15 metres above ground level and caught them, not on glass slides on the ground as Stepanov had done, but, because they were interested in diseases of fruit trees, on vertical sticky slides placed on three vertical posts at 1·5, 3·0, and 5·1 metres down wind from the source, and at thirteen heights above ground at each distance; seven tests were done at wind speeds ranging from 1·7 to 7·2 metres per second. Other tests measured horizontal dispersion. Wilson and Baker calculated the standard deviation of the spores deposited at each distance in each test, and from their values of σ we can now estimate the parameters for Sutton's equation. In a few individual experiments their values obtained for m lie outside the limits of 1·24 and 2·00 postulated by Sutton. But their mean values are within the range. Though their values for C_z agree well with Sutton's findings, their values for C_y are higher and agree with those of Gregory, Longhurst & Sreeramulu (1961).

At Waltair, India, Sreeramulu & Ramalingam (1961) used slide traps at ground level, as in Stepanov's earlier tests (Figure 10), and also obtained values for C and m. The twelve tests with *Lycopodium* embraced a variety of conditions by night as well as day, and in some tests the considerably smaller spores of *Podaxis* (14×11 μm) were also liberated. Values for C varied over a wide range, but m agreed closely with the results obtained in England, and again supported Sutton's theory over the distances of 30 to 45 metres tested.

Evidently for microbiological work we must use large values for C_y, perhaps because we are concerned with longer periods than Sutton. We shall therefore choose $C_y = 0·8$ (metre)$^{1/8}$, and $C_z = 0·12$ (metre)$^{1/8}$, as standard in Chapter XVI where deposition gradients are considered in detail.

COMPARISON OF THEORIES OF SCHMIDT AND SUTTON

According to Schmidt's theory, $\sigma^2 = 2At/\rho$, so, because $t = x/u$, we have $\log \sigma = \frac{1}{2}\log x + \frac{1}{2}\log (2A/\rho u)$. If this relation holds true in field tests, plotting experimental data for $\log \sigma$ against $\log x$ should give a line of slope $\tan^{-1} \frac{1}{2}$, that is 26°34'. However according to Sutton's theory $\sigma^2 = \frac{1}{2}C^2 x^m$, therefore $\log \sigma = m/2.\log x + \frac{1}{2} \log (\frac{1}{2}C^2)$. If this holds true, plotting the observed values of $\log \sigma$ against $\log x$ should give a line of slope: $\tan^{-1} m/2$. For values of Sutton's m between 1·75 and 2·00, the line should slope at some angle between 40°36' and 45°.

Field tests with *Lycopodium* spores liberated over short grass by Gregory, Longhurst & Sreeramulu (1961) allow the theories of Schmidt and of Sutton to be compared directly. Spore cloud concentrations were measured near

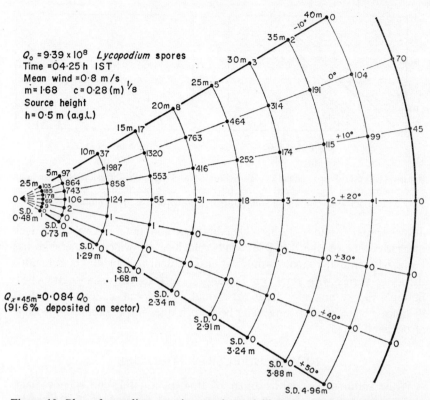

$Q_0 = 9.39 \times 10^8$ *Lycopodium* spores
Time $= 04.25$ h IST
Mean wind $= 0.8$ m/s
$\dot{m} = 1.68$ $c = 0.28$ (m)$^{1/8}$
Source height
$h = 0.5$ m (a.g.l.)

$Q_{x=45m} = 0.084\ Q_0$
(91.6% deposited on sector)

Figure 10. Plan of sampling area in experimental liberation of *Lycopodium* spores at Waltair, India (Sreeramulu & Ramalingam, 1961), showing numbers of spores trapped and standard deviations at various positions down wind of a point source. Q_0 = number of particles liberated; Q_x = total number of particles estimated to remain in suspension in spore cloud; a.g.l. = height above ground level.

ground level at distances up to 10 metres simultaneously at 24 points. Results plotted in Figure 11a show the lines sloping at angles varying between 40° and 46°. This is incompatible with Schmidt's theory which requires a slope of 26°34'. Furthermore, if log σ is plotted against log t (calculated from the mean wind speed and distance), according to Schmidt's theory, σ should be the same after a given *time* whatever the wind speed, but this is not so (Figure 11b). On Sutton's theory at a given distance log σ varies only over a comparatively narrow range of values depending on the parameter m. The results of these experiments (which were the first to make a direct volumetric measurement of change in χ with distance) are compatible with Sutton's theory, which requires a slope of 40°36' for $m = 1.75$, and 45° for $m = 2.00$.

In biological applications we are usually interested in the relation between

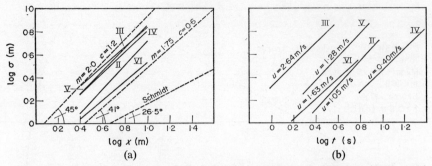

Figure 11. Test of agreement of W. Schmidt's and Sutton's diffusion theories with experiments using *Lycopodium* spores liberated over grass field at Imperial College Field Station, Ascot, England. (Gregory, Longhurst & Sreeramulu, 1961.)

diffusion and distance, rather than diffusion and time. As we often lack measurements of the variable wind velocities in which dispersion has occurred, Schmidt's theory would be inconvenient to handle. On Schmidt's theory σ varies with time; on Sutton's σ varies with distance travelled. Sutton's theory not only fits experimental results well, at least over short distances, but it is also convenient because it does not require a knowledge of the wind speed.

NEWER DIFFUSION THEORIES

While Sutton's classical treatment of atmospheric diffusion is convenient, and satisfactory over short distances, its general applicability is limited by its very simplicity which neglects certain factors which assume increasing importance, and modify the dispersal pattern, as the distance travelled increases beyond a kilometre or so. Discrepancies between theory and practice become apparent during field tests.

For distances over a kilometre or so, theory should take into account the following factors.

(1) The increase of wind speed and decrease of turbulence with height above the ground.

(2) Different values for n_y and n_z (where $n =$ Sutton's $1-m$); also different values for C_y and C_z.

(3) The effect of wind shear. Tyldesley (1967) in a computer study of eddy diffusion suggested that because wind speed increases with height the top of the diffusing cloud will travel faster than, and diffuse vertically downwards in advance of the lower portions. This will result in a longer but more dilute cloud, which takes longer to pass a given point than calculated from the original theory. Maximum concentration will be decreased, but the 'area dose' (p. 88) will ultimately be unaltered.

(4) When the cloud has travelled a long distance and expanded, too few of the larger eddies that dominate further expansion may have occurred in a limited period of time to have smoothed the concentration, and the distribution of concentrations may be unpredictably patchy.

(5) Sutton's theory also assumes that the diffusing cloud retains its initial number of particles, Q_0, in suspension. This is valid for smoke and gases, but microbial clouds lose material by deposition. Discussion of this limitation is postponed until Chapter XVI.

(6) The diffusion parameters, C and m, used by Sutton, apply to thermally neutral, overcast weather, and parameters are needed for a wider range of meteorological conditions.

Pasquill (1961, 1962a) devised a convenient method of predicting aerosol concentrations at various points down wind, for use when a measure of wind speed and gustiness is available or when values for these can be assumed. The method uses readings from an anemometer and a recording wind direction vane, and assumes decreasing values for the standard deviation of lateral wind direction as conditions change from extremely unstable conditions, through six categories, through neutral to moderately stable conditions. If gustiness is not measured, stability conditions can be judged from ordinary meteorological observations. The method has agreed well when tested against results from diffusion experiments in the field. It is useful in predicting maximum concentration *during* the course of an atmospheric pollutant incident, either accidental or deliberate. For avoidance of allergens the method could be used to calculate maximum possible concentrations. For use in plant pathology an estimate of mean deposition is required, and this is still perhaps most easily obtained from other modifications of Sutton's theory.

For an account of recent work on atmospheric diffusion, in theory and from experiments with various tracers, the authoritative volume on *Meteorology and Atomic Energy* (United States Atomic Energy Commission, 1968) should be studied.

BROOKHAVEN NATIONAL LABORATORY STUDIES ON AMBROSIA POLLEN

Pollen of ragweed (*Ambrosia* spp.) is an important aeroallergen in North America, and has served as a convenient natural material for a long series of dispersion studies by the Meteorology Group at Brookhaven, working with the New York State Museum and Science Service. Two main series of tests were done in the field: (1) using small area sources consisting of groups of ragweed plants (either treated to flower early, or flowering at normal season); and (2) using point sources liberating stained ragweed pollen from aqueous suspensions. Sources, approximately 1·5 metres above ground level, were surrounded with arrays of air-sampling devices, down wind in open country, to measure concentration of pollen at various points in the x, y and z directions, and also deposition onto glass slides on the ground. All experiments were supported by ample meteorological records. (Other kinds of pollen were also used, and diffusion into forest was also studied.) Dispersion patterns were conveniently plotted as 'isopleths' (lines of equal concentration) of the diffusing cloud or plume.

TABLE XII

CONCENTRATION OF *Ambrosia* POLLEN AT APPROXIMATELY 60 METRES FROM SOURCE, EXPRESSED AS PERCENTAGE OF CONCENTRATION AT 1 METRE DOWN WIND OF EDGE OF SOURCE. (ALL MEASUREMENTS ON AXIS PLUME AT SOURCE HEIGHT.)

(Raynor *et al.*, 1970)

Source diameter (*m*)	Distance from edge of source (*m*)	Concentration (% of χ at 1 m)
Point	68·6	0·3
5·5	65·8	2·7
9·1	64·0	5·0
18·3	59·4	9·2
27·4	54·9	12·5

Based on average results from numerous tests (Raynor, *et al.*, 1970), decrease of concentration of *Ambrosia* pollen on the axis of the plume at source height is shown in Table XII to vary with the diameter of the source.

The relative concentration, at any given position on the plume axis, increased as the wind speed increased – an effect attributed to decreasing lateral spread coupled with decreasing loss by deposition as wind speed increased.

In these tests, vertical standard deviation, σ_z, was only about one-tenth that of the lateral standard deviation, σ_y, from a low level source (about 1·5 m above ground level) even in unstable conditions – a result widely different from the ratio of $\sigma_y/\sigma_z = \pm 2$ appropriate to emission from a chimney. This characteristic favours local deposition of pollen, thus increasing the chance of local pollination, and decreasing the amount of pollen reaching distant stigmas and noses.

Other data from the Brookhaven workers showed that loss of material from the pollen cloud by deposition to ground was many times faster than could be accounted for by terminal velocity of the pollen grain, yet long distance travel is still possible for a fraction of the emission, about 1% of which still remained airborne after travelling 1 kilometre.

Table XII shows that horizontal diffusion, aided by deposition, reduces the concentration of a pollen cloud emitted by a point source to about one hundredth of its source concentration within 60 metres of travel down wind. Nevertheless a 'tail of the distribution' amounting to 1% of Q_0 was estimated to remain airborne after travelling one kilometre, consequently there is, at any point in the ragweed geographical area, a 'background concentration' made up of the tails of distant sources, which is super-imposed on the effects of a local source. Raynor, Ogden & Hayes (1968) measured background concentration of *Ambrosia* pollen at points up wind of their experimental sources, and then plotted isopleths to show at what distance down wind from the experimental source its contribution equalled selected multiples of background.

Mean distances down wind at which local concentration equalled background concentration varied in different seasons from about 110 metres to 200 metres (with maximum distances in individual tests several times these values). Effects of local sources (varying in size from 200 to 2000 m²), became negligible in comparison with background at distances of 500 to 1000 metres.

The concentration of particles derived from a small source decreases rapidly with distance down wind. Because a pollen cloud is rapidly diluted, an individual forest tree will tend to be crossed by its neighbours (Strand, 1957) which are favourably placed to compete for space on the stigma. But longer distances are needed to protect against a pathogenic or allergenic source because in this context competitors are not protective.

VII

Deposition Processes

We have considered airborne microbes as diffusing clouds. Before we can discuss ways in which they are deposited in the complex environment out of doors, we must deal with deposition processes under simplified, ideal conditions. The word 'deposition' is used in a general sense to include all processes by which airborne particles are transferred from aerial suspension to the surface of a liquid or solid. One form of deposition, the impaction of droplets or particles on surfaces, has been studied extensively, both theoretically and in wind-tunnel experiments. Impaction is highly relevant to the problems of spore deposition in nature and of the sampling techniques which form the topics of Chapters VII and IX.

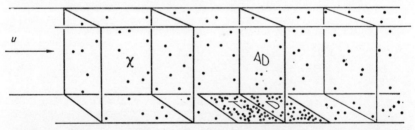

Figure 12. Diagram showing relation between concentration (χ = number of spores per unit volume of air); u = wind speed; A.D. = area dose (number of spores passing through 'frame' of unit area); and T.D. = trap dose (number of spores deposited on unit area of surface).

The relation between concentration of the spore cloud, χ, and deposition on the surface, T.D. (= trap dose), over which the spore cloud travels, is illustrated in Figure 12, together with the concept of 'area dose' (A.D. = the number of particles flowing through an imaginary frame of unit area cross-section at right angles to the direction of the wind). Concentration of the cloud (χ = number of spores per cubic metre) is the more fundamental measurement, and the one of greatest interest to the allergist, whose patients inhale volumes of air. The trap dose, which measures deposition on a surface, is of more interest to plant pathologists, plant breeders and pollen analysts. The area dose is a useful concept in passing from one measurement to the other. For a given concentration of particles per unit volume of air, the area dose must increase with wind speed, but whether the trap dose will be similarly affected is a matter for experiment. With a continuous source emitting during a limited time, the area dose will be the same as if the same total quantity of particles, Q_0, had been liberated in a number of successive instantaneous puffs arriving in a series of greatly fluctuating concentrations.

We can conveniently express the percentage trapping efficiency of a surface as:

$$E = \frac{\text{Trap dose/cm}^2}{\text{Area dose/cm}^2} \times 100$$

This convention expresses the efficiency with which a surface clears the spore cloud to a height of one centimetre above the surface. Deposition on a surface takes place in several ways, including impaction and sedimentation, for sedimentation seldom acts alone.

MECHANISM OF IMPACTION

When a bluff object such as a cylinder is placed in wind, the oncoming airstream must flow around the obstruction, but airborne particles will be carried some distance towards it by their own momentum before they are in turn deflected by the wind flowing round the obstacle (Figure 13).

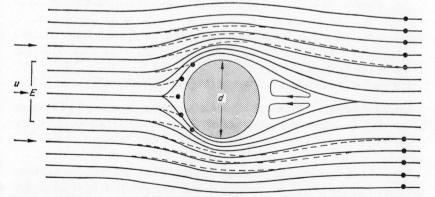

Figure 13. Diagram showing streamlines of air and particle trajectories around a cylindrical obstruction (vertical cylinder seen in plan). E = streamlines carrying spores towards cylinder; d = diameter of cylinder; arrows on left show direction of wind of velocity = u.

If all those particles were impacted, whose trajectories in the free wind stream would have passed through the obstruction, impaction efficiency would be 100%, but apparently in practice it never is. The distance travelled by the moving particle towards the cylinder before being deflected by the air streamlines flowing around the cylinder is related to both the momentum of the particle and the size of the object disturbing the airflow. Another effect, collection by direct interception, becomes important when the diameter of the particle is an appreciable fraction of the diameter of the cylinder, and with particles in the range of 1 μm and less, Brownian diffusion may play a major role in deposition.

The general principles of impaction were made clear by Sell (1931) in connection with dust filtration. From aerodynamical considerations Sell postulated that the efficiency of impaction, E, is related to a non-dimensional constant, which may be written in a form convenient here as: $k = v_s u / \frac{1}{2} dg$, where: v_s = speed of fall of particle in still air; u = speed of wind; d = diameter of cylinder, strip, etc; and g = acceleration due to gravity.

Sell derived the relation between E and k by observing the trajectories of uniform droplets of Indian ink on paper bisecting a vertical cylinder in a small wind-tunnel. It is not clear from Sell's paper what range of conditions he tested. Figure 14 plots experimental values for the relation between E and k obtained in the wind-tunnel tests described in the following section, compared with theoretical values calculated by Langmuir & Blodgett (1949), using a refinement of Sell's method. More extensive data of this relation are given by May & Clifford (1967).

Two useful concepts in this connexion are 'stop distance', λ, and 'relaxation time', τ (Chamberlain, 1967a).

Stop distance. If a spore is given an initial horizontal velocity, v, (as in ejection from an ascus, p. 47, or blown by the wind) the distance it travels in still air before being brought to rest by air friction under Stokes's law depends on the kinetic energy of the particle and on air drag. This stop distance, $\lambda = (v.v_s)/g$.

Sell's impaction parameter, $k = \lambda/d$ (May & Clifford, 1967).

Relaxation time. The time it takes for a particle, as a result of friction, to acquire the speed and direction of the surrounding air is called the relaxation time, $\tau = v_s/g$. For a 10 μg spore of unit density, $\tau = 3\cdot1\times10^{-4}$ seconds.

Efficiency of impaction on spheres, cylinders, strips and discs has attracted two types of investigators, theoretical and experimental, and large discrepancies have appeared between calculations from different workers and again between theory and experiment. Deposition on cylinders has been studied by Chamberlain (1967a), May & Clifford (1967) and Starr (1967b).

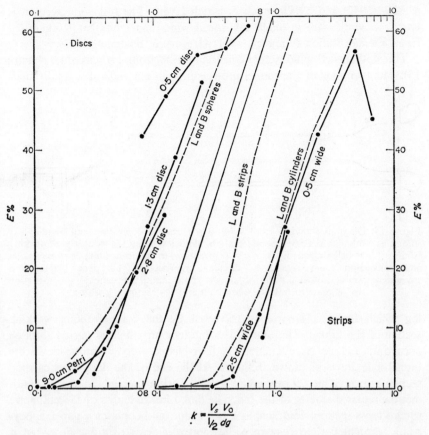

Figure 14. Observed relation between $E\%$ and $k = v_s u / \frac{1}{2} dg$. Solid lines from Gregory & Stedman (1953). Broken lines = values for spheres, strips and cylinders as predicted by Langmuir & Blodgett (1949). (Reproduced by permission from *Annals of Applied Biology*.)

Deposition mechanisms other than impaction still await a coherent theory, and we must rely on experimental values.

WIND-TUNNEL STUDY OF IMPACTION

Recent experiments on impaction on cylinders agree reasonably well with the theory, e.g. May & Clifford (1967), using an improved wind-tunnel and small liquid drops, closely approached theoretical values.

For understanding of spore deposition in the field and in sampling devices an experimental study of efficiency of various geometric surfaces in trapping real, solid particles is required, and as a first approach much information can be obtained from wind-tunnel tests. The properties of small wind-tunnels for use with animal pathogens are described by Druett

E

& May (1952) and by Dimmick & Hatch (1969). The following account is based on work with a small, low-speed wind-tunnel built at Rothamsted Experimental Station (Gregory, 1951; Gregory & Stedman, 1953).

The Rothamsted wind-tunnel consists of a horizontal square duct (Figure 15). The two ends of the tunnel project through the ends of a small build-

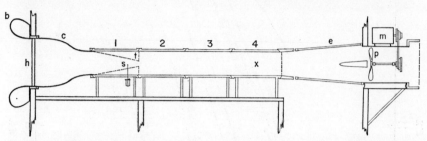

Figure 15. Diagram showing small wind-tunnel used in deposition and mouldy hay studies at Rothamsted Experimental Station, elevation view. 1–4 = 'Perspex' working sections; b = bell-shaped intake; c = contraction to smooth flow; e = expansion and conversion from square to circular cross-section; h = paper honeycomb straightener; m = motor; p = propeller; s = spore input; t = removable constriction to generate turbulence when required; x = trapping position.

ing which forms a laboratory traversed by the 2·8-metre-long working section of the tunnel. The tunnel uses outdoor air which is passed through once only and not re-circulated. A four-bladed wooden propeller in the tunnel exit draws air down the duct. At the end of the duct an expansion section converts the 29-centimetre-square working section to the 46-centimetre-circular diameter of the fan. The flared intake end is of 51 centimetre square cross-section, contracts to a bell shape, and contains a paper honeycomb 'straightener' to remove eddies and produce streamline flow; but if turbulent flow is desired a constriction is inserted in the first part of the working section (indicated by dotted lines at t in Figure 15).

Spores under test are injected (or otherwise liberated) usually on the tunnel axis near the constriction. Spore trapping equipment, plants or other objects for testing deposition, can be inserted farther down wind through removable panels in the walls of the working section. Wind speeds of from 0·5 to nearly 10 metres per second are obtained by changing pulleys on the belt drive between constant-speed electric motor and the fan, and by inserting screens across the tunnel to increase its resistance. The slower wind speeds are best obtained by increasing the resistance of the tunnel, rather than by slowing the fan, because outdoor wind movement disturbs the flow less when the total tunnel resistance is large than when it is small.

Most tests of spore dispersal or deposition in the wind-tunnel involve knowledge of the time–mean spore concentration of the air. The Cascade Impactor (K. R. May, 1956; see also Chapter IX), operated isokinetically (i.e. with the orifice facing the wind and with suction adjusted to draw air in through the orifice at the same speed as the wind) is taken as standard to estimate the mean number of particles per cubic metre of air during the period of the experiment. With this information, and knowing the wind speed, we can calculate the area dose (A.D. $= \chi ut$). The trapping efficiency of any surface exposed to the spore cloud in the wind-tunnel can then be determined by estimating the number of particles collected per square centimetre of trapping surface, and expressing it as a percentage of the area dose.

Most work was done with spores of *Lycopodium clavatum* which are convenient for experimental work. They can be bought easily; they separate readily from one another when blown into the air. They are not smooth spheres, but their mean diameter is about 32 μm, density 1·175, and terminal velocity estimated variously from 1·76 to 2·14 cm/s; the number per gramme is 9·39 to 9·4×10⁷.

Liberation of one million *Lycopodium* spores at a point on the central axis of the tunnel produced a conical cloud. At a sampling point on the central axis of the tunnel, 1·4 metres down wind, the area dose was about 4500 spores per square centimetre under turbulent conditions at wind speeds of from 5·57 to 9·7 metres per second. Under streamline conditions the dispersal cone was visibly narrower and the area dose nearly double at these wind speeds. At 1·1 metre per second, however, streamline conditions gave a much lower area dose because the cloud was displaced downwards under gravity. On the whole, efficiency of impaction was not much affected by whether the flow was turbulent or streamlined.

IMPACTION ON CYLINDERS

Efficiency of impaction on vertical cylinders is *increased* by: (1) increasing wind speed; (2) increasing the mass of the particle (by increase in size or density); (3) making the cylinder sticky; and (4) decreasing the diameter of the cylinder (except that large spores such as *Lycopodium* tend to blow off narrow sticky cylinders at high wind speeds, whereas small spores, such as those of *Ustilago perennans*, do not). As will be shown later, similar relations hold for impaction on surfaces of other shapes, such as spheres, discs, strips (ribbon) and, in the field, leaves, stems and stigmas. The value for E thus depends on shape of surface. The order of decreasing efficiency is probably: cylinders and spheres, strips and discs (May & Clifford, 1967). The way efficiency, E, for trapping *Lycopodium* varies with wind speed and cylinder diameter is shown in Figure 16.

Wind-tunnel results confirm the theoretical conclusion of C. N. Davies & Peetz (1956), that under extreme conditions efficiency of impaction can be zero; for example, in the Rothamsted tunnel the very small spores of

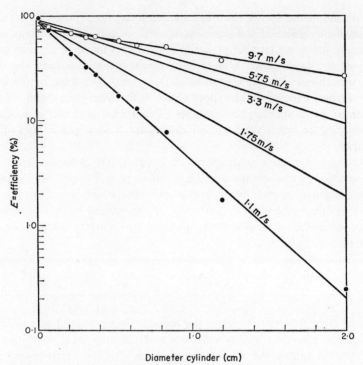

Figure 16. Linear regression of log E on cylinder diameter at five wind speeds; experimental values shown for 9·7 and 1·1 m/s only (Gregory, 1951; reproduced by permission from *Annals of Applied Biology*).

the giant puff-ball, *Calvatia (Lycoperdon) gigantea*, were not impacted under any of the conditions tested. The slightly larger spores of the smut fungus, *Ustilago perennans*, however, showed appreciable efficiencies, but probably because they often travel on the wind in clumps which effectively behave as rather heavier particles than their component single spores.

The deposit is not uniform over the cylinder, but is densest on the part of the circumference projecting farthest forward into the wind; this is the so-called stagnation line. The density of the deposit decreases towards the sides of the cylinder, and there is a spore-free zone on the shoulders where the airflow is more or less tangential to the surface. Usually there is no deposit at the back or down-wind side of the cylinder, but with *Lycopodium* spores (32 μm diameter) a narrow line of deposit has been observed down the back of sticky cylinders of less than 1 mm diameter at wind speeds of 1 metre per second or less. Deposit on the side opposite to the oncoming wind is negligible or zero under most conditions tested. The angle subtended by the deposit on the up-wind side was less than 180° in both turbulent and streamline wind. Other things being equal, the angle subtended at the leading edge of the cylinder by the deposit was increased by increasing the wind speed and by decreasing the cylinder diameter; increasing efficiency of the whole cylinder evidently runs parallel with increases in the angle subtended by the deposit. The low efficiency of wide cylinders and slow winds shows both as a narrow trace and thinner deposit per unit area.

The values for E shown in Figure 16 are lower than the optimum obtained by May & Clifford (1967) using liquid drops in an improved, streamlined wind tunnel, probably because *Lycopodium* spores tend to bounce off even from a sticky surface. Chamberlain (1967a) stresses the much poorer retention from dry surfaces such as leaves, which, however, is increased when wetted by rain.

The decrease of efficiency with increasing cylinder diameter was first noticed in field tests. Per unit length, a large cylinder 12 cm in diameter, may collect no more pollen grains or spores than a cylinder 1 cm in diameter, and per unit area of surface it may collect many fewer (Gregory, 1951).

IMPACTION ON A ROTATING STICKY CYLINDER

Rotating the cylinder at a peripheral speed comparable with the wind speed would be expected to reduce the thickness of the boundary layer on the cylinder surface, to induce the well-known Magnus effect, and to produce a local rotation of air around the cylinder itself. It was not obvious whether these effects would alter impaction efficiency, so Professor J. R. D. Francis suggested tests and I am indebted to him for advice and the loan of equipment.

The tests were done in the Rothamsted wind-tunnel, with a 15·5-centimetre square constriction up wind to generate turbulence. *Lycopodium* spores (32 μm diameter) were blown in at a point on the axis 18 cm up wind of the constriction, and, after a diffusion path 63 cm long, the spore cloud reached the trapping section where a stationary and a rotating cylinder, each 0·5 cm in diameter, were exposed simultaneously. The rotating cylinder, which was placed vertically across the axis of the tunnel, consisted of a steel rod mounted in a ball-race secured flush with the floor of the tunnel at one end, and connected by a sleeve of stout rubber tubing to the spindle of a 225-watt 'Universal' motor mounted on the roof of the tunnel. The stationary cylinder was 2 cm to one side and 2 cm up wind of the rotatable cylinder, and both carried adhesive coatings of cellulose film which, after exposure, were removed and scanned under the microscope to measure the deposit. Speed of rotation of the cylinder was controlled by a 'Variac' transformer and was measured by a stroboscopic lamp. The range of speeds that could be tested was limited by the steel rod which began to bend at speeds above 7000 r.p.m.

With a wind of 1·1 metres per second, efficiency was substantially unchanged until the peripheral speed of rotation attained approximately the speed of the wind; at higher rates of rotation the efficiency decreased rapidly, reaching zero before the peripheral speed had reached twice that of the wind. At a lower wind speed of 0·68 metre per second, efficiency of the cylinder, when rotating at 0·4 times the wind speed, fell to 53% of the stationary cylinder, and to 24% at 0·82 times the wind speed. These results suggest that the centrifugal effect is sufficient to decrease impaction to zero. The phenomenon would be interesting to explore at a wide range of cylinder diameters and wind speeds (the work of Brun *et al.*, 1955 refers to much greater wind speeds than we tested).

IMPACTION ON PLANE SURFACES AT VARIOUS ANGLES TO WIND

The interest in deposition on narrow horizontal and vertical sticky strips lies in the widespread use of microscope slides for routine trapping of fungus spores and pollen.

The theory of gravity deposition assumes that the air flowing past the surface contains a large population of particles distributed at random. The particles fall at their terminal velocity, v_s cm/s, and the wind blows horizontally at u cm/s. A plane surface of area 1 cm² faces the wind, making an angle Θ with the horizontal plane, as shown in Figure 17. Then only particles contained in the rectangular skew prism ABCDEFGH have trajectories in the free air which would carry them to the surface during time t seconds. The volume of this prism is given by: $V = t(u \sin \Theta + v_s \cos \Theta) \text{cm}^3$.

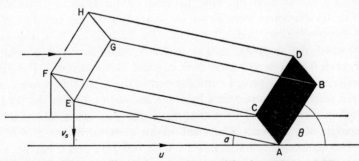

Figure 17. Diagram illustrating gravity particle deposition. ABCD = trap surface; u = wind speed; v_s = terminal velocity of particle; ABCDEFGH = rectangular skew prism containing airborne particles whose trajectories would bring them to rest on trap surface; Θ = presentation angle; tan $a = v_s/u$.

If $\Theta = 90°$, then $v_s \cos \Theta = 0$, and hence the number of particles with trajectories cutting a vertical surface should be independent of the terminal velocity of the particles, but will depend on the wind-run. (Deposition will in practice be reduced below this value because the airstream is deflected by the surface itself, and the efficiency of collection depends on u, as explained in the previous section.)

If $\Theta = 0°$, then $u \sin \Theta = 0$, and deposition under the influence of gravity should depend on the terminal velocity, v_s, so that for a horizontal trap the volume of air sampled should be independent of wind speed and should depend only on the terminal velocity of the particles. From a cloud of uniform concentration, most trajectories should pass through a surface inclined at an angle Θ, when $\tan \Theta = u/v_s$ ($\Theta = 45°$ when $v_s = u$).

Assuming the time–mean density of the spore-cloud = χ spores per cubic metre, it will be apparent that the area dose A.D. = χu on this gravity deposition theory, and that trap dose T.D. = χv_s, from which, if the deposition is by gravity as assumed, the expected area dose will give a convenient test of the validity of the theory.

Our wind-tunnel experiments show that deposition on a horizontal flat surface is a fairly complex process depending on several factors besides the simple resultant of gravity and wind assumed in the diagram (Figure 17). The surface studied in greatest detail has been the $76 \times 25 \times 1.3$ mm glass microscope slide, as this has been extensively used in routine spore trapping. Experiments with other plane surface traps are reported by Gregory & Stedman (1953) in more detail than given here.

The slide was placed with its long axis at right-angles to the wind and held by clips placed at the two ends to avoid disturbing the airflow. The slide surface was orientated at various angles to the wind in different test runs, the convention adopted for descriptive purposes being: presentation angle $0°$ = parallel with the wind; $45°$ when the leading edge was lower than the trailing edge; and $90°$ at right angles to the wind. (Presentation angle so defined differs from the aeronautical 'angle of incidence' in which at $45°$, for example, the leading edge is higher than the trailing edge.)

The effect of gravity was studied in two sets of experiments. In one set, with the long axis of the slide vertical (parallel with the z axis) and the surface making various angles with the x,z plane, the effect of gravity on deposition must be neutral. In the other set with the long axis of the slide horizontal (parallel with the y axis) and the surface making various angles with the x,y plane, the effect of gravity must be positive at angles from $0°$ up to less than $90°$, neutral at $90°$, and negative at angles greater than $90°$ and up to $180°$ (these gravity conditions are denoted by the symbols: g_+, g_0, and g_- respectively). Angles greater than $180°$ represent the back of the slide.

Preliminary tests showed that trapping efficiency varied in different parts of the slide. Accordingly the slide was divided into five ½-cm zones, denoted: A, B, C, D, and E respectively from the leading to the trailing edge.

Results of the main series of tests are plotted in Figure 18, where the efficiency of deposition expected on the gravity theory at 0° for each wind speed, taking v_s for *Lycopodium* as 1·76 cm/s, is indicated by dotted lines. Observed values less than $E = 0·1\%$ are unreliable, but are plotted to show the trend. Zero and values below 0·01 are all plotted as 0·01% as they cannot be distinguished with the data available.

The curves obtained probably result from the interaction of several mechanisms: sedimentation, impaction, turbulence and edge effects. In certain sets of conditions, one or other of the mechanisms can be found acting singly; but for the most part deposition is interpreted as resulting from the simultaneous action of several mechanisms.

DEPOSITION ON HORIZONTAL SLIDES

(1) *Deposition by sedimentation*, under the influence of gravity alone, is seen on the upper surface of a horizontal slide at the lowest wind speed tested (conditions denoted by: 0°, 0·5 m/s, g_+). Here deposition over the slide as a whole was very close to the expected value predicted by the gravity theory ($E = v_s/u \times 100 = 1·76/50 \times 100 = 3·5\%$), but even at this low wind speed the bluff edge of the slide, 1·3 mm thick, caused some edge shadow, shown as a decreased deposit just behind the leading (up wind) edge. That deposition is solely caused by gravity is shown by the absence of deposit on the under side of the horizontal slide, or on either side of a vertical slide held parallel with the wind (0°, 0·5 m/s, g_- and g_0).

At wind speeds more usual outdoors, say between 1·0 and 2·0 m/s, a surprising effect developed in these wind-tunnel experiments. With gravity positive, the bluff edge of the slide produced an edge shadow deflecting a large proportion of the approaching spores; at 1·1 m/s the part of the surface just behind the leading edge was almost free of deposit, and efficiency reached about 50% of the expected value only on the rearmost zone. At 1·7 m/s efficiency was almost zero over the whole slide. With gravity neutral or negative, efficiencies were also almost zero.

(2) *Turbulent deposition*. As the wind speed was raised still further, the efficiency of the horizontal slide recovered; but deposition cannot have been due to gravity sedimentation, because at 9·5 m/s the amount deposited was almost the same on the under side (g_-) as on the upper side (g_+) of a horizontal slide, and also on the two sides of a vertical slide held parallel with the wind (g_0). At this speed turbulent deposition is seen in its almost pure condition. As indicated below, there is some evidence that this deposi-

tion may result from turbulence generated by the bluff edge of the slide itself, though C. N. Davies (1960) considers that it is a wind-tunnel artefact.

At 0° (horizontal slide) and higher wind speeds, although deposition was turbulent, gravity appeared to interact with the process in some way that is at present obscure. With gravity neutral (g_0) and wind speeds of 5·7 and 9·5 m/s, deposition was greater at the leading and trailing edges than in the middle of the slide. With gravity negative (g_-) at 9·5 m/s, however, the leading edge showed an anomaly, having a deposit 8 to 10 times that found with gravity positive. With gravity negative at 5·7 and 3·2 m/s, deposition behind the leading edge was negligible.

DEPOSITION OF *Lycopodium* SPORES ON INCLINED PLANE SURFACES

Deposition on a horizontal microscope slide is best regarded as a special case of deposition on an inclined plane with a presentation angle 0°. A number of possible angles ranging from 0°, through 90° (vertical slide) to 180°, were tested in the turbulent wind-tunnel, and results are also shown in Figure 18.

(1) *Impaction*. Deposition by impaction should be zero at 0°, but it would be expected to occur to some extent at all other presentation angles, as at these angles the surface subtends the oncoming airstream. With the slide vertical (90°), the gravity effect should be neutral, and at low wind speeds of 0·5 to 1·1 m/s the slide would not be expected to generate turbulence. Deposition under these conditions should be due to impaction only. In the tests, as wind speed was increased (90°, 1·7 to 9·5 m/s), deposition increased over the whole surface and was more uniformly distributed; but even at the highest speed tested the deposit at the margin exceeded that at the middle. In this respect impaction on a plane surface contrasts strikingly with that on cylinders, where the centre of the trace is always denser than the edges.

(2) *Edge drift*. The effect of the bluff edge of the slide in 'shading' the leading edge has been referred to above. Behind the edge shadow, a region of greater deposit caused by an edge drift might be expected. Figure 19 shows that at 0° this edge drift fell behind the trailing edge, but that when the slide was inclined at 15° or 30° to the wind, the edge drift impacted on the slide. This is shown by the deposit on the leading edge which greatly exceeded the expected value on slides inclined over the range 15° to 60°, 1·7 to 9·5 m/s.

(3) *Mixed effects*. Over most of the range of zones, presentation angles and wind speeds, the deposition was from a mixture of two or more mechanisms whose relative importance can be roughly assessed from the empirical results shown in Figure 18.

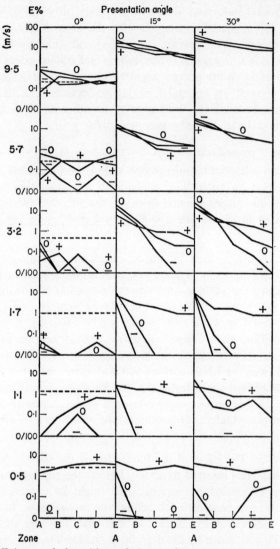

Figure 18. Efficiency of deposition of *Lycopodium* spores on zones across glass microscope slide at presentation angles from 0° (left-hand side of left-hand page) to 90° (right-hand side of right-hand page), as observed in wind-tunnel experiments.

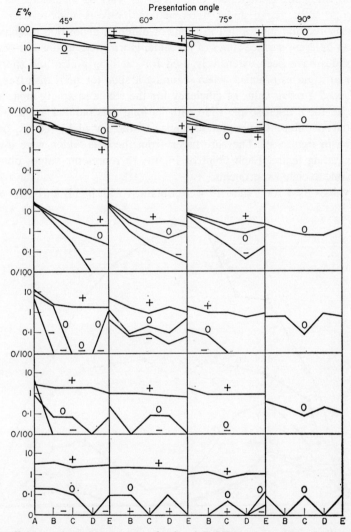

$E\%$ = efficiency (percentage area dose); A,B,C,D,E = successive half-centimetre zones across slide from leading edge (A) to trailing edge (E); + = gravity positive; 0 = gravity neutral; − = gravity negative. (From Gregory & Stedman, 1953; reproduced by permission from *Annals of Applied Biology*.)

MEAN DEPOSIT ON INCLINED SLIDES

In the foregoing paragraphs, efficiencies of various arrangements of a microscope slide, acting under different conditions as a spore trap, have been measured and interpreted in terms of different deposition mechanisms acting on different surface zones of the slide. Horizontal, inclined or vertical sticky slides have been extensively used for catching pollen and spores in outdoor air, and in practice, when scanning a slide (or petri dish trap), we usually need a mean value of efficiency for the whole sampling area. Mean efficiencies for a microscopic slide with its long axis parallel to the y axis, orientated at different presentation angles to the x,y plane (i.e. g_+, or g_-), are given in Figure 19. They are taken from the data which were used in the preceeding figure. Each point of Figure 19 represents values obtained in from one to eight experiments.

The values used for Figure 19 were obtained with highly turbulent wind.

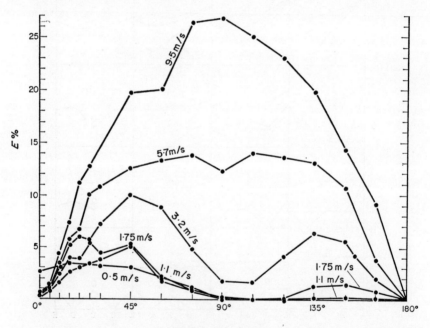

Figure 19. Mean efficiency of deposition of *Lycopodium* spores on glass microscope slide (all zones) at presentation angles of 0° to 180°. E = efficiency expressed as percentage area dose. (From Gregory & Stedman, 1953; reproduced by permission from *Annals of Applied Biology*.)

Partially streamlining the flow, by removing the turbulence-generating obstruction from the tunnel, had little effect on the deposit – except with presentation angles between 0° and 10°, and with wind speeds below 5 m/s, when efficiency of deposition was reduced, being least with the horizontal slide (0°).

Efficiency of deposition on the back of the microscope slide was usually less than 1%.

Decreasing the width of the slide from the customary 2·5 to a mere 0·5 cm, increased efficiency most at the slowest wind speeds. This narrow trap was most efficient at 5·5 m/s and 90° presentation angle (vertical), and a fall in efficiency at 9·5 m/s and 90° was comparable with the anomalous reduction in efficiency with very narrow cylinders at higher wind speeds (Gregory, 1951).

Lycopodium DEPOSITION ON PLANE SURFACES: SUMMARY

Results of deposition tests on plane surfaces with the Rothamsted wind-tunnel have been critically examined and partially explained by C. N. Davies (1960). He concluded that deposition on the horizontal slide was by gravity unless the wind was fast enough to denude, by adhesion or bounce-off at the leading edge of the slide, the layer of air passing close to the slide surface. The equal deposition on the upper and lower surface observed at high speeds is attributed to eddy impaction arising as an artefact in the wind-tunnel. However, this explanation overlooks the fact that deposition of pollen on downward facing surfaces outdoors is known from the work of Durham (1944, p. 233), and has been confirmed by outdoor observations at Rothamsted. Davies's general conclusion is acceptable: the average deposition of *Lycopodium* spores over a large area of smooth surface can be calculated from the terminal velocity, and while effects due to impaction and turbulence may affect distribution locally they cannot affect the total rate. Nevertheless, because of these local effects the stationary microscope slide and petri dish are clearly poor tools for quantitative air sampling.

DEPOSITION OF OTHER SPORES ON INCLINED PLANE SURFACES

Lycopodium was studied in greatest detail because it is easily handled and has relatively large spores. Less detailed tests were also made with the smaller spores of *Calvatia (Lycoperdon) gigantea*, with *Erysiphe graminis* conidia, and with spores of *Ustilago perennans* (Table XIII). In general it may be said that impaction efficiency on a microscope slide held at right-angles to the wind is low, but might be as high as 25% with grass pollen or rust uredospores in winds of 9 m/s.

EFFECT OF THICKNESS OF SLIDE

Deposition on a horizontal or inclined surface is evidently complex and may be disturbed by edge-shadowing. Tests were therefore made with both

TABLE XIII

EFFICIENCY $(E\%)$ OF DEPOSITION ON INCLINED SLIDES IN TURBULENT WIND-TUNNEL

(Gregory & Stedman, *unpublished*)

Presentation angle	Wind speed (m/s)					
	9·5	5·5	3·2	1·7	1·1	0·5
Lycoperdon (calvatia) giganteum (c. 4 μ)						
0°	0·07	1·1	0·6	1·2	5·6	3·3
90°	0	0	0	0	0	0
180°	0·04	1·1	1·7	1·1	6·1	2·2
Ustilago perennans (c. 6 to 8 μ)						
0°	0·2	0·2	0·2	0·1	0·02	0·09
45°	0·3	0·4	0·3	0·04	0·1	0·1
90°	0·3	0·2	0·6	0·2	0·01	0·03
135°	0·4	0·2	0·4	0·09	0·04	0·04
180°	0·2	0·2	0·2	0·02	0·02	0
Erysiphe graminis (conidia c. 25 × 12μ)						
0°	0·25	0·13	0	0	0·05	0·92
45°	7·7	3·70	1·48	0·93	0·86	1·39
90°	2·70	0·77	0·06	0	0·07	0
135°	4·10	1·83	0·18	0	0	0
180°	0·72	0·19	0	0	0	0
225°	0·28	0·78	0·31	0·22	0·25	1·16
270°	0·39	0·14	0·12	0	0	0
315°	0	0	0	0	0·07	0
360°	0·72	0·19	0	0	0	0

thicker and thinner sticky slides, and also with thick plates, 10 cm wide having a double-bevelled edge. Bluff or bevelled edges were not sticky.

With a horizontal plate 6·4 mm thick, the edge-effect was present, and at medium wind speeds edge-shadow became pronounced. At 1·1 m/s nearly the whole surface was in the shadow of the leading edge, and at 3·2 and 5·5 m/s there was almost no deposit on the slide. At 9·5 m/s, however, there was some turbulent deposition on both upper and lower surfaces, both in the turbulent and streamlined wind-tunnel.

When the leading edge of the plate, 6·4 mm thick, was sharpened with a double bevel to form a 45° edge facing the wind (as used by Landahl & Herrmann, 1949), a very different effect was observed. In a wind of 0·5 m/s there was a very even deposit over the whole upper surface, but at 1·1 m/s an edge-shadow developed and spread across the surface as the wind speed was increased, until at 5·7 m/s deposition was negligible. There was no recovery by turbulent deposition at higher wind speeds, and at 9·5 m/s the thick bevelled trap underestimated spore concentration by a factor of 200 times.

Thin horizontal surfaces, by contrast, gave efficiencies much nearer to the expected values for gravity sedimentation. However, edge-shadow and turbulent deposition occurred – even with an edge 0·016 cm thick (a microscope cover-glass). A double-edged 'wafer' safety razor blade gave uniform deposits on the upper surface, which were close to expected values – except at 9·5 m/s, where the deposit was three times that expected.

ORIENTATION OF SPORES

Lycopodium spores showed different orientation in different parts of the deposit. Gregory (1951) stated that, on the stagnation zone up wind of a vertical cylinder, the spores lie with the rounded distal surface uppermost, and that spores settling in air under the influence of gravity come to rest in the same position. Further observation shows that this statement was incorrect, and that in the stagnation zone or its equivalent the spores lie with the rounded distal surface touching the cylinder. Evidently the spore becomes orientated with the point trailing as it moves through the air. Orientation with the point up wind is therefore characteristic of *Lycopodium* in the stagnation zone.

Microscopic observation of deposits showed that, as was expected, when the glass slide was horizontal the stagnation zone was on the edge, and when the slide was vertical this zone was in the middle of the slide. At intermediate presentation angles, seen most clearly at angles near 90°, the stagnation zone shifted, at 115° occupying zone B, and reaching zone A at about 120° at the higher wind speeds. At lower wind speeds orientation was less definite.

DEPOSITION ON 9 CM DIAMETER PETRI DISH

The petri dish trap, extensively used in aerobiological mould surveys, was tested horizontally after pouring with 15 cm³ of 2% water agar (tests showed that deposition and retention on this medium were similar to those on glycerine jelly). Mean deposition efficiencies (per cent A.D.) for 1 cm square zones on the agar surface are given by Gregory & Stedman (1953). At all wind speeds, narrow edge drifts occurred behind the rim of the leading edge, and in front of the rim of the trailing edge. Efficiency at 0·5 m/s was low, but at 1·1 and 1·7 m/s efficiency was high, apparently because of the large contribution to the total made by the front and back edge drifts. At 3·2 m/s and above, efficiency fell off substantially below expectation – apparently because the sampling surface was almost entirely shadowed by the 1 cm high rim of the dish. Somewhat similar effects were noted by Chamberlain (1966a).

Effects produced by the rim of the petri dish were nearly eliminated by placing the dish at the bottom of a metal cylinder 13 cm deep and 11·5 cm in diameter, sunk below a horizontal flat surface consisting of a square cardboard platform cutting the central axis of the wind-tunnel. The cardboard fitted flush with the mouth of the cylinder and extended 11 cm up and down wind. Practical use has been made of this principle by Tauber (1965).

The petri dish was also tested in a vertical position. At 9·4 and 5·5 m/s, the deposit was four times as great in the central zone of 2·5 cm radius as it was in the peripheral centimetre around the rim. At and below 3·2 m/s the difference was reversed, with nearly 75% more spores just inside the rim of the dish than elsewhere.

RETENTION AND BLOW-OFF FROM CLEAN SURFACES

Experiments showed that there is no appreciable loss of *Lycopodium* spores from the deposit on the surface of a slide with a sticky coating of glycerine jelly at any of the wind speeds tested. Blow-off from a non-sticky glass surface, however, depended on the wind speed and the angle of incidence of the wind. Clean microscope slides were placed in a spore cloud at 0·5 m/s to obtain a deposit, and were then placed successively in winds of increasing speed. Tests were made at angles of 0°, 45° and 90°. The percentage of the original deposit that was retained after 1 minute at each wind speed was estimated by counting. At the highest wind speed, 9·5 m/s, slight traces of grease on the slide greatly increased retention when the slide was horizontal, and unless the surface was carefully cleaned before use, erratic results were obtained under these conditions.

With the slide vertical, blow-off was nearly linear with wind speed, 98% being retained at 1·1 m/s, and 60% at 9·5 m/s. Blow-off was least at 45°, and at this angle retention was 100% at wind speeds up to 5·5 m/s, and 95% was retained even at 9·5 m/s. By contrast, blow-off was greatest with the surface horizontal, when 70% was retained at 1·7 m/s, and only 26% at 9·5 m/s. These results illustrate the way the laminar boundary layer acts as a dust trap (p. 33) but the actual values probably have little application to plant surfaces. The same problem has been studied in the form of blow-off of ragweed *(Ambrosia)* pollen from glass slides by Dingle et al. (1959, p. 376), who found that successively faster wind may shear off layers from pollen clusters, in a wind-tunnel.

Blow-off and bounce-off of solid particles such as large pollen and spores decrease the actual trapping efficiencies of surfaces. Values nearer to ideal are attained with the liquid drop tests by May & Clifford (1967), who showed that at higher values of E and k there was no significant difference between the collecting efficiencies of spheres, strips and cylinders, although with $k < 0·4$ strips were more efficient than cylinders, which in turn were more efficient collectors than spheres. Flat discs were markedly more efficient than other shapes.

Bounce-off. Chamberlain (1967a) used radioactive tagging of *Lycopodium* spores (32 μm diameter) and ragweed pollen (19 μm diameter) to measure efficiency of deposition on beech *(Fagus sylvatica)* twigs (0·65 cm diameter) in a wind-tunnel at 5 m/s, and found that the stickiness of the surface had a major effect on E. On twigs coated with a stopcock grease, both species of particle showed efficiencies similar to a sticky cylinder used as a standard of reference (E = approximately 35%). Twigs wetted with water were only slightly less efficient (E = 29%); whereas with dry twigs E = 4% for *Lycopodium* spores and 0·9% for ragweed pollen.

The inefficiency of dry twigs is attributed to bounce-off because re-exposure to wind of 5 m/s failed to blow off a further fraction.

In other tests, Chamberlain prepared artificial plastic 'grass leaves' to which he affixed segments cut from real leaves of grass, plantain and clover, and pieces of dry filter paper, in order to explore the effects of leaf surface texture on deposition.

With respect to *Lycopodium* spores and ragweed pollen, real dry leaf surfaces were only about one-tenth to one-half as efficient as sticky plastic 'leaves'. But with respect to polystyrene spheres, 5 μm diameter, and still more with droplets of tricresyl phosphate, 1 μm in diameter, efficiency of real dry leaves was *greater* than smooth sticky plastic – an effect attributed to the hairiness of natural surfaces.

DEPOSITION AND RETENTION ON PLANT LEAVES

The tests described in the preceeding sections on artificial surfaces give information on principles of particle deposition from the air, and are useful in evaluating apparatus for sampling airborne particles. We now need to ask how relevant this work is to spore deposition on plant surfaces which, though not sticky, are rough and flap in the wind *(see* Evans, 1961).

To imitate natural conditions, shoots of potato *(Solanum tuberosum)* with rough leaves, and broad bean *(Vicia faba)* with smooth leaves, were placed in the turbulent wind tunnel at Rothamsted and exposed to clouds of *Lycopodium* spores in the usual manner. The petiole of the leaf was clamped and leaflets allowed to flap freely, trailing the leaf tip down wind. After exposure, the leaf surfaces were examined under the microscope and the deposit was counted on zones across the tip, middle and base of the lamina (upper and lower surfaces), the deposition efficiencies being calculated (Table XIV).

Considerable differences from deposition on rigid, sticky horizontal

TABLE XIV

EFFICIENCY ($E\%$) OF DEPOSITION OF *Lycopodium* SPORES ON UPPER AND LOWER SURFACES OF POTATO AND BROAD-BEAN LEAFLETS IN TURBULENT WIND-TUNNEL

(Gregory & Stedman, *unpublished*)

Part of leaf		Wind velocity (m/s)						
		9·4	5·3	3·0	1·6	1·3	1·1	0·6
Potato leaflet								
Tip	(upper)	0·0	0·16	0·17	0·68	—	0·42	0·0
	(under)	0·0	0·06	0·04	0·19	—	0·89	0·0
Middle		0·06	0·24	0·30	1·34	—	1·62	3·20
		0·0	0·0	0·02	0·28	—	0·0	0·0
Base		0·0	0·12	0·47	1·10	—	8·7	5·8
		0·04	0·0	0·0	1·06	—	0·0	0·0
Broad-bean leaflet								
Tip	(upper)	0·06	0·0	0·15	—	1·4	2·82	5·20
	(under)	0·0	0·0	0·0	—	0·0	0·02	0·0
Middle		0·0	0·0	0·46	—	2·0	3·52	6·0
		0·0	0·0	0·0	—	0·0	0·06	0·0
Base		0·0	0·0	0·35	—	2·2	1·31	2·0
		0·0	0·0	0·02	—	0·03	0·05	0·0
Horizontal slide		0·42	0·32	0·05	0·09	—	0·29	0·6
		0·36	0·05	0·02	0·01	—	0·004	—
Theoretical		0·20	0·34	0·57	1·05	1·35	1·6	3·45
(sedimentation)		0·0	0·0	0·0	0·0	0·0	0·0	0·0

slides are apparent. Turbulent deposition either failed to develop at the higher wind speeds, or the spores were bounced off or blown off again in the wind. Deposit on the under sides of the leaves was small at all wind speeds, but on the upper side it was up to 5% of the area dose at winds of 0·6 to 1·0 m/s – both on potato leaves and on the smoother broad-bean leaves. At these low wind speeds the deposit was similar to that expected from sedimentation under gravity; but potato leaves had more spores at the base, and broad bean leaflets more at the tip. At higher wind speeds efficiencies were low.

More relevant to plant pathogenic fungi is the work of Carter (1965) with *Eutypa armeniacae* on apricot shoots. *E. armeniacae* discharges its ascospores in clumps of eight (= octads), forming wind-dispersed particles averaging $8·1 \times 4·4$ μm, which are deposited by impaction on leaves, stems and petioles of apricots (when deposited on wet surfaces the octads separate into single ascospores which can be redistributed by rain-splash).

Efficiencies for leaves and stems in wind-tunnel tests at the Waite Agricultural Research Institute, Adelaide were less than 1% at all wind speeds tested (Table XV), but for petioles efficiencies reached between 2 and 3% at 3·5 and 5·0 m/s. The total number of ascospores deposited on a defoliated shoot was only 3·5% of that on a similar leafy shoot. Blow-off of spores, once deposited on a dry surface, was negligible. Although deposition efficiency at a single encounter was small, the large area of tree exposed, giving multiple encounters, must make an orchard a highly efficient air filter.

TABLE XV

MEAN EFFICIENCY OF DEPOSITION FOR *Eutypa armeniacae* OCTADS ON APRICOT SHOOTS IN WIND-TUNNEL TESTS

(Carter, 1965)

| Wind speed (m/s) | Efficiency (E per cent) | | |
	Leaves	Petioles	Stems
1·0	0·09	0·93	0·15
2·0	0·10	1·59	0·30
3·5	0·36	2·76	0·73
5·0	0·41	2·57	0·89

There is reasonable agreement between observation and the theory of impaction, but other forms of deposition are complex and impure. Efficiency of a collecting surface as an impactor spore trap increases as the wind speed and particle size increase, and as the size of the collecting surface decreases. Deposition on horizontal trap surfaces and leaves can be predicted in terms of sedimentation under gravity at low wind speeds, but edge effects can easily predominate as wind speed increases; at 5 to 10 m/s turbulence may result in deposition upwards against gravity.

Deposition on leaves and other natural surfaces is decreased by bounce-off and blow-off of large spores when the surface is dry, and increased by impaction of small spores on hairs and microscopic projections in dry air.

VIII

Natural Deposition

Having become airborne and been transported by the wind, the spore will eventually quit the turbulent layers of the atmosphere, re-cross the boundary layer, and come to rest in the still layer of air on a solid or liquid surface, which may or may not prove favourable for growth. Some characters of spores seem to have been evolved in response to problems of take-off, while others may well have evolved as adaptations for deposition.

Little was known about deposition processes until the development of wind-tunnel techniques – originally for research in aerodynamics – made it possible to experiment on the behaviour of spores in controlled winds in the laboratory. The principal methods of spore deposition in nature can now be tentatively suggested: impaction, sedimentation, boundary layer exchange, turbulent deposition, rain-washing, and electrostatic deposition.

RELATION BETWEEN CONCENTRATION AND DEPOSITION

The relation between χ, the cloud concentration, and d, the surface deposition, has been little studied, although it is a relation of considerable biological importance – for instance in pollination, and in epidemics of plant diseases and their control by protectant dusts and sprays. With wind flowing across a smooth surface, it should be possible to calculate deposition directly from a knowledge of concentration, wind speed and terminal velocity (Figure 17). However, it is not obvious what factors dominate the situation under the complex surface conditions obtaining in nature, and so the problem must be approached experimentally. With some insight, outlined in Chapter VII into deposition under the relatively simple conditions in a wind-tunnel, we can start to analyse factors controlling deposition in nature. In wind-tunnel studies, and in calibrating spore traps, it is convenient to use percentage efficiency of deposition, $E = $ (trap dose/area dose) $\times 100$.

110

For field conditions, two expressions have been used for deposition on the ground. The 'deposition coefficient', $p = d/n$ (where d = number of spores deposited per square centimetre of surface, and n = number of spores passing through the cubic centimetre above the surface), measures the thickness of the slice of cloud cleared of particles by travelling over unit length of ground surface (Gregory, 1945). Under a given set of conditions, p is assumed to depend only on concentration, though it is probably affected by wind speed, turbulence and other factors. The 'velocity of deposition', introduced by Chamberlain (1956), was defined as:

$$v_g = \frac{\text{amount deposited per cm}^2 \text{ of surface per second}}{\text{volumetric concentration per cm}^3 \text{ above surface}}.$$

If p is assumed to be independent of wind speed, it will be apparent that, in our notation: $p = v_g/u$. Numerically p is identical with v_g in a wind of approximately 1 m/s (see Chamberlain, 1967a).

It might have been assumed that the rate of deposition would be given by Stokes's law, giving a sedimentation velocity v_s. But the factors p, or v_g, were deliberately chosen because sedimentation under gravity is obviously only one of many deposition mechanisms. Consequently the deposition coefficient was designed to express the thickness of the spore cloud cleared by all mechanisms including terminal velocity. The thickness of cloud cleared by deposition is assumed to have its concentration restored during continued travel over the surface by downward eddy diffusion as well as by terminal velocity. This treatment was justified, for example, by experiments with ragweed pollen at Brookhaven by Raynor et al. (1970, p. 893) who obtained a mean value over 22 tests of $v_g = 5\cdot05$ cm/s, compared with $v_s = 1\cdot56$ cm/s.

MEASUREMENT OF p AND v_g

The first attempt to evaluate p by Gregory (1945) was based on experiments by Stepanov (1935), whose results were tested against Sutton's (1932) eddy diffusion theory.

Sutton's formulae had been developed for calculating the concentration of a cloud of particles whose deposition was negligible – the number of particles in the cloud, Q_0, remaining constant throughout the diffusion. In our problem, although the effect of gravity on dispersion can be neglected, the quantity of spores remaining in suspension is steadily diminishing owing to a relatively large deposition from that part of the cloud which is in contact with the ground, so that Q_x, the total quantity remaining suspended in the cloud when its centre has moved to a distance, x, is less than the original Q_0. It has been shown (Gregory, 1945, Appendix by Margaret F.

Gregory) that Q_x will decrease exponentially with increasing distance, according to the equation:

$$Q_x = Q_0 \exp\left[-\frac{2\,p\,x^{(1-\frac{1}{2}m)}}{\sqrt{(\pi)}\,C\,(1-\frac{1}{2}m)}\right].$$

Values of Q_x and d for two values of the parameter m were calculated and it became possible to test the theory against Stepanov's results. In each of his experiments the total number of spores liberated differed, so the data had first to be put on a comparable basis by equating the mean deposition, d, at 5 metres to 100%, and then expressing the deposition observed at greater distances by relative percentages. The logarithms of the observed relative depositions were plotted against the logarithms of the distances in centimetres. The expected depositions when $p = 0.05$, 0.025, and zero, respectively, were also plotted, and the line calculated for $p = 0.05$ was seen to approach most nearly to the observed values (Gregory, 1945). A deposition coefficient of $p = 0.05$ means that, in travelling across 1 cm^2 of surface, the entire cloud would deposit a quantity of spores approximately equivalent to the number contained in a slice 0.5 mm thick through the axial plane of the cloud. This value of p was estimated from Stepanov's experiments in winds of about 1 m/s (Chapter VII).

Deposition was measured experimentally by Gregory et al. (1961) in field tests at the Imperial College Field Station, Ascot. Spores of *Lycopodium* and the much smaller spores of a bracket fungus, *Ganoderma applanatum*, were liberated a short distance above ground level in a field of short, rough grass. Cloud concentration and deposition, measured at a number of positions simultaneously, enabled p or v_g to be estimated at various distances up to 20 metres from the point source (Table XVI). (Two additional estimates with *Lycopodium* spores at 1 metre from the source on a smooth lawn at Rothamsted Experimental Station, gave $p = 0.05$ and 0.09, and $v_g = 3.1$ and 5.3 cm/s, respectively.) A remarkable phenomenon, shown in Table XVI, is that both v_g and p vary with distance, decreasing with distance from the source, until at 10 and 20 metres values for v_g similar to those for v_s are reached. (On theoretical grounds, Schrödter (1954, 1960) deduced that the 'probable final velocity' of a spore is only half the terminal velocity: a result at variance with observation where any deviation is in the opposite direction.)

Ground deposition conforms better to a sedimentation theory than does deposition on the horizontal microscope slide in the wind-tunnel recorded in Chapter VII. For comparison with ground deposition, Table XVII shows wind-tunnel values, calculated in terms of p and v_g, for deposition on a horizontal slide.

For very small spores in the 1 μm range deposition by diffusion becomes

TABLE XVI

DEPOSITION OF SPORES ON GROUND, FROM SIMULTANEOUS MEASUREMENTS OF χ AND d BY VISUAL COUNTS UNDER MICROSCOPE
(Gregory, Longhurst & Sreeramulu, 1961, and unpublished)

Height of liberation	Mean wind speed m/s	2·5 m p	2·5 m v_g cm/s	5·0 m p	5·0 m v_g cm/s	10·0 m p	10·0 m v_g cm/s	20·0 m p	20·0 m v_g cm/s
Lycopodium spores									
1·0 m	1·98	0·01	2·3	0·02	3·6	0·02	3·4	—	—
	0·83	0·02	1·2	0·05	3·8	0·06	5·1	—	—
	0·62	—	—	0·09	5·5	0·05	2·9	0·03	1·7
	0·40	0·44	17·6	0·05	2·1	0·02	1·0	—	—
0·25 m	1·28	0·05	6·9	0·03	4·4	0·02	2·7	—	—
	1·10	0·07	7·2	0·01	1·5	0·02	1·7	—	—
	1·09	0·13	13·7	0·04	4·5	0·02	1·8	—	—
	Mean:	0·12	8·16	0·042	3·64	0·029	2·8	0·03	1·7

Ganoderma spores (distances: 1·0 m, 1·5 m, 2·5 m, 5·0 m)

Height	Mean wind m/s	1·0 m p	1·0 m v_g	1·5 m p	1·5 m v_g	2·5 m p	2·5 m v_g	5·0 m p	5·0 m v_g
0·25 m	1·70	—	—	—	—	0·003	0·44	0·0008	0·14
	0·91	0·022	1·97	0·006	0·56	0·005	0·46	0·004	0·34
	0·76	0·015	1·13	0·006	0·43	0·003	0·24	0·001	0·08
	0·61	0·014	0·83	0·010	0·61	0·006	0·34	0·005	0·29
	Mean:	0·017	1·31	0·007	0·53	0·004	0·37	0·003	0·21

Summary

Lycopodium	Mean of all values p	v_g	Mean at selected distance p	v_g
c. 32 μ dia. $v_s = 1·76$ cm/s	0·057	4·35	0·03	2·7 at 10 metres
Ganoderma c. 10×6 μ $v_s = 0·18$ cm/s	0·007	0·56	0·004	0·37 at 5 metres

TABLE XVII

OBSERVED VALUES OF p AND v_g FOR *Lycopodium* SPORES ON HORIZONTAL MICROSCOPE SLIDE IN WIND-TUNNEL (TURBULENT AND STREAMLINED CONDITIONS)

	950	575	320	170	110	50
Turbulent wind						
p	0·0043	0·0029	0·0005	0·0009	0·0050	0·026
v_g cm/s	4·10	1·7	0·10	0·15	0·55	1·3
Streamlined wind						
p	0·0042	0·0013	0·00008	0·00002	0·0004	—
v_g cm/s	4·0	0·75	0·025	0·004	0·004	—
p expected for $v_g = 1·75$ cm/s	0·00185	0·003	0·0055	0·01	0·0016	0·035

equally important with impaction (C. N. Davies, 1966). Wells & Chamberlain (1967) show experimentally that particles in the same size range are deposited on a rough surface with v_g several orders of magnitude greater than their terminal velocity, suggesting that hairs on leaves and other surface roughnesses on a small scale, increase microbial deposition from the wind. Studying particles ranging from 0·4 to 1·2 μm diameter, Peirson & Cambray (1965) found dry deposition rates ranging from 0·25 to 2·0 cm/s, respectively, values much greater than Stokes's law would predict. Perhaps in nature there is a lower limit of 0·25 cm/s, even for the small spores of Actinomycetes.

Deposits from artificially liberated spore clouds on glass surfaces can be recorded easily, but on natural surfaces in the field deposited spores are elusive. Chamberlain (1966a, 1967a) overcame this difficulty by radioactive tagging of *Lycopodium* spores with [131]iodine (half-life 8 days). In addition, thorium B (half-life 10·6 hours) was used in wind-tunnel experiments. In the field tests tagged *Lycopodium* spores were liberated at a height of 50 cm above ground level and sampled at various points down wind, both on sticky cylinders to measure cloud concentration (wind speed being known), and on filter paper in petri dishes among the grass, also on the grass itself, the underlying herbage and the soil. Measurable activity was detected at up to 100 metres from the source. At low wind speeds v_g for *Lycopodium* spores to dry grass and soil approximated to v_s (about 2 cm/s), but at high wind speeds impaction gained importance with $v_g = 3$–4 cm/s, in spite of loss of spores from bounce off leaves. With grass wetted by rain a value of $v_g = 6·8$ cm/s was recorded.

In the wind-tunnel tests with *Lycopodium* spores and *Ambrosia* pollen depositing on real grass, small particles in the 0·1 to 5·0 μm range were less affected by stickiness because leaf hairs and other irregularities assumed greater importance. With particles larger than 20 μm, v_s and v_g were similar, but for smaller spores other components of v_g predominated, and Chamberlain observed $v_g/v_s = \pm 10$ for particles in the 1–5 μm range. (Below 1 μm, v_g actually increased again.)

The increased values for p or v_g near the artificial source recorded by Gregory *et al.* (1961) were not observed by Chamberlain, who considers the effect may have been an artefact due either to clumping of spores, bulk sedimentation (Chapter II, p. 25), or excessive turbulent deposition where the cloud first approaches the ground.

Chamberlain (1967a) breaks down v_g into the sum of two components: $v_g = v_s + v_T$ (where v_T is the additional apparent velocity of deposition due to impaction). At very low wind speeds, v_T is small compared with v_s, but at moderate or high wind speeds it may be comparable or greater.

LOSS BY DEPOSITION FROM SPORE CLOUD

To judge from outdoor values of p and v_g, a spore cloud liberated near ground level must lose an important fraction of its original number by deposition on the ground during the early stages of its diffusion. Relevant field measurements are scanty, but we have attempted to make minimum estimates of total ground deposit from the experiments of Stepanov (1935), Gregory et al. (1961) and Sreeramulu & Ramalingam (1961), in all of which Q_0 was estimated and where deposition was measured at several distances along lines radiating from the point of dispersion. Each experimental value was plotted to scale on graph paper, points were joined, and the total number of spores deposited within the sampling area was estimated by integrating areas under the curves (Table XVIII). It is clear that spores must also have been deposited on each side of the area sampled, and the values are therefore underestimates; they were made without extrapolation, except that, where the number of radii on each side of the axis of dispersion was unequal, a symmetrical dispersion was assumed. With liberation at 0·25 m above ground level, from 14 to 24% of the *Lycopodium* spores liberated were deposited within 10 metres radius in winds of little over 1 m/s. With the much smaller spores of *Podaxis*, only about 1% were deposited within 10 metres (Table XVIII).

We have no estimates of recovery on test areas for distances greater than 10–45 m shown in Table XVIII. But calculations by Chamberlain (1966a, p. 67) suggest a greater flight range for pollen than the experiments quoted.

DEPOSITION MECHANISMS OUTDOORS

BOUNDARY LAYER EXCHANGE

Observations on deposition outdoors are still meagre, but values for v_g obtained with spores of two widely different sizes and fall velocities, *Lycopodium clavatum* (1·75 cm/s) and *Ganoderma applanatum* (0·18 cm/s), suggest the following picture of deposition.

Deposition on the ground tends to remove spores from the base of the spore cloud. The concentration of the cloud near ground level is then restored by diffusion – either horizontal diffusion from a nearby source, or downward vertical diffusion from a reservoir of particles overhead, when the atmosphere itself is acting as a source. Diffusion thus brings particles down to the boundary layer, and here they settle out mainly under gravity. This is why, when v_g is estimated from χ as measured just above ground level, it numerically approaches the Stokesian terminal velocity. This process, which has been called 'boundary layer exchange', continually re-

TABLE XVIII

PERCENTAGE OF TOTAL SPORES (Q_0) LIBERATED NEAR GROUND-LEVEL THAT
WERE ESTIMATED TO HAVE BEEN DEPOSITED ON GROUND IN OPEN AIR TESTS

	Wind speed	Height of liberation	Dimensions of sector of annulus		Estimated proportion of spores recoverd on test area
			Distance limits	Angle	
	(m/s)	(m)	(m)	(deg)	$(\%)$
	Stepanov				*Tilletia caries*
	0·5–4·0	0·8–1·2	5–20	105	11·2
	2·3–3·0	1·5	5–40	105	8·55
	Gregory, Longhurst & Sreeramulu				*Lycopodium*
	1·01	0·25	2·5–10	120	24·4
	0·40	1·0	2·5–10	100	18·1
	1·28	0·25	2·5–10	120	14·5
	0·83	1·0	2·5–10	120	13·5
	Sreeramulu & Ramalingam				*Lycopodium* *Podaxis*
II	2·1	0·5	2·5–30	140	11·4 —
III	2·2	0·5	2·5–30	140	12·1 —
IV	0·8 (Cold night)	0·5	2·5–30	140	45·5 —
V	2·3	0·5	2·5–30	140	10·4 —
VI	0·8 (Cold night)	0·5	2·5–45	60	91·6 —
VII	3·2	0·5	2·5–35	80	11·7 0·48
VIII	1·1 (Dawn)	0·5	2·5–30	70	30·0 1·6
IX	4·3	0·5	5–30	80	8·1 0·3
XII	1·8 (Before dawn)	0·5	2·5–30	60	9·5 —
XIII	4·4	0·5	2·5–30	80	5·4 —

plenishes the ground layer of still air in which even minute spores can
sediment-out.

The demarcation between the laminar surface layer and the turbulent
windstream is not sharp. From time to time spore-bearing eddies break
into the laminar layer which is being cleared by sedimentation, removing
spore-free air and leaving in exchange small volumes of spore-laden air.
These spores sediment at terminal velocity under the influence of gravity
and are soon out of the reach of further eddies. In boundary layer exchange,
turbulence has a great effect in bringing spores down to the layer where
they can slowly sediment under the influence of gravity.

SEDIMENTATION

As we have seen, the effect of gravity is usually negligible on vertical
distribution in the atmosphere. Still air, as Tyndall (1881) found long ago,
soon becomes purified because microbes settle out from it under gravity –
a process we call sedimentation. Outdoors the air is almost never still,
except within about a millimetre of the surface. In the turbulent air layer

above the ground, the effects of gravity are slight and difficult to demonstrate; but Rempe (1937) demonstrated gravitational sorting of different species of tree pollen near Göttingen (*see* p. 28). And Hirst *et al.* (1967) observed gravitational sorting of pollens and spores during travel in the upper air over the North Sea (p. 275). Under stable conditions on clear nights, when air at ground level is cooled by radiation, the laminar layer may extend to a height of several metres. This zone may be nearly cleared of pollen by sedimentation, while above it a much larger spore and pollen concentration may be retained in the turbulent layer.

Sedimentation is mainly observed in the laminar layer which normally extends only a few millimetres above the surface, but under exceptional conditions at night it may reach up to several metres. Wind-tunnel experiments confirm that the effect of sedimentation is slight at wind speeds of 2 m/s and upwards.

IMPACTION

When a small surface, such as a leaf or twig, projects into the wind, spores may be impacted on its windward side. Wind-tunnel experiments confirm that impaction is inefficient when small spores approach large obstructions at low wind speeds. Conversely, impaction is more efficient when large spores are blown towards small objects at high wind speeds. So it seems that large spores, in addition to carrying a bigger food reserve, have the advantage of a favourable size for impaction on surfaces. Dry-spored, airborne leaf pathogens usually have comparatively large spores (e.g. uredospores, aecidiospores, *Phytophthora*, *Helminthosporium*, etc).

On the other hand, dry-spored soil inhabitants are characterized by small spores, unsuitable for impaction (e.g. *Penicillium* and *Aspergillus*). Among vegetation, where the wind speed normally reaches an upper limit of about 2 m/s, spores of *Lycoperdon perlatum* 4–5 μm in diameter would not be impacted at all, even on objects as narrow as 1 mm; and it can be calculated that spores of this size would require a wind of about 25 m/s to be impacted with only 10% efficiency on a blade of grass. Evidently we must look to processes other than impaction to deposit minute spores, such as those of the puff-balls, earth-stars, *Ustilago*, and the common moulds. The loose smuts of cereals, *Ustilago* spp., with spores in the 7–9 μm range, would not be impacted efficiently on leaves and stems; but on narrow surfaces such as glumes and stigmas of grasses, they would be expected to reach an efficiency as high as 50 to 75%.

Agaricus (Psalliota) campestris has spores about 7×6 μm which should be near the lower limit of size for impaction on grass leaves or stems at the limiting wind speed in closed vegetation. Uredospores of *Puccinia graminis* and conidia of *Erysiphe graminis* would impact on a wheat leaf with efficiencies near 40–60%. *Botrytis polyblastis*, a leaf pathogen of *Narcissus*, with spores up to 90 μm diameter, comes into the same group of plant pathogens that are relatively efficiently impacted on leaves and stems.

The difficult problem of impaction on objects as wide as tree trunks seems to have been solved by some lichens. *Pertusaria pertusa*, as observed by the author, can shoot single ascospores, measuring some 150×50 μm to a horizontal distance of 40 mm. Extrapolation from wind-tunnel data suggests that, once launched into a wind of only 2 m/s, these giant spores (nearly the largest in the fungi) would impact on a tree trunk of 20 cm diameter with about 50% efficiency.

Although a high impaction efficiency may be necessary for fungi which attack leaves and stems, it may be positively disadvantageous for spores produced among dense vegetation. Johnstone *et al.* (1949) point out that the ability of an airborne particle to penetrate close vegetation is the inverse of impaction efficiency. In close vegetation a high impaction efficiency would reduce the chances of a spore getting very far from its point of liberation. It is possible that the 10 μm-diameter spore represents a working compromise between the conflicting requirements of dispersal and deposition, evolved under the normal range of winds encountered among vegetation. The large-spored leaf and stem pathogens appear to be specialized *impactors*, whereas the minute-spored puff-balls and moulds appear to be specialized *penetrators* – perhaps normally deposited by processes other than impaction.

TURBULENT DEPOSITION

Turbulent deposition has been observed in the wind-tunnel and on artificial spore traps in the open air. Spore-laden air flowing over horizontal surfaces will deposit spores at rates greater than those calculated for sedimentation under the influence of gravity. In the wind-tunnel turbulent deposition increases with increasing wind speed, and at 5 to 10 m/s, deposition may be as great on the under side of a horizontal surface as it is on the upper side – an effect which clearly cannot be caused either by impaction or sedimentation. The effect has also been noted on pollen traps by Durham (1944, p. 233), who found the catch on the lower surface of a horizontal slide to be as high as 50% of the upper surface. Rishbeth (1959) found that spores of *Fomes annosus* and *Peniophora gigantea* can be deposited on the under surface of pine stem sections exposed a few metres above ground.

Turbulent deposition in the open air is also illustrated by experiments with *Lycopodium* spores liberated just above short grass at Rothamsted Experimental Station (Table XIX). Deposition was recorded on the upper and lower surfaces of horizontal sticky plates held clear of supports on long pins projecting from a wire frame. Catches shown in Table XIX are for traps at the same level as the spores were liberated ($h+0$), and at 25 cm above this level ($h+25$) in a down wind direction. The dimensions of the traps were $18 \times 18 \times 4$ mm in the x, y and z directions respectively.

The greater deposit on the under surfaces of traps at 25 cm above the level of emission, together with the unexpectedly large deposition on ground near the source as shown in Table XVI, is evidence of factors near the source of the cloud which still need to be elucidated.

TABLE XIX

NUMBER OF *Lycopodium* SPORES DEPOSITED ON UPPER AND LOWER SURFACES OF HORIZONTAL TRAPS, $Q_0 = 10^8$ SPORES LIBERATED NEAR GROUND LEVEL

(Gregory, *unpublished*)

	Height of trap (cm)	Distance down wind of source (cm)							
		25	50	75	100	105	130	155	180
11 February 1948									
	$h + 25$	12	5	4	3	—	—	—	—
		19	19	8	15				
	$h + 0$	7640	1915	659	353	—	—	—	—
		208	238	88	27				
30 April 1948									
	$h + 25$	0	4	4	7	30	0	0	4
		0	11	4	22	55	0	0	60
	$h + 0$	1000	3500	730	520	540	320	120	166
		18	0	0	7	0	7	4	4

Conditions	Wind speed at 2 metres (u)	Height of liberation (h)
11 February 1948	3·7 m/s	5 cm
30 April 1948	3·9 m/s	7 cm

ELECTROSTATIC DEPOSITION

The basidiospores of the cultivated mushroom and some other fungi were shown by Buller (1909) to carry small electric charges when falling in air. Little is known about the phenomenon, and its effect on spore movement is probably negligible, except when the spore is within about a millimetre of another body. The origin of the charge, its effect over very short distances, and its relation to the vertical potential gradient in the atmosphere, which is said to average about 150 volts per metre, might repay future investigation.

Ingold (1957) suggested that in polyporous fungi, with long and narrow vertical hymenial tubes, electrostatic forces may keep a basidiospore in the middle of the tube, preventing deposition on the walls of the tube, while the spore is falling slowly under gravity. Taggart *et al.* (1964), finding that slightly tilting the tubes inhibited spore discharge, considered that such a mechanism is inoperative. Gregory (1957) showed that basidiospores of *Ganoderma applanatum* usually carry a positive charge when allowed to fall between two condenser plates which are charged + and − 200 volts with respect to earth potential. In one test the average deflection of the spores from the vertical was 12°, indicating a velocity component towards the plate of 0·036 cm/s in a field of 400 volts per cm. Swinbank *et al.* (1964) found that the basidiospores of *Merulius (Serpula) lacrymans* carry negative charges averaging $(1·35 \pm 0·12) \times 10^{-8}$ electrostatic units, which they considered could not significantly affect the movement of spores in relation to the hymenial surface.

The ascospores of *Lophodermium* [*pinastri*], the cause of pine leaf-cast disease, are shot into the air from fruit-bodies on fallen pine needles. Rack (1959) using a potential difference of 75 volts in an apparatus resembling ours, but without earthing the battery, claimed that immediately on ejection from the ascus the spores carried a negative charge, but that, at a height of 1·0 to 1·6 metres above ground level, from 93 to 99% of the spores carry a positive charge. He concluded that this interesting phenomenon might aid adhesion of spores to healthy pine needles on the tree.

In the normal, fine-weather electrical field of 1 volt/cm over a flat surface with the Earth negatively charged, a *Ganoderma* spore would gain a negligible 0·05% of its terminal velocity. But with smaller microbes the effect may not be negligible. Ranz & Johnstone (1952) showed that aerosol particles 0·5 μm in diameter could easily carry charges by which their deposition or suspension would be controlled by the Earth's field rather than by gravity. A particle near a projection on the Earth's surface might be attracted or repelled by a force ten times greater than that of gravity.

The charges observed on basidiospores and ascospores are presumably acquired during the liberation process. Later in their aerial transport the initial charge may very well be masked by capture of atmospheric ions which are normally present in the air (*see* Krueger *et al.*, 1969). Little is known of such effects however.

MINOR DEPOSITION MECHANISMS

It is also known that small particles tend to move down a temperature gradient (Cawood, 1936; Watson, 1936), and that forces exist which cause particles, at least up to 2 μm in diameter, to be repelled by a hot surface and attracted by a cold one. Smoke particles and *Lycopodium* spores are also known to move away from light in the phenomenon of 'photophoresis' (Whytlaw-Gray & Patterson, 1932).

RAIN-WASHING (SCRUBBING, RAIN-OUT, WASH-OUT)

Natural raindrops vary in size up to a maximum diameter of about 5 mm, above which they become unstable and break up during fall into two smaller drops. They have terminal velocities of fall ranging from 2 to 9 m/s (Best, 1950; see also Gunn & Kinzer, 1949).

The pick-up of small spheres in the path of falling raindrops was studied theoretically by Langmuir (1948) in connection with artificial rain-making in Hawaii. To judge from Langmuir's figures, the minute spores of *Lyco-perdon* and of the soil-inhabiting Penicillia would fail to be collected at all by drops much below 1 mm in diameter, and efficiency of collection would rise to a maximum of about 15% with droplets about 2 mm in diameter, decreasing again with still larger drops. Basidiospores of *Agaricus* (*Psalliota*) *campestris* should begin to be collected by raindrops of 2 mm in diameter, and decrease slightly with drops of larger diameters. Spores of *Tilletia caries*, uredospores of *Puccinia*, and conidia of *Erysiphe graminis*, could be collected by any possible raindrop, and collection would reach a maximum of about 80% efficiency with drops 2·8 mm in diameter.

The optimum size of spore for deposition in rain varies with the size of prevalent raindrops. As we have seen, spores of *Lycoperdon perlatum* would not be collected by drops less than 1 mm in diameter, and it has been shown that drops of less than this diameter do not operate the bellows mechanism of this species (cf. p. 46, and Gregory, 1949). According to Langmuir, collection efficiency is at a maximum with drops of about 2 mm in diameter for all spore sizes, and is then about 25% for spores 4 μm in diameter, and 80 to 90% for spores 20 to 30 μm in diameter.

Nothing is known of the pick-up of non-spherical spores and how properties other than size and terminal velocity affect collection by rain-drops. The spores of *Ustilago perennans*, although not easily wetted by water, are evidently readily collected by rain. The work of Burges (1950) suggests that, when non-wettable spores reach the ground, they remain in the upper layers of the soil. For the leaf and stem parasites, removal from the upper air is possibly unfavourable, while for small-spored soil inhabi-tants as well as for smuts and other fungi that infect seedlings, it may be the normal method whereby they come to ground.

Figures 20 and 21, compiled from data of Best (1950) and others, show the proportion of raindrops of various sizes occurring in natural rainfall, and the collection efficiencies of such drops (*E%*) as given by Langmuir (1948). However, Starr & Mason (1966) found experimentally that the maximal drop size for collection of *Lycoperdon*, *Ustilago nuda*, and pollen

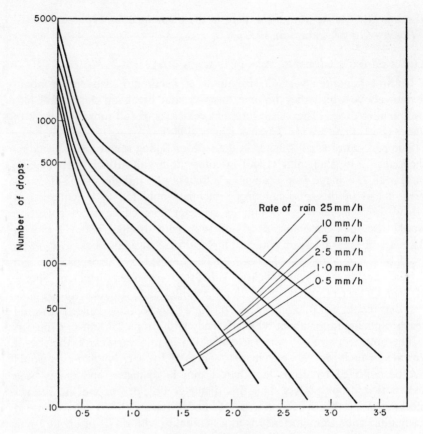

Figure 20. Numbers of raindrops of various diameters passing through a horizontal square metre 'frame' during rainfall of different intensities, from 0·5 to 25 mm/h. (Smoothed curves from data of Best, 1950.)

of paper mulberry, was by water drops rather smaller (800 μm diameter) than calculated by Langmuir. On the basis of these experiments, Starr (1967a) estimated that 75% of *Lycoperdon* and 20% of paper mulberry pollen remain in the air after 2 hours of rainfall of 1 mm per hour. Whereas with 5 mm of rain per hour, about 35% of *Lycoperdon* would remain airborne after 2 hours.

It might be supposed that hydrophilic particles would be more easily picked up by falling raindrops than hydrophobic ones. However, McCully *et al.* (1956), R. R. Davies (1961), and McDonald (1962, 1964) all indicate that both wettable and non-wettable particles are readily collected by falling water drops.

Chamberlain (1956) used the results of work by Best (1950) to calculate

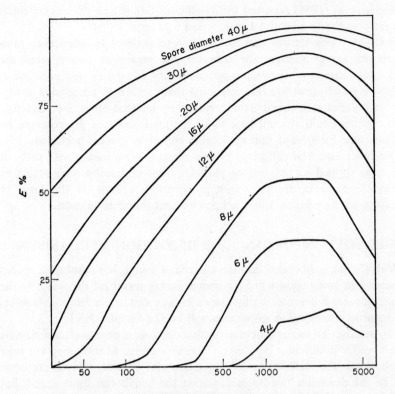

Figure 21. Collection efficiencies with which raindrops of diameters from 0·03 to 5·0 mm collect spores of from 4 to 40 μm diameter. (Smoothed curves from values computed by Langmuir, 1948.)

that, if other causes of diminution of the spore cloud are neglected, rainfall at the rate of 1 mm per hour would deplete an aerosol of *Lycopodium* spores according to the equation. $\chi_t = \chi_0 \exp(-3\cdot3 \times 10^{-4} t)$. Tsis implies that after one hour the concentration will be reduced to one-third of its initial value. Chamberlain applied the concept of v_g to rain wash-out, and showed that, the greater the height of the spore cloud through which rain falls, the greater the rate of wash-out. The study was continued experimentally at Harwell by F. G. May (1958), who found good agreement with Chamberlain's theoretical values when tested with *Lycopodium* spores marked with [131]iodine. When testing thundery showers, he also found good agreement with values calculated by Langmuir (1948) and by Mason (1957), but the predominantly smaller and more variable drop sizes in frontal rain removed particles faster than had been predicted by theory.

F

McCully *et al.* (1956) reported highly efficient removal of fluorescent dusts by rain from the air in both laboratory and field tests.

F. G. May pointed out that, in addition to removal by impaction, raindrops can act as electrostatic collectors by means of two effects: the coulombic attraction between oppositely-charged spores and raindrops, and by induced attraction. He estimated that capture by coulombic attraction in thunder rain might equal capture by induced attraction, and that both together could amount to 20% of the total wash-out. By contrast, he suggests that for frontal rain coulombic collection can be neglected.

One other aspect of collection by rain remains to be mentioned: rain-out is a term applied to removal of particles from the air by their acting as condensation nuclei around which precipitation collects in the form of raindrops or snowflakes. Little is known about this phenomenon.

RELATIVE IMPORTANCE OF DEPOSITION MECHANISMS

With Q_0 spores liberated into the air from a source, the number of spores remaining in suspension will be reduced during travel on the wind by the action of various deposition processes and, at a distance x from the source, the number remaining in suspension will be Q_x, Chapter XVI.

The relative importance of the various deposition processes can now be seen to differ in different positions. Close to a source liberating spores near ground-level, the mechanisms of impaction and boundary layer exchange will be the dominant factors in depleting the windborne spore cloud, but in calm airs sedimentation will play a leading role. Near the source, rain-wash will have a relatively slight effect, because the height and breadth of the cloud are small. But as the cloud diffuses to its maximum height, rain will have an increasingly large effect. With the very small particles pushed up into the stratosphere by megaton bombs and falling back into the troposphere, rain-wash with rain-out appear to be the most important mechanisms for removal. The level of ^{90}strontium in soil at Antofagasta, Chile, in rain-free desert, is only 1% of that general for places of the same latitude with normal rainfall (Libby, 1956).

Vegetation outdoors constitutes an important agent for clearing the atmosphere of suspended microbes. Neuberger *et al.* (1967) found that at a distance of 100 metres within a dense stand of coniferous trees, 80% of the ragweed pollen entering the forest had been removed by filtration on vegetation, which had also captured 34% of the much smaller Aitken nuclei. Deciduous trees were less effective, and the amount of pollen increased relatively with afternoon turbulence.

Local irregularities of terrain may alter the total deposition on an area; for example, Heise & Heise (1957) found that the warm air rising from a city prevented much of the deposition of ragweed pollen and *Alternaria* spores.

Measurements of deposits on leaves are few, but Rishbeth (1959) used surface microflora of conifer needles and birch leaves to measure infection potential of forest contamination with spores of *Fomes annosus* and *Peniophora gigantea*. Preece (1963) scanned apple leaves microscopically to map deposited spores. Evans (1961) mapped the local irregularities in fungi causing lesions on banana leaves.

Working in Danish forests, Tauber (1967) measured heavy deposits of pollen on branches and leaves which effectively filtered the air in the trunk space. Moreover when subjected to rain, substantial percentages of the pollen could be refloated and carried by droplets.

Rain-washing of air, when it occurs, rapidly ends the process of spore dispersal, and seems to be effective even with small spores which are deposited only slowly and ineffectively by other processes.

Besides washing spores out of the air, raindrops may possibly serve to transport spores from one atmospheric level to another: (1) by the drop entraining particles in its wake for a time, during its fall (Magarvey & Hoskins, 1968); and (2) by evaporation of the drop before reaching ground level, leaving its captured particles again in suspension in the air lower down. If carried up again in a thermal, this could be a mechanism by which spores become concentrated at the base of a cumulus cloud.

IX

Air Sampling Technique

Knowledge of the air spora has depended on developing sound techniques for air sampling. Direct microscopic examination of spores suspended in air is scarcely ever practicable. All convenient methods depend on apparatus to remove the spores to a surface where they can be examined – either directly by microscopy or after growth in culture.

Until recently, sampling was largely instinctive. Early methods of air sampling were summarized by Cunningham (1873). Developments during the next seventy years were reviewed by the Committee on Apparatus in Aerobiology of the National Research Council, Washington, D.C. (1941) and again by duBuy et al. (1945) of the United States Public Health Service.

The basic principles for efficiently removing microbes from moving air for enumeration have been enunciated by K. R. May. In 'isokinetic sampling' (a term introduced by Druett, 1942), air is drawn in, through a feathered orifice facing the wind, at the same speed as the wind, followed by acceleration to a velocity adequate to deposit the entrained microbes on a retaining surface (May, 1945). In 'stagnation point sampling' (May, 1966, 1967) the wind is brought to rest by a baffle and the sample withdrawn from almost still air.

In light of the discussion of deposition mechanisms in the last two chapters, sampling methods can be reviewed once more. The account is aided by modern reviews by Wolf et al. (1959), Zhukova (1963), May (1967), Akers & Won (1969) and R. R. Davies (1971).

The various sampling techniques now possible have different advantages and limitations. The following questions have to be answered before a technique is chosen for a particular job. Is an assessment of the total air spora wanted, or is the study concerned only with a few groups or a single species? Is a continuous day and night record needed, or will short and regular or even occasional samples suffice? If changes in the air spora are being studied, what time intervals are necessary: Would a 24-hour mean

126

suffice, or must an accuracy of ± 1 hour be attained? Or again, is an instantaneous cut-off needed to give a time discrimination of minutes or seconds? The size of the sample must be decided upon, and the choice made between the volumetric and deposition methods. Allergists are normally interested in the number of particles (dead or alive) in a given volume of air, whereas plant pathologists and plant breeders are more interested in the deposition of viable spores or pollen grains.

Biological air-sampling requires apparatus, and many questions must be asked about this. What are its capital and running costs? (It is uneconomic to use the time of a trained scientist with a good microscope on inefficient sampling methods.) Does it need a power supply? Must the equipment be portable? Is it to be operated by a skilled staff or must it be robust and foolproof? Does its efficiency vary with wind speed?

When the organism has been caught it has to be identified, but is this to be done visually or by cultural methods? Visual methods (using various forms of microscopy) should give the complete picture of the air spora, with spore aggregates as well as inorganic particles, whereas cultural methods allow the most accurate taxonomic determinations – but they will show only viable microbes that can grow on the media chosen. Visual pollen identification can be carried to species level in many genera: but whereas spores of a few pathogenic fungi can be recognized with certainty, the identification of spores and cells of bacteria, Actinomycetes and most of the smaller fungi, is tentative in the extreme (*see* Appendix I, p. 299).

To economize description in the following account, the various kinds of apparatus for sampling the air spora are grouped according to the physical processes by which they remove particles from the air and deposit them on surfaces for examination. Many of the techniques can be adapted to give either a deposit for visual examination under the microscope or cultures on artificial media. The basic types of equipment tend to proliferate variants with minor modifications.

GRAVITY SEDIMENTATION METHODS

SEDIMENTATION FROM STILL AIR

A simple box, developed by Alvarez & Castro (1952) for the study of airborne fungi, had two hinged sides and a covered tray at the bottom for inserting a microscope slide or a petri dish. To take an air sample, the two hinged sides were raised horizontally and wind was allowed to blow through the box. Closing the hinges trapped a boxful of air and the entrapped spores then sedimented under gravity. In theory the result is not affected by wind velocity or particle size. Sampling is discontinuous, only a small volume of air being sampled at a time, and there will be losses on the walls and roof due to convection and diffusion (*see* Tyndall, 1881; Green & Lane, 1957, p. 229). An improved form of this is described by Ogden, Raynor & Hayes (1971).

SEDIMENTATION FROM WIND

The method of examining the deposits on a freely exposed, horizontal surface, such as a glass miscroscope slide, was used by Pouchet in his controversy with Pasteur, but it can be traced back to van Leeuwenhoek. The method is more correctly described as 'dust examination'; it fails to examine air quantitatively, and its defects will be apparent from Chapters VII and VIII.

(i) *The 'gravity slide'* has been the routine method for investigating the pollen and spore content of the air since the early days of hay fever studies. Scheppegrell (1922) used ordinary 76×25 mm microscope slides exposed without protection from rain. Most workers, however, e.g. Blackley (1873), Wodehouse (1945), Durham (1946), Hyde & Williams (1950), and Hyre (1950), have exposed the slides horizontally, with the sticky side facing upwards in some form of shelter, open to the wind, but giving protection from rain.

The gravity slide is cheap, simple, and operates continually, but has serious defects as a quantitative method of air sampling outdoors, giving a highly distorted picture of the air spora because it preferentially selects the larger particles. Terminal velocity increases as the square of the effective particle radius (Chapter II), and terminal velocities are such that under ideal conditions, while, for example, a slide is receiving the puff-ball spores from a layer of air 0·5 mm thick, it also receives rust uredospores from a column 10 mm thick, and grass pollen from a column 50 mm thick. A volume of air containing puff-ball spores and grass pollen in equal concentration would be recorded by the gravity slide as having 100 times as many pollen grains as puff-ball spores. To correct this distortion Scheppegrell (1922) tried to calculate the volumetric concentrations from gravity-slide deposits using a formula based on particle *diameter*, which was later corrected by Cocke (1937) to particle *radius*. (This easy error recurs in the literature regularly.)

Because of edge effects, and turbulent deposition at high wind speeds (Chapter VII), the gravity-slide catch is very difficult to interpret; but in spite of these defects it has been widely used and has contributed much knowledge of the air spora.

(ii) *The gravity petri dish.* A few workers have exposed petri dishes horizontally in outside air to provide pollen deposits for visual examination (Hesselman, 1919; Ludi & Vareschi, 1936).

More commonly, petri dishes of sterile medium are exposed in the open air for periods of from 1 to 10 minutes to investigate the cultivable bacterial or mould flora of the atmosphere. (Some workers expose empty sterile dishes and add the medium on returning to the laboratory.) Indoors the method is subject to distortion owing to the sedimentation rate. Outdoors, however, the method is also subject to aerodynamic effects from the edge of the dish – unless the dish is sunk below a hole in a plane surface, in which case the observed catch is close to expectation (Gregory & Stedman, 1953, p. 666): an arrangement adopted for palynology in the Tauber trap (Tauber, 1965). For sampling out-of-doors the gravity petri dish has been used in tests by: P. F. Frankland & Hart (1887), Saito (1904, 1908 and 1922), Zobell & Mathews (1936), Dye & Vernon (1952), Menna (1955), Richards (1954a and b), Werff (1958), and many others.

Apart from convenience and economy, the method is valued for the precision with which the catch can be identified in resulting cultures. Its defects are: sensitivity to particle size, wind speed, and aerodynamic effects (p. 105); also the small volume of air sampled intermittently. Continuous sampling is impracticable and diurnal changes of the air spora are not revealed. Its restriction to viable and cultivable particles may be developed with advantage to a high degree of selectivity when sampling is aimed at a limited group of organisms. With pleomorphic fungi, positive growth in culture may still leave the origin in doubt; a *Phoma* or a *Fusarium* culture might have arisen from either a conidium or an ascospore – a detail which it is sometimes important to resolve.

(iii) *Conical funnels* have been exposed for trapping conidia of downy mildew of the vine *(Plasmopara viticola)* by Savulescu (1941), for forecasting mildew outbreaks in Rumania.

SEDIMENTATION FROM ARTIFICIALLY MOVING AIR

Hesse's method of sedimentation in long horizontal tubes has been described in Chapter I. Cocke (1938) used the same principle for visual microscopic airborne pollen. A small chamber was lined top and bottom with eight glass slides, leaving a passage of only 1–2 mm between floor and roof. With 1·4 m³ of air per 24 hours drawn between the slides, Cocke reported good agreement with the gravity slides exposed simultaneously outdoors.

Similar in principle are the funnel device (Hollaender & Dalla Valle, 1939), in which air enters the stem of an inverted conical funnel suspended 3 mm above a petri dish of medium, and the bottle device of Scharf (in duBuy *et al.*, 1945), in which air plays from the end of a tube on the surface of a culture medium in a horizontal, medical-flat bottle. As the volume of air sampled in a given time is often inconveniently small, and they are not easily made quantitative, none of these methods have passed into routine use.

INERTIAL METHODS

In the 'inertial methods' the particles may be retained on filters, on flat surfaces or in liquids. The air sample may be drawn through a jet or tube, or may be spun to induce centrifugal separation. Alternatively the apparatus may move the trapping surface through the air, as in the whirling arm devices, or in sampling from aircraft.

IMPACTION USING WIND MOVEMENT

(1) *Vertical and inclined sticky microscope slides* have been used for catching pollen in hay fever research, and for fungus spores in cereal rust studies. Blackley (1873) exposed four slides at a time, facing different points of the compass; but commonly a single slide is exposed for 24 hours in a pivotted vane shelter which swings to face the wind: cf. Craigie (1945), Clark (1951), Mehta (1952). Slides were fixed in trees, in positions comparable with leaves, in order to trap spores of *Hemileia vastatrix*, a rust fungus devastating coffee, in Ceylon by Ward (1882) and in India by Mayne (1932; and cf. Zwert & Lewis, 1963).

Various methods are used to test for the spore liberation period of a green plant or fungus, but this is a different problem from air sampling (Pettersson, 1940; Oort, 1952; Hopkins, 1959).

Vertical slide traps have been sent aloft fixed to kites, first apparently by Blackley (1873). A variety of such devices were used by Mehta (1933, 1940, 1952) in his extensive studies of rust dissemination in India. During 1930–32 he used free hydrogen balloons to carry vertical slides, the cylinder containing the slide being opened and closed again by burning fuses after about 5 minutes at the right altitude (Chatterjee, 1931); but this method was discontinued in favour of sticky slides attached to kites which could be kept aloft in the wind for 2 or 3 hours. It is now obvious that when carried by a free balloon (contrary to a kite or captive balloon), even if the slide faced the wind, its trapping efficiency would be low as there would be little or no movement of the slide relative to the air.

Sampling from aircraft in the early days was by sticky slide or petri dish, exposed at right angles to the line of flight. The wooden paddle and bottle devized by Stakman *et al.* (1923) was also used by Mehta (1952) and others, and, at aircraft speeds, should be a reasonably efficient collector except for the smallest spores. Newman (1948) claimed an increase in efficiency by placing a leading wire in front of the slide to break the stagnation zone when sampling from aircraft.

(2) *Vertical cylinder.* A removable sticky coating applied to a vertical cylinder was apparently first used by Rempe (1937) for studying airborne pollen at Göttingen. To measure pollen 'drift', cellulose film coated with petroleum jelly was wrapped around the surface of small brass tubes, 14 mm wide by 45 mm long. These tubes were hung vertically from trees or stakes at the required height. Rempe compared pollen 'drift' with pollen 'deposition' (which he measured on horizontal slides 1 cm above ground), and found that drift was usually greater than deposition. At 100 metres distance from a *Corylus* bush, and at a height of 3 metres, the drift was 25 times as great as the ground deposition. Day and night values for a whole month were obtained on the level roof of the Göttingen Botanical Institute. The vertical sticky cylinder (0·5 cm in diameter) has also been used by Turner (1956) for field studies on cereal powdery mildew *(Erysiphe graminis).*

Sticky-surface traps have usually been scanned visually but they can be used to provide cultures: thus Martin (1943) showed that spores can be picked off a trap slide.

Sticky gravity slides, vertical slides and cylinders, are all convenient and cheap. Their defects are (1) theoretically a zero catch of large particles in still air, and of small spores even at ordinary wind speeds; and (2) very great changes in efficiency with alterations in wind speed (*see* Figure 19).

Landahl & Herrmann (1949) claimed that the amount of aerosol deposited on a vertical slide does not change much if it is orientated at various angles within 30° of the wind. Our results indicate that with *Lycopodium* spores this is true only at wind speeds of about 5 m/s. Rotating the slide from 90° to 60° increases the efficiency by about 10 times at 1·1 and 0·75 m/s, and by about 5 times at 3·2 m/s, whereas at 9·5 m/s, the efficiency is decreased by one-quarter.

The use of the 7·6×2·5 cm microscope slide at angles intermediate between 0° and 90° would increase trapping efficiency at the wind speeds normally experienced in the open air near the ground. Hyre (1950) observed that a slide with a presentation angle of 45°, exposed in the field, caught more sporangia of *Pseudoperonospora cubensis* than either a vertical or a horizontal slide at wind speeds of up to 2·4 m/s – a result easily explicable from wind-tunnel work (cf. Figure 19). It ought to be possible to orientate a slide at 5° to the oncoming wind which should give an efficiency varying only between 0·6 and 1·4% over the range of 1 to 10 m/s; or at 10°, when its efficiency would vary between 1·5 and 4·5% over the same range of wind speeds. An alternative approach would be to make the slide tilt, according to the wind's force, over the angles of 10° to 30° as the wind speed decreases from 10 metres to 1 m/s, over which range it would have a nearly constant efficiency of 4% with *Lycopodium.*

The 45° inclined slide has been widely used in plant pathology, e.g. the Canadian wheat rust studies (Craigie, 1945), and Hyre (1949) on *Phytophthora infestans*, and by Jones & Newall (1946) on pollen.

When used vertically the 76×25 mm microscope slide, widely used in plant pathology, is an impactor trap; it suffers from the defects of being relatively inefficient as an impactor at low wind speeds and of being highly sensitive to changes in wind speed. To avoid wrong conclusions and to be able to translate deposit into time–mean concentrations, it is necessary to know the wind speed under which deposition took place and to make the necessary correction. This may be illustrated by considering the deposit received from a cloud containing 10 000 *Lycopodium* spores per cubic metre, using wind-tunnel data for two wind speeds. At 1·0 m/s impaction at 5% efficiency would deposit about 20 spores per square centimetre per hour. With an increase of wind speed to 9·4 metres per second, the deposit would increase to about 9000 spores per square centimetre per hour – a 450 times increase in deposit without change in the number of spores per cubic metre of air. A further source of error is that at low wind speeds the efficiency is low, and as the catch is therefore small it has to be multiplied by a large factor, and the error of estimation becomes great.

(3) *Aeroconiscopes*, used first by medical workers and later by plant pathologists, are now mainly of historical interest. They seem to have been used first by Salisbury (1866) in the Mississippi Valley, but were developed more fully by Maddox (1870, 1871) and Airy (1874) in England, and by Cunningham (1873) in India (Chapter I). (The word 'aeroscope' has been used with some ambiguity, referring either to this type of instrument or to bubblers of the type described by Rettger (*see* p. 135).

Aeroconiscopes of the Maddox and Cunningham type (Figure 2) have been used in plant pathological investigations by Christoff (1934) in Bulgaria, and by A. A. Shitikova-Russakova (*see* Stepanov, 1935) in Russia. Such wind-operated aeroconiscopes or aeroscopes are essentially qualitative and there is no information on their efficiency. Finding an organism means that it was certainly present in the air, but gives no information about its numbers. As with many trapping methods, failure to detect an organism does not necessarily mean that it was absent.

FORCED AIR-FLOW IMPACTORS

Samplers through which air is drawn by pumps, fans or aspirators can be made relatively independent of changes in wind speed and differences in particle size, and they can accordingly give a volumetric reading under field conditions.

For sampling clouds of inorganic particles, which are often present in high concentrations in closed spaces, various forms of koniometer have been developed (e.g. the Kotze koniometer, the Owen jet dust counter, and the Aitken nucleus counter). None of these is well suited to the requirements of aerobiology; because of the small volume sampled they are suitable only for high concentrations, and their aerodynamic features make isokinetic sampling impossible (also, aggregates are shattered by a high speed of impact). Their efficiency has been tested by C. N. Davies *et al.* (1951), and their advantages and drawbacks are well summarized by Green & Lane (1957). Many of the devices extensively used in biological work are satisfactory for sampling aerosols consisting of single bacterial cells or mould spores in *still* air, but fall far short of the ideal isokinetic sampling when used for bacteria carried on 'rafts', and larger spores and pollen grains in *moving* air.

The main errors in impactor traps operated by suction are: (1) *intake errors* due to failure of the spore to enter the orifice; and (2) *retention errors* due to failure to deposit spores in the correct place, either because they are lost on the walls of the trap or because they pass right through the trap. Splash-borne spores, travelling mostly in droplets of 150–600 μm diameter, are not yet satisfactorily collected by any existing equipment. Future development of 'stagnation point' sampling could overcome some of these difficulties; meanwhile some of the errors to which suction impactor traps are liable have been explored. Failure to sample isokinetically is a fault shared by many instruments: the factors to be considered in terms of intake efficiency are: terminal velocity and stop distance of the particles, size and orientation of the sampling head and intake orifice, the suction rate and the wind speed.

C. N. Davies (1968) made a theoretical study of the intake efficiency of spherical unit-density particles, over 1 μm diameter, entering three kinds of sampling head: (1) a 'small' thin-walled tube (a 'small' tube being defined as a tube whose efficiency is unaltered by orientation); (2) a large thin-walled tube; and (3) a bulky head containing a small orifice. The same general problem was studied in wind-tunnel experiments at Brookhaven by Raynor (1970), who measured intake efficiency of filter heads, using particles of 0·68 μm diameter (uranine crystals), 6 μm (*Ustilago* spores), and 20 μm (*Ambrosia* pollen). According to wind speed, particle size, suction rate and orientation of filter head, observed intake efficiencies varied from under 1% to over 100%. In the absence of a reliable theory, sampling devices need experimental calibration.

(i) *Sieving filters.* Drawing air through a filter with pores too small for the organism sought to penetrate, is a simple technique which is, however, relatively little used as it is difficult to obtain a sufficiently high rate of airflow through small pores. Filter paper was used by Frey & Keitt (1925) for *Venturia* spores, by Chamberlain (1956) in wind-tunnel tests with *Lycopodium*, by Gordon & Cupp (1953) for *Histoplasma*, and for pollen collection on Atlantic liners by Erdtman (1937). Membrane filters of re-precipitated cellulose can now be bought with pores from 0·1 to 14 μm in diameter which allow flow rates adequate for some air sampling (First & Silverman, 1953; Goetz, 1953; Haas, 1956). After exposure these membranes can be placed directly on the surface of solid media in a petri dish; or they can be shaken in water and the suspension plated out; or they can be mounted as a transparency and examined directly under the microscope.

(ii) *Impaction filters* differ from sieving filters in that they consist of a deep layer of fibres or granules separated by relatively wide air spaces. Airborne particles are subjected to repeated encounters, and are impacted in the foremost layers of the filter substance. Pasteur's air filter (*see* p. 5) was packed with a nitrocellulose plug which, after sampling, was dissolved in an alcohol–ether mixture and the particles were examined microscopically. As this treatment killed the microbes trapped, P. F. Frankland (1887) substituted powdered glass which was washed, diluted and plated-out after exposure; Rosebury (1947) used a cotton plug. Because of the difficulty of washing organisms off a solid filter, many workers have preferred a completely soluble filter such as sodium suphate (Miquel, 1890). Buller & Lowe (1911) substituted a plug of powdered sugar; Richards (1955) intro-duced the sodium alginate–wool filter, and others have used ammonium alginate, which is soluble in Ringer's solution or sodium hexametaphosphate. Greene *et al.* (1964) used polyurethane foam filters for stratospheric sampling from balloons: large volumes could easily be aspirated from the material and the entrapped microbes readily removed by washing. Timmons *et al.* (1966) used a soluble gelatine foam filter for sampling bacteria and moulds from aircraft.

For many years the standard technique for trapping airborne microbes was that of Petri (1888) or modifications of it (e.g. Weinzerl & Fos, 1910). Air was drawn through a sterilized tube 9 cm long and 1·6 cm wide, con-taining two columns of sand separated by wire gauze. The two sand filters were plated-out separately, the rear portion acting as a control. Equipment for exposing a set of five such tubes from aircraft during flight is described by Overeem (1936).

To study fungus pathogens in coniferous forests, Rishbeth (1958, 1959) exposed squares of butter muslin on a wire frame to the wind, commonly at 2 metres above ground, or on a moving vehicle. Plating of the washings on freshly-cut slices of pine trunks gave a sensitive and extremely selective method of detecting spores of two species of fungi, *Fomes annosus* and *Peniophora gigantea*.

(iii) *Liquid scrubbing devices* remove particles suspended in air, as it bubbles through a liquid, by various combinations of impaction, sedimentation and diffusion. A simple 'aeroscope' bubbler devized by Rettger (1910), consisting of an inlet tube ending in a submerged perforated bulb appears to be the prototype of a practicable method. Another simple form is described by Gilbert (1950). Improved forms of bubbler were described by Wheeler *et al.* (1941), by Moulton *et al.* (1943) who added an atomizer stage before the bubbler, and by duBuy *et al.* (1945).

(iv) *Impingers* differ somewhat in principle from bubblers. A small flask carries a wide inlet tube; the inner end of this tube is fused to a short length of capillary tubing which dips (in the original form) at least 5 mm below the liquid but remains at least 4 mm above the bottom of the flask. The capillary tube acts as a limiting orifice, controlling the flow rate under suction (e.g. *see* Rosebury, 1947). This type of impinger has been found to have a high retention efficiency in work on experimental airborne infection of animals. If used to sample clouds containing large particles, however, substantial wall-loss may occur at the bend of the inlet tube. This can be prevented by adding the pre-impinger, originally devised by K. R. May & Druett (1953) to collect particles over 8 μm diameter as a separate fraction representing material not penetrating to the lower respiratory tract during inhalation. K. R. May & Harper (1957) raised the end of the capillary *above* the level of the collecting fluid to decrease damage to microbes on impingement. In the form of the all-glass capillary impinger this type of instrument has been a standard through the 1960's for aerobiological investigation.

The multistage liquid impinger (K. R. May, 1966) was devised to separate the collected particles into three fractions approximately corresponding to the sizes retained in the upper respiratory tract, in the bronchi and bronchioles, and penetrating to the alveoli of the lung, respectively. The gentle impingement possible with the multistage liquid impinger also minimizes damage to microbes and improves collection efficiency (Plate 4).

(v) *Centrifugal samplers.* Centrifugal impaction was used in the Wells (1933) air centrifuge, in which 30 to 50 litres of air per minute are drawn through an agar-lined glass cylinder rotating at 3500 to 4000 r.p.m.; this makes the air rotate and throws all suspended particles on to the wall of the cylinder by centrifugal force. After exposure the cylinder is incubated, and colonies developing on the agar-coated walls give a direct volumetric reading of the organisms present – so far as they are cultivable on the medium used. This apparatus was used extensively for routine bacterial sampling (Wells, 1955). Its disadvantages include the difficulty of examining the catch inside the curved tube, and a deposition efficiency falling from 100% with 2·3 μm particles to 50% with 0·77 μm particles (Phelps & Buchbinder, 1941).

A more elaborate development is the 'conifuge', designed for microscopic study of size-distribution in particulate clouds (*see* Green & Lane, 1957). It appears to be satisfactory for comparing terminal velocities of spherical with irregularly-shaped particles, but has the disadvantage that it samples only 25 cm³ of air per minute.

Different in appearance, but essentially similar in principle, are the cyclone dust collectors, extensively used in industry for removing dust from the air, and which have the advantage of allowing a large through-put with a small pressure drop. Small cyclone dust collectors have been used for fungal spores by Tervet (1950), Tervet *et al.* (1951), Ogawa & English (1955), Cherry & Peet (1966). (For the design of cyclones, *see* C. N. Davies, 1952.)

For field sampling where microbial concentration is low and large quantities of air have to be screened, Errington & Powell (1969) devised a robust cyclone separator, sampling up to 350 litres of air per minute, and concentrating the contained microbes into a few cm³ of fluid by means of an irrigating system which continually washes the internal walls of the collector.

(vi) *Impactors (historical).* Impactor samplers use suction from a pump to accelerate air through an orifice to a speed at which deposition efficiency becomes high (80–100%) and airborne particles are collected on a sticky, solid surface. The prototypes, such as the Owen jet dust counter, the Aitken nuclei counter and various forms of koniometer (*see* p. 132), are inconvenient for microbial work, but Durham (1947) found such pumps effective for spot samples of pollen clouds in high concentration.

(vii) *The slit sampler* (Bourdillon *et al.*, 1941) was designed primarily for indoor studies of bacteria. A stream of air is drawn through a narrow slit placed just above the surface of sterile medium in a slowly rotating petri dish. Particles, spores, etc, are impacted on the medium and, after a few minutes' running time, the petri dish is removed and incubated so that colonies may develop and be counted. The 'duplex radial-jet air sampler' of Luckiesh *et al.* (1949) is similar in principal and allows sterilization by autoclaving. Retention in the slit sampler is good; identification is made in culture; the apparatus is suitable for occasional or regular samples of short duration, but not for continuous use. It is best used indoors as it is not suited for isokinetic sampling, and collection errors would be high when sampling large particles in a wind. Lidwell (1959) designed a variant of the slit sampler for grading particles carrying bacteria by size.

A device, similar in principle but using a round jet impacting microbes on agar covering a rotating drum, is the Monitor for airborne bacteria of Andersen & Andersen (1962). A record lasting up to 28 hours can be obtained on a trace 12·3 metres long.

(viii) *The Cascade Impactor* (K. R. May, 1945) is a highly efficient suction trap. Used isokinetically it can serve as a standard of reference against which other devices can be calibrated, although a simple feathered sampling head with a membrane or other filter is sometimes preferred on account of freedom from wall losses. The cascade impactor consists of a folded tube through which the air to be sampled is drawn by suction. During passage, the air is accelerated through a series of four progressively narrower jets, and airborne particles are impacted on sticky slides placed close behind these jets. The first jet, which is the intake orifice, faces the wind. In the original model, at the sampling rate of 17·5 litres per minute, the speeds through the four jets are: 2·2, 10·2, 20·4 and 34 m/s, permitting 50% of the particles whose diameters are 12, 4, 2 and 1·1 μm to penetrate the respective jets. Later models give good retention down to 0·5 μm.

The cascade impactor thus fractionates the catch into four sizes, and it avoids breaking down spore clumps as do the liquid impingers. It is ideal for short sampling periods with visual counting under the microscope, but if used for sampling the air spora outdoors in summer in England the deposit becomes unmanageably dense after an hour or two. To overcome this kind of difficulty and to give time-discrimination, a version with moving slides was developed by K. R. May (1956). A five-stage impactor was developed by Wilcox (1953), and another version was designed for simplicity in manufacture by Lipmann (1961). Wall loss and other errors affecting calibration of the cascade impactor are discussed by Mercer (1963, 1965).

(ix) *The automatic volumetric spore trap* (Hirst, 1952) is a power-driven trap designed for operating continuously in the field (Figure 22). It consists essentially of a single impactor slit (this has the same dimensions as the second jet of the cascade impactor), behind which is placed a sticky microscope slide moving at 2 mm per hour. In the course of twenty-four hours a trace is deposited in a band 48 mm long. This can be scanned under the microscope longitudinally to obtain a daily mean, or it can be scanned transversely at intervals of, say, 4 mm to get a reading corresponding to the air spora current every two hours throughout the day and night.

The Hirst trap is ideal for survey work, and it is a useful compromise over many conflicting requirements. It is power driven; sampling is approximately isokinetic; identification is visual, under the microscope, and time discrimination is obtained to ±1 hour or better. The advantages of the Hirst trap are its robustness, simplicity, and continuous operation with a

minimum of servicing. Its disadvantages are capital cost, power requirement, and its unsuitability both for identification in culture and for trapping splash-dispersed spores. The original version of the Hirst trap (made
by C. F. Casella & Co, London, N.1, England), with its slide changed
daily, has an alternative version (made by Burkard Manufacturing Co,
Rickmansworth, Herts, England) which collects the deposit on a plastic

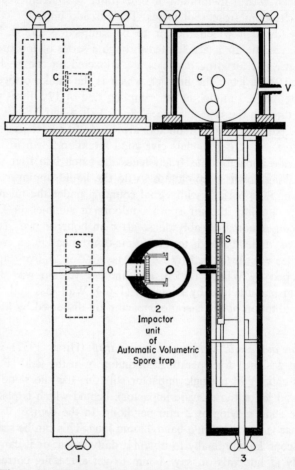

Figure 22. Diagram of the Hirst volumetric suction trap:
1 = elevation facing wind; 2 = plan of section through
orifice; 3 = elevation side view; S = sticky microscope
slide; C = clockwork drive for raising slide; O = orifice
facing wind, with slide in holder moving up past orifice;
V = connection to vacuum pump. (Wind-vane and cap
to protect orifice from rain are not shown.) (Illustration
by permission of J. M. Hirst.)

band on a drum rotating once in 7 days, so that the trap will run for a week without servicing. Both versions have proved satisfactory for continuous operation over years, provided an extra glass-wool dust filter is inserted immediately before the vacuum pump to protect it from wear. The Kramer-Collins spore sampler (Kramer & Pady, 1966) is based on the third stage of the cascade impactor (a 14×0.75 mm slit) in front of a discontinuously moving microscope slide on which 24 bands are deposited, 2 mm apart, in an attempt to obtain accurate time discrimination. Efficiency is reported as 100% at winds of about 1 m/s, but efficiency declines as wind speed increases because sampling is not isokinetic.

Scanning volumetric trap slides under the microscope is tedious unless the deposit is dense. Hirst (1959) reports a comparison of mean density of spore deposit on horizontal slides, vertical slides, and Hirst trap slides (Table XX), exposed during four summer months in the open air. Small spores, which almost entirely escape the vertical cylinder and horizontal slide, are caught in rewarding numbers in the Hirst trap deposits.

TABLE XX

AVERAGE DENSITY OF SPORE DEPOSIT FROM THREE TYPES OF TRAP 2 m ABOVE GROUND, HARPENDEN, ENGLAND, JUNE TO SEPTEMBER 1951
(Hirst, 1959)

	Spores per cm²			Ratio	
	a	b	c	b/a	c/a
	Hirst spore trap	Vertical cylinder 5 mm diameter	Horizontal slide		
Smuts	621	15	3	0·024	0·005
Alternaria	156	58	3	0·372	0·016
Cladosporium	8930	376	59	0·042	0·007
Erysiphe	100	69	2	0·690	0·018
Pollen $< 20 \mu$	206	377	8	1·830	0·038
Pollen $> 20 \mu$	181	490	13	2·707	0·072

Other continuously operating suction traps, deriving from the cascade impactor, include variants by: Voisey & Bassett, 1961; Husain, 1963; Schenck, 1964; Davis & Sechler, 1966; Roelfs et al., 1970; and Tilak & Kulkarni, 1970. Portable variants for taking small samples in the field include: a light-weight portable trap (Gregory, 1954); and a trap for sampling the air spora of pastures (Brook, 1959).

(x) *The Andersen Sampler* can be regarded as deriving from the so-called 'sieve device' of duBuy & Crisp (1944) and the cascade impactor (Andersen, 1958). After entering a circular orifice, air is drawn through a series of six circular plates, each perforated with 400 holes through which spores are impacted directly onto sterile medium in petri dishes (Figure 23). Succeeding plates in the series have progressively smaller holes; the largest particles are deposited in the first dish and the smallest in the last; different media are used for the different size fractions. Air is sampled at the rate of 28·3 litres per minute; wall losses are claimed to be negligible, and retention is said to be 100%, even for single bacterial cells. In our work the Andersen sampler has proved very convenient and has given good results with bacteria, Actinomycetes and moulds, though for particles

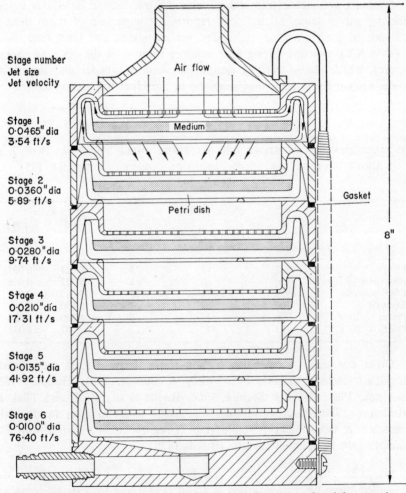

Figure 23. Diagram of six-stage Andersen sampler. (Reproduced by permission of Dr A. A. Andersen from the *Journal of Bacteriology*, **76**, 1958.)

larger than about 8–10 μm the addition of some form of pre-impinger seems desirable for avoiding wall losses on the front of the first plate. Sayer *et al.* (1969) found the Andersen sampler more efficient than the gravity petri dish for enumerating airborne fungi. K. R. May (1964, 1967) modified the pattern of the holes to avoid wall losses with larger particles, and recommended sampling without the nose cone of the original model; further, if used in a wind, the intake should be fitted with a large 'stagnation point' shield.

Reference here must be made to the use of the animal lung as a sampling device (*see also* Chapter XV). Lurie & Way (1957) injected macerated lung tissue of various animals intraperitoneally into mice, from whose livers and spleens two dermatophytes (*Trichophyton mentagrophytes* and *Microsporum gypseum*) were subsequently isolated in culture.

(xi) *Whirling arm.* The principle of moving an object through air on a rotating arm has been developed in aerodynamics laboratories as an alternative to the wind-tunnel. As a device for air sampling it has been developed in the United States – beginning with the 'airwhip' of Durham (1947), who used a 36 inch (92 cm) aluminium rod to swing a forward-facing, sticky glass slide in a circle at 100 r.p.m. Near a stand of flowering ragweed *(Ambrosia)*, Durham recorded a maximum pollen concentration of about 10 million per cubic metre of air.

A high-speed whirling arm trap was developed by Perkins (1957) as the so-called 'Rotorod sampler'. It has been used in plant pathology by Asai (1960), and has been somewhat modified by Harrington *et al.* (1959). The original rotorod sampler consists of a length of 1/16th inch-square (1·6 mm) cross-section brass rod, bent to form a vertical U-shaped collector with arms 6 cm high, 8 cm apart, and fixed to the shaft of a miniature electric motor running at 2520 r.p.m. over the range 9 to 15 volts from dry batteries. The sticky arms effectively sample air at 120 litres per minute. Rotating at a peripheral speed of about 10 m/s, high collecting efficiency would be expected for pollen, and for spores down to about 12 μm diameter; for smaller spores, efficiency would fall below 50%. Carter (1961) measured the efficiency of the rotorod sampler with spores of *Mycosphaerella pinodes* ascospores (12–15×6–8 μm), using a cascade impactor as standard, and recorded 50% efficiency in a wind of 1–2 m/s, and about 90% in wind of 4–5 m/s. Modifications include, instead of square brass rod, inclined microscope slides as collectors for work with rice blast disease (Suzuki, 1965a); the rotoslide sampler for airborne pollen by Ogden & Raynor (1967) collects on the *edge* of a rotating microscope slide, and to protect the collecting edges when not collecting during intermittent working, Raynor & Ogden (1970) devised a centrifugally operating 'swing-shield'. In its various forms the rotorod is convenient for occasional use, but storing the deposits obtained for future reference and identification is inconvenient.

THERMAL PRECIPITATION

A hot body placed in a dust-laden atmosphere produces a dust-free space around itself (Watson, 1936). This well-known phenomenon has been used in the thermal precipitator, in which the dust-laden airstream flows slowly past a wire heated electrically to 100° C above the ambient temperature, depositing the dust particles on glass slips for examination. It has been little used for aerobiological work, but is highly efficient for sub-microscopic particles, and larger ones up to 5 μm diameter. It is most suitable for use when the particles are in high concentration, as the volume of air sampled is only about 7 cm³ per minute (*see* Green & Lane, 1957).

ELECTROSTATIC PRECIPITATION

The movement of charged particles in an electrical field is widely used in industry to extract dust from air because the pressure drop imposed by the requisite apparatus is small, even with high rates of air flow, Berry (1941) recognized that an efficient microbial sampling method could be developed on this principle, and the General Electric electrostatic air sampler was devised by Luckiesh *et al.* (1949). Petri dishes of culture medium are placed on flat metal plates (electrodes) which are oppositely charged to 7000 volts from a half-wave rectifier. Air enters through the apex of a fairly flat inverted metal cone extending to near the edge of each dish, and each cone carries a charge opposite to that of the electrode under its dish. A pump draws air at 14 litres per minute over each dish, and particles move in the electrostatic field and are deposited on the agar surface.

When *Escherichia coli* in aqueous suspension was atomized into a room, the dish on the positive electrode collected nearly ten times as many cells as that on the negative. With naturally-occurring airborne bacteria, 30% more were deposited over the *negative* electrode. Presumably each dish collects a proportion of the uncharged particles by gravity and impaction, as well as collecting the charged particles moving towards it. The positions and dimensions of the upper electrodes have been decided by empirical test, and may need modifying for mould spores and pollen. For naturally airborne bacteria the concentration, based on the sum of the counts of the two dishes, was slightly greater than simultaneous tests with the 'duplex radial-jet sampler'. How far particles are charged by ions after entering the apparatus is not yet known. The apparatus was used in aerobiological work by Kelly *et al.* (1951), and later by S. M. Pady and his colleagues in Kansas.

Contrasting with the small-volume intake of the General Electric electrostatic sampler, the large-volume sampler, designed by Litton Systems Incorporated, has a through-put of up to 10 cubic metres per minute passing through a corona discharge, and precipitating entrained particles on a rotating disc which is washed with a small quantity of re-circulating collecting fluid. Gerone *et al.* (1966) collected in a medium containing serum and antibiotics in order to study room air contaminated by human volunteers with coxsackie virus.

COMPARISON OF METHODS

Under simple conditions it is not difficult to define an absolute standard for air sampling. With non-aggregated spores of one species liberated in a wind-tunnel, isokinetic sampling through a feathered orifice facing up wind collecting into a suitable membrane filter, cascade impactor, or liquid impinger, with precautions against overloading, should give a reliable visual estimate of the number of particles in a measured volume of air. The cascade impactor, catching on a thick layer of soft adhesive, tends to reveal spore clumps intact, and, if this feature is undesirable, the liquid impinger should be used to break aggregrates. The more varied the population in species, size, and state of aggregation, the harder it becomes to devise equipment to measure microbial concentration in air. Particles over 10 μm in diameter must be sampled directly and cannot be ducted around corners on the way to the apparatus without heavy wall losses.

Air hygiene in bacteriology has been mainly a study of air within buildings, and its equipment has therefore been developed for sampling from still or slowly moving air. Aerodynamic effects have been neglected, despite the fact that bacteria are carried on 'rafts' or spray droplets of greatly varying size, so that while efficiency of retention has been achieved, efficiency of collection has been neglected. Most devices, such as the slit sampler and the electrostatic avoid this difficulty in still air by pointing the orifice upwards, but this makes them unsuitable for use in moving air.

In outdoor aerobiology the sizes of pollen grains and fungus spores (*see* Appendix I), and the variability of wind speeds, has focused attention on collection efficiency. Results from the various 'surface' traps, depending on natural deposition processes, are difficult to translate into volumetric results. Over short periods outdoors while wind velocity is constant, or in a wind-tunnel, a vertical strip or cylinder can be used to estimate concentration provided the wind speed and the deposition efficiencies of the particles concerned are known. If the particle's terminal velocity is known, fairly close estimates can also be made using theoretical formulae (cf. Davies & Peetz, 1956). Most data from surface traps, such as the gravity slide and petri dish counts, cannot be translated into concentration but merely measure surface deposition. Only the vast differences in natural concentrations that occur at different times and places make it possible to infer changes from deposition records. Nevertheless Hyde (1959a) showed that in general and over a long period, gravity slide sampling indicated the same qualitative composition and seasonal variation of the pollen cloud over south Wales as did the Hirst trap: as an exception, the Hirst trap revealed that the abundance of nettle *(Urtica)* pollen had been greatly underestimated by the gravity slide method.

Hayes (1969) compared pollen counts by the gravity slide Durham sampler (used as standard for pollen allergy in the United States) with the count from an intermittent rotorod, over a period of five months at Albany, New York State. No way was found for converting gravity slide counts to measure airborne concentration.

Continuous records provided by the Hirst trap have proved highly illuminating in mycology, plant pathology and allergy – even though the results are still limited to visual identification. Formidable problems are posed in developing convenient, continuous sampling in culture (*see* Miquel & Benoist, 1890; Andersen & Andersen, 1962) to record concentrations of viable organisms. Problems include selectivity of media, temperature and aeration during incubation, and competitive and antibiotic interactions between organisms present. Highly selective sampling in culture seems easier than non-selective methods. Air sampling, indoors or out, can have either of two aims: (1) to attain the broadest knowledge of the whole range of organisms in the air, which requires the most complete and undistorted sampling method possible; or (2) to obtain detailed knowledge about a single species or group, which may require highly selective methods.

Air sampling has been successful in revealing the diversity of organisms forming the air spora, in defining conditions for the outbreak of epidemics of some plant diseases, and in measuring dispersal gradients of spore concentration (*see* Chapter XVI). So far it has proved disappointing as a routine measure for forecasting outbreaks of crop disease, because existing methods are mostly insensitive to small concentrations of inoculum in the air (Table XXI).

As pointed out by Hirst (1959): 'No trap is likely to detect spores as sensitively as an acre of susceptible crop in weather favourable to infection. Thus epidemics may be started by spore concentrations which traps will not reveal, so we must define the value of "nil catches". With volumetric traps this can be done by calculating the "detection threshold", or concentration at which one spore should appear in the area scanned for each

TABLE XXI

ESTIMATED DETECTION THRESHOLDS OF CONCENTRATION (SPORES PER CUBIC
METRE OF AIR) OF HIRST TRAP AND STICKY MICROSCOPE SLIDE INCLINED
AT 45°, ASSUMING EXPOSURE FOR ONE HOUR AND COMPLETE COUNT OF
28 mm² OF SPORE DEPOSIT

(Hirst & Stedman, 1961)

| | Wind speed (m/s) | | | | | |
	0·5	1·1	1·75	3·2	5·5	9·5
Microscope slide inclined at 45°						
Lycopodium	941	260	150	40	20	8
Erysiphe	2 000	1 500	860	340	70	10
Ustilago	28 000	13 000	10 000	1400	640	490
Hirst trap						
Lycopodium	2	2	2	3	3	—
Ustilago	2	2	3	2	2	—

sample. In our routine scanning of hourly samples from the Hirst trap the detection threshold is less than 10 spores per cubic metre of air. This high value explains why spore traps are of little practical use in forecasting epidemics of potato blight which start from minute local sources, but are valuable for apple scab or black rust, with which initial spore concentrations may be high because of sudden liberation of accumulated spores or the arrival of a spore-laden air mass.'

Before using an air sampler, its performance should be explored experimentally. Consistency of performance alone is an unsatisfactory criterion, because a trap may be consistently misleading. However, the standard of accuracy required will depend on the purpose of the operation. To measure a gradient greater accuracy may be required than in routine surveys, where fluctuations are on a logarithmic scale, and errors due to failure to match suction rate with wind speed can be tolerated.

X

The Air Spora Near the Earth's Surface

Ultimately we hope to attain an undistorted picture of the ambient outdoor air spora, studied in its own right as a natural phenomenon. All air-sampling methods are more or less selective. This chapter deals with the concentrations of microbes in suspension in air near the ground, that is, within the laminar and turbulent boundary layers ordinarily inhabited by man, animals and plants. The account is based on the limited amount of information obtained by volumetric air sampling with reasonably efficient apparatus. No attempt will be made to summarize the extensive results from gravity slide and petri dish traps, as these are already covered by excellent summaries by Feinberg *et al*. (1946), Maunsell (1954a), Werff (1958), Morrow *et al*. (1964) and others. But the results of long-term sampling with such surface traps will be drawn upon for supplementary information when required.

COMPOSITION OF THE AIR SPORA

Some 1200 species of bacteria and Actinomycetes are recognized. Other spore-producing organisms include perhaps 40 000 species of fungi, numerous mosses, liverworts, ferns and their allies, and more than 100 000 species of pollen-producing flowering plants of which about 10% are wind pollinated. (Of the Protista [Protozoa] able to enter the air spora, our information is very meagre and unsystematic.)

A taxonomist, having in mind the twenty-five volumes of Saccardo's *Sylloge Fungorum*, or the many volumes and supplements of the *Index Kewensis*, may wonder what useful statements can possibly be made about the air spora where most of the fungus or other plant bodies, whose characters could aid identification, are lacking. Fortunately, as a cursory microscopic examination of the deposit from an impactor sampler shows, the potentially airborne microbes are not all equally common in the air. One sample is normally dominated by one or two types of spore, with

146

several other types in fair abundance, and many more encountered in ones and twos only. The frequency of distribution of individuals of different species in an air spora resembles the series of the logarithmic and log-normal types discussed by Fisher *et al.* (1943), and by Williams (1947, 1960). Investigation will doubtless show that different air sporas have different 'diversities'.

In practice the problem of recording an air spora requires the recognition of a number of categories for the organisms most commonly present in the sample, and inclusion of a miscellaneous group, which may ultimately contain from 10 to 15% of the total, for microbes not further categorized. Pollens of flowering plants can often be identified to the species level, and so can a few fungus spores – especially of the Uredineae and some other plant pathogens (Appendix I, p. 299). In samples of the outdoor air, bacteria can seldom even be recognized visually as such, let alone identified, and the only sampling devices suitable for their study involve making cultures. In practice the categories adopted are of varying degrees of arbitrariness; but the different categories behave so differently that it would be intolerable to have no way of referring to them.

TAXONOMIC GROUPS NEEDING STUDY IN THE AIR SPORA

We now have some knowledge of the occurrence of bacteria, fungi and pollens as components of the outdoor air spora, but there are some groups whose presence is obvious enough yet about which we have scarcely any quantitative data. Thus I know of no attempts at continuous sampling to assess the concentration of Actinomycetes, and of moss and liverwort spores in the atmosphere.

(i) *Protista* [*protozoa*]. For these we have the estimate of Miquel (1883) of an average of 0·1 to 0·2 airborne protozoan 'eggs' per cubic metre at the Observatoire Montsouris, Paris. Using a Pasteur-type filter, Pusch-karew (1913) sampled near ground level on the right bank of the Neckar down stream from Heidelberg. In forty-nine tests, on different occasions and at different times of the day, his catch averaged 2·5 protozoan cysts per cubic metre of air. His catch included species of *Amoeba, Bodo, Monas, Calpoda*, etc. Curiously enough, he concluded that this concentration was too small to account for the observed almost world-wide uniformity of species of freshwater protozoa, and that other dispersal routes must be important (as no doubt they are). Schlichting (1969) summarizes these and later records, and the study evidently awaits convenient techniques.

(ii) *Algae*. Microscopic terrestrial and freshwater algae occur in the air, but have been little studied. A few samples were taken on the roof of buildings at Leiden by Overeem (1937), using the 'standard aeroscope' and

Rettger bubbler. At least forty algae were obtained from a total of 20 cubic metres of air, including: *Chlorococcum, Chlorella, Pleurococcus, Stichococcus,* and *Navicula.*

The occurrence of blue-green algae resembling species of *Gloeocapsa* or other members of the Chroococcaceae was recorded by Gregory *et al.* (1955) from continuous sampling with a Hirst trap at Thorney Island in Chichester Harbour, England, from 30 June to 13 July, 1954. Concentrations averaged 110 colonies per cubic metre of air (averaging 8 cells per clump). Diurnal periodicity showed maximum numbers near midnight (210 colonies per cubic metre) and a minimum in the morning (30 colonies). Similar, but fewer colonies were found regularly in London and Rothamsted. They showed no pronounced seasonal trend according to Hamilton (1959), who also reported the rare occurrence of diatoms and desmids. Earlier records have been compiled and new ones added by Schlichting (1969) and by Brown *et al.* (1969). Evidently microscopic algae are widely prevalent in the atmosphere in numbers varying from a few to a few hundred per cubic metre, and occasionally they may form a heavy deposit on the ground (D.S.I.R., 1931).

(iii)   *Ferns.* For ferns the reports are few. At Rothamsted, with no large areas of bracken within several kilometres, and only small quantities within 1 km, spores of the *Pteridium* type occurred frequently in warm dry weather from late July to mid September. They averaged 4 per cubic metre, with a maximum daily concentration of 36 per cubic metre (Gregory & Hirst, 1957; Hamilton, 1959).

MIQUEL'S WORK ON BACTERIA AND MOULDS

Recognizing the paucity of information on airborne microbes, Pierre Miquel made daily counts in Paris during the last quarter of the nineteenth century (cf. p. 8). Miquel's 'contribution to the microscopic flora of the air' is probably the most sustained series of volumetric measurements of the microbial population of the air ever attempted. Daily observations in the Parc Montsouris, about 5 km south of the centre of Paris, served him as a standard for comparison with the polluted air in a densely populated city. The bacteria of the outdoor air were classified in the following percentages as: *Micrococcus* 66, *Bacillus* 25, *Bacterium* 6, *Vibrio* 1–2.

Miquel (1899) shows a seasonal variation in total bacterial and mould concentrations (Table XXII). Most of the samples were taken with a form of the Pasteur trap (*see* p. 5), using a sterile plug of powdered anhydrous sodium sulphate as a filter. This was dissolved after exposure and inoculated into flasks of filtered saline beef extract. At the Parc Montsouris bacteria were nearly three times as numerous in summer as in winter, but moulds fluctuated rather less. Near the Hôtel de Ville in the centre of Paris, bacteria showed a similar seasonal variation but were $2\frac{1}{2}$ times as many as in the Parc; moulds were 10 times as numerous but showed little seasonal variation. At first Miquel argued that, as only one-tenth could have been blown in from the country to the centre of the city, the rest must have come from houses. But after the year 1881 he noted a steady annual decline and he attributed this to improved street cleaning and washing to lay dust which, we may suppose, consisted largely of soil enriched with horse droppings. Data are also given for a narrow, unhygienic street, and for one of the main sewers of Paris. The air of sewers was no more highly contaminated than the outside air, and was often surprisingly pure (Chapter XIV).

TABLE XXII

MEANS OF MONTHLY MEAN NUMBERS OF BACTERIA AND MOULDS PER CUBIC METRE OF OUTDOOR AIR IN PARIS (Miquel, 1899), IN CULTURE IN NEUTRAL BEEF BROTH

Month	Parc Montsouris (16- and 9-year means, respectively)		Near Hôtel de Ville, place Saint-Gervais (1888–1897)		Passage Saint-Pierre 1897–1898 (Mean)		Main sewer Blvd. Sebastopol (1891–1897)	
	Bacteria	Moulds	Bacteria	Moulds	Bacteria	Moulds	Bacteria	Moulds
January	198	160	3 840	1555	6 610	1665	2670	4535
February	148	110	3 475	1375	3 265	1790	3095	1965
March	209	155	4 995	1290	2 790	1630	2555	2485
April	362	140	8 260	2445	11 710	1885	3875	6290
May	295	230	8 725	1560	4 910	1650	3845	1865
June	355	222	10 830	1835	5 015	2630	2705	2360
July	464	205	12 040	2590	5 930	4235	4460	3490
August	450	270	10 300	2450	4 265	2770	4645	3195
September	395	215	9 920	2435	5 545	1735	3630	1845
October	260	228	7 160	2445	7 900	2165	3965	4135
November	195	240	5 845	2175	4 735	2270	3800	5210
December	167	166	4 365	2005	4 015	1390	6750	2560
Average	290	195	7 480	2015	5 555	2150	3835	3330

EFFECT OF RAIN

The numbers of bacteria in summer thus averaged several hundred per cubic metre, and were reduced in a few hours by rain to a mere 20 to 30, but they increased again as the ground dried. Surprisingly enough, bacterial numbers often increased after snowfall. The numbers increased with increasing wind speed and remained high during a drought, unless it was prolonged. To Miquel it was clear that rain had a complex action: air that contained many bacteria after a fine, dry spell of weather was rapidly purified by rain, but often during a spell of humid weather the fall of rain would contaminate the air more than it purified it – possibly because raindrops collected bacteria in their fall towards the ground and, by evaporating before reaching the soil, added their collection to the air near ground level. We may also suspect that bacteria were put into the air by rainsplash. The first rain after drought might contain 200 000 bacteria per litre instead of the average number of 3380 per litre (*see* Chapter XIII).

Slowly Miquel came to the conclusion that the source of most outdoor airborne bacteria is the surface of the ground, whence they are picked up with dry soil particles by wind – a conclusion still acceptable 90 years later.

Moulds reacted differently to the fall of rain. With the onset of rain, the air was at first purified; but when rainy periods lasted for some days, the number of mould spores in the air at Montsouris often increased remarkably, even reaching 95 000 to 120 000 per cubic metre. In dry weather coloured spores abounded. Re-invasion of the air after rain was mostly by colourless organisms, which Miquel diagnosed, probably incorrectly, as immature spores. Rain and warmth increased the atmospheric spore content, though this might decrease during high winds because their extra lifting power did not compensate for their power of desiccating and killing.

CIRCADIAN PERIODICITY

(i) *Periodicity in numbers of bacteria* outdoors could be studied only on 40 selected days of the year because, by Miquel's methods, for hourly studies, from 600 to 700 culture flasks had to be handled in one day. The fullest data are probably those of 1882 to 1884, which showed continual change from hour to hour of the day in relation to changes in meteorological factors that have not yet been unravelled.

Bacterial numbers showed circadian periodicities differing between Montsouris and the centre of Paris. At Montsouris there were two daily maxima at 08.00 hours and 20.00 hours, and two minima at 02.00 hours and 14.00 hours respectively (Figure 24). In the centre of Paris, however, there tended to be a single maximum at 14.00 hours and a minimum at 02.00 hours during most of the year, but in autumn the double peaks tended to occur in central Paris as at Montsouris.

This daily variation was shown to hold irrespective of wind direction; an effect that was probably in part attributable to mechanical causes such as traffic and sweeping of streets. Furthermore the peaks occurred also in rainy weather so long as not more than 2–3 mm of rain fell in 24 hours. Miquel also showed that the outside changes soon penetrated into the rooms of buildings unless they were exceptionally well sealed.

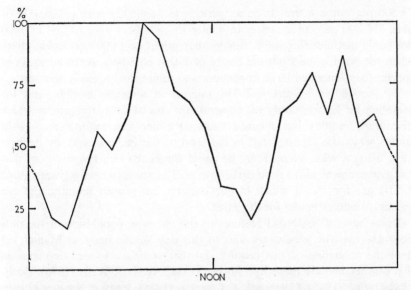

Figure 24. Circadian periodicity of total numbers of bacteria in air at the Observatoire Montsouris, Paris, based on hourly readings between March 1882 and September 1884 (Miquel, 1886).

(ii) *Periodicity in numbers of mould spores* resembled that of bacteria. Trapping by impaction on a moving slide which gave him hourly readings, Miquel found that moulds had two maxima at 08.00 and 20.00 hours. These maxima were independent of wind velocity and fluctuated much more than did bacterial counts. Miquel then tried 15-minute sampling periods and discovered the important principle that hourly values are merely a smoothing of still more rapid fluctuations; he says (transl.) 'what I wish to establish by all these examples is the variability of the nature of the organisms living in the atmosphere'.

RELATIVE NUMBERS OF BACTERIA AND MOULDS

Miquel had started with the aim of describing the cryptogamic flora of the atmosphere, and in the earlier years of his work he reported much larger numbers of moulds than bacteria.

In 1879 Miquel was assessing mould spores visually by a continuously operated, aspirated aeroscope at 2 metres above ground level at the centre of a lawn in the Parc. He caught microbes at 100 times the rate of the non-aspirated aeroscopes of Maddox and Cunningham, and he concluded that less than 10% of the organisms seen visually would grow in culture. The numbers of germs (principally mould spores as shown by his drawings) in the air at the Parc Montsouris during continuous sampling in 1878 averaged

28 500 per cubic metre. In rainy periods in June they rose to 100 000 or even 200 000 per cubic metre. In winter the numbers were as low as 1000 per cubic metre during snow, though they might be 14 000 per cubic metre when the wind came from the centre of Paris. Numbers increased again in spring; they remained large in summer and diminished again in autumn.

By the 1890's Miquel had lost interest in airborne moulds, and for sampling air he constantly recommends the use of sugar-free media which discourage moulds but enhance bacterial counts. Henceforth the mould counts which he reported fell to the level of the bacteria and, by deliberately using a selective medium, he could forget the rich fungus spora that had embarrassed him in the earlier years. The values already given (Table XXII) are for media which favour bacteria but repress moulds, and are certainly underestimates for the latter.

Knowledge of the broad features of the bacterial population of the outdoor air near the ground remains to this day substantially as Miquel left it at the beginning of this century. Further comparable measurements in this century include those by: Forbes (1924), Wells & Wells (1936), Buchbinder *et al.* (1945), Colebrook & Cawston (1948), Pady & Kramer (1967). On the whole the topic has been neglected and, significantly, the American Association for the Advancement of Science's book *Aerobiology* (Moulton, 1942) has no chapter on bacteria in outdoor air over land.

RECENT STUDIES OF FUNGI AND POLLEN

The pollen and fungus components of the outdoor air spora have attracted much attention in this century and the use of volumetric sampling equipment in their study has been highly illuminating. Quantitative visual counting of spores from 3 μm upwards, confirms Miquel's impression of recurrent 'tides' of spore concentration; but different groups of microbes are now known to have separate 'tidal waves', and the composition and concentration of the air spora varies enormously with place, season, time of day, weather and human activity.

Continuous surveys with volumetric samplers have been reported from many regions, but much of the work is unpublished. A few of the longer series are listed below with citation of recent key papers, as a clue to finding other publications in series.

Rothamsted Experimental Station, Harpenden, England. For example: Gregory & Hirst (1957); Hirst & Stedman (1961); Gregory & Henden (1967).

Wright-Fleming Institute, St Mary's Hospital, London, England. For example: Hamilton (1959); R. R. Davies *et al.* (1963).

St David's Hospital, Cardiff, Wales. For example: Hyde (1959a); Adams 1964); Adams *et al.* (1968).

Imperial College Field Station, Ascot, England. For example: M. E. Lacey (1962); Sreeramulu (1963, 1964).

Andhra University, Waltair, India. For example: Sreeramulu & Ramalingam (1966).

Kansas State University, Manhattan, Kansas, U.S.A. For example: Pady *et al.* (1969).

Jamaica (Banana plantations), West Indies. For example: Meredith (1962a, b).

University of Nebraska, Lincoln, Nebraska, U.S.A. For example: Meredith (1967).

University College, Cardiff, Wales. For example: Harvey (1970).

THE AIR SPORA AT TWO METRES ABOVE GROUND LEVEL

Continuous records in a mixed agricultural environment were obtained during the summer of 1952 by Gregory & Hirst (1957), using the Hirst automatic volumetric spore trap. The mean spore concentration at 2 metres above ground level over the period 1 June to 25 October was 12 500 spores per cubic metre. These were grouped visually into twenty-five categories. The commonest spore type was *Cladosporium* (probably mainly *C. herbarum*), which accounted for 47% of the total. The second commonest were classified as hyaline basidiospores and made up 31% of the season's catch; most of these were probably spores of species of *Sporobolomyces*, with spores of *Tilletiopsis* adding another 0·56%. Coloured basidiospores of mushrooms and toadstools (agarics, boleti and bracket-fungi) amounted to 3·3% of the season's total. Pollen made up 1% of the total. Conidia of powdery mildews (Erysiphaceae), 'brand spores' or 'chlamydospores' of *Ustilago* species, and conidia of *Alternaria*, amounted to between 1 and 2% each. All other particles recognizable as spores of micro-organisms were put into an 'unclassified' category, totalling 8% of the season's catch, and included many organisms which, although abundant in soils, form only an insignificant fraction of the summer outdoor air spora (for example, *Penicillium*, *Aspergillus*, and various Mucoraceae). Bacteria and Actinomycetes are not revealed by this method, which is efficient only for particles over 3 μm in diameter.

In Britain the attempts to get a relatively undistorted picture of the outdoor air spora have demonstrated beyond doubt that *Cladosporium* and *Sporobolomyces* predominate, followed by the hyaline and coloured basidiospores of the mushrooms and toadstools. Fewer in number, but not necessarily less in total volume, are the pollens, *Alternaria*, ascospores, and the large-spored plant-pathogenic fungi. Under ordinary conditions, splash-borne types seem not to amount to more than a few per cent of the total air spora. Many other types are found also, but they are infrequent, except in special localities or under special circumstances.

The importance of basidiospores from agarics and bracket-fungi, and especially the mirror yeasts, as components of the air spora remained unrealized until demonstrated by continuous volumetric sampling (Gregory & Hirst, 1952, 1957). It is remarkable that even the existence of the Sporobolomycetes was unrecognized until 1930. Basidiospores are not efficiently caught by surface traps, and confirmation of their frequency in the atmosphere (which was doubted at first) has been slow in forthcoming. However, Hyde & Adams (1960) report that at Cardiff over the whole year 1958 the basidiospore types collectively averaged 1059 out of the average fungus spore count of 2164 per cubic metre of air. Furthermore, estimating *volume* instead of number, they showed that basidiospores came second only to grass pollen. Daily estimates with a slit sampler for one year at Manhattan, Kansas, gave the numbers of basidiospores as 24·3% of all spores caught – second only to *Cladosporium* (Kramer *et al.* 1959a, b).

Spores of *Cladosporium*, and a few other types notably *Ustilago*, often occur in groups of several spores in a clump. This is clearly due to failure to separate at time of liberation, rather than to secondary aggregation during wind transport. Basidiospores, which are normally launched separately, seldom occur in groups on the spore trap slide: probably they are kept apart by the presence of like charges at the time of liberation.

THE AIR SPORA AT OTHER HEIGHTS NEAR THE GROUND

In general spore concentration is greater at positions nearer the ground than the standard sampling height of 2 metres, and is less at greater heights. Using a Hirst trap at 24 metres in a lattice tower at Rothamsted, the average spore concentration was 81·5% of that at 2 metres, though some spores characteristic of the night air spora were actually commoner at the higher level (Gregory & Hirst, 1957).

Tests at heights below 2 metres with a portable suction trap in the New Forest, England, showed a general decrease of concentration with height (Table XXIII). The difference was greatest at night, when the total spore concentrations were smallest. By way of exception, *Cladosporium* numbers were reversed at 13.00 hours: this is taken to mean that sources of *Cladosporium* were not present close to the trap and that the air nearest the ground was being depleted of this organism in passage over the surface.

In his studies of the 'phyllosphere' of cereal leaves, Last (1955) sampled air among wheat plants at 11, 46, and 80 cm above ground level and found, 237 000, 170 000, and 41 000 spores of *Sporobolomyces* per cubic metre, respectively.

TABLE XXIII

TOTAL NUMBER OF POLLEN GRAINS AND SPORES PER CUBIC METRE OF AIR IN OAK–BIRCH WOOD, NEW FOREST, ENGLAND, 23 JULY 1953

(Gregory, 1954)

	Height above ground level (cm)		
	7	30	120
05·00 hours GMT	20 600	19 000	7 250
13·00 hours GMT	31 300	24 200	20 300

The air spora at heights greater than in these examples is dealt with in Chapter XII.

CIRCADIAN PERIODICITY OF THE FINE-WEATHER SPORA

Concentrations of spores of a single species, or a group of related species, often show 'diurnal' rhythms (more appropriately called circadian rhythms), comparable with those observed by Miquel (*see* p. 150). This effect was studied in detail by Hirst (1953), whose mode of presenting the results has proved a convenient model for the many subsequent workers who have followed this lead. Hourly or two-hourly concentrations are obtained on as many rain-free days as possible. At regular intervals throughout the day, the mean spore concentration is plotted as a percentage of the maximum. Geometric means are preferred to arithmetic means, and separate curves can be drawn for different weather types. Examples are shown in Figure 25.

Interpretation of these periodicities is not always clear. The rhythmic variations in *spore liberation*, referred to briefly in Chapter IV, clearly contribute to changes in *airborne concentration*, but other important factors affect concentration. Supply of particles from the source is controlled by geographical distribution and seasonal growth cycles of organisms producing pollen or spores. Even if the number *liberated* per hour remains constant throughout day and night, *concentration* will tend to be greater at night because on average the spore cloud will suffer less dilution owing to the combined effect of slower winds, decreased turbulence, absence of convection and presence of a temperature inversion at night. Superimposed on these will be the effects of external factors on spore liberation, including light and darkness, temperature, wind, humidity changes and water supply, as well as endogenous rhythms.

Circadian rhythms in fine weather have been identified for many components of the air spora by investigators in various countries. Information on a particular organism can be traced through abstracts. The periodicities described are average effects, not necessarily discernible in one 24-hour period. Further, times of maximum and minimum concentration may differ in different localities.

G

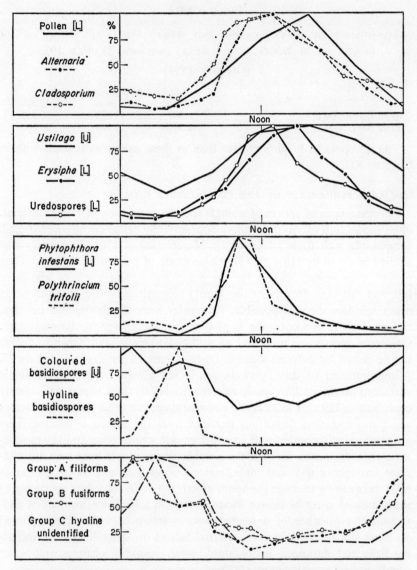

Figure 25. Mean circadian periodicity curves of thirteen spore groups expressed as percentage of the peak geometric mean concentration. (From Hirst trap records at Rothamsted Experimental Station, summer 1952.) The symbols L and U refer to corrections applied to the data, based on wind-tunnel measurements of efficiency of the Hirst trap, made with *Lycopodium* and *Ustilago* spores respectively. (Reproduced by permission of J. M. Hirst from the *Transactions of the British Mycological Society*, 1953.)

Several patterns of periodicity in fine weather are now evident. The scheme which follows differs slightly from groupings described by Meredith (1962b), and Sreeramulu & Ramalingam (1966).

(i) *Night pattern* (Group III of Meredith; Type A of Sreeramulu & Ramalingam). These occur in greatest concentration sometime between sunset and sunrise, and the determinants are probably varied and complex. Maxima usually occur after midnight. This group includes all typical basidiospores including *Sporobolomyces*, but in cooler weather coloured basidiospores may reach a maximum in the afternoon (Gregory & Stedman, 1958). Also included are many ascospores; also *Entomophthora* (Wilding, 1970), and *Podaxis* (Sreeramulu & Seshvataram, 1962).

(ii) *Post-dawn pattern* (Group I of Meredith; Type B of Sreeramulu & Ramalingam). A group which occurs in the early part of the daylight hours, probably in direct relation to liberation mechanisms operated by decreasing vapour pressure, and characterized by a rapid decline after the peak concentration is reached. Here belong *Phytophthora infestans* and *Polythrincium trifolii* (Hirst, 1953); *Cordana musae, Deightoniella torulosa, Nigrospora, Zygosporium, Zygophialia,* and *Corynespora* (Meredith, 1962b).

(iii) *Middle-day pattern* (Group II of Meredith; Type C of Sreeramulu & Ramalingam). Organisms of this group have maxima during what is typically the warmest part of the day, with greatest wind speed and turbulence. Included here are: *Cladosporium, Alternaria, Ustilago, Erysiphe* and many pollens (Hirst, 1953); *Trichocomis, Phaeotrichocomis padwickii,* and some *Cercospora* spp. (Sreeramulu & Ramalingam, 1966); *Curvularia, Tetraploa, Memnoniella, Periconiella* (Meredith, 1962b).

(iv) *Double-peak pattern* (Type D of Sreeramulu & Ramalingam). The double-peaked rhythm for bacterial concentrations, found by Miquel at Montsouris (p. 151), with two daytime maxima, remains an unexplained phenomenon worth re-examining. Similar patterns are reported for *Cladosporium* (Rich & Waggoner, 1962; Pawsey, 1964), and for *Curvularia, Helminthosporium* spp., *Tetraploa, Cercospora* spp., and *Aspergillus* (Sreeramulu & Ramalingam, 1966).

All these circadian periodicities are based on average catches for a number of days. Weather on a particular day may disturb the normal rhythm: for example, *Sporobolomyces* may persist to 10.00 hours or later; anthesis of grasses may be suppressed in dull weather. In the dry season in Nigeria, certain typically afternoon types may occur in the forenoon (Cammack, 1955). With some organisms the rise from a low concentration is often steep whereas the subsequent decrease is relatively slow; with other organisms such as *Sporobolomyces*, and the post-dawn group, the reverse is often true.

WEATHER EFFECTS

Atmospheric spore concentration fluctuates with changes in weather. It also fluctuates for biological reasons such as growth and differentiation of spore or pollen producing organs. Studies reported by Hirst (1953) show that the pollens, and spores of *Cladosporium, Erysiphe, Alternaria,* smuts, and rusts (which together form the main components of the middle-day fine-weather air spora), are mostly removed from the air by prolonged rain which, however, soon puts into the air a characteristic damp-air spora. Temporarily, with the start of heavy rain, there may be a marked increase in some of these types resulting from the puff and tap mechanism (Hirst & Stedman, 1963).

Fluctuation is a marked characteristic of the fine-weather air spora, but some types depend on rain to get into the air and occur in high concentration only after measurable rainfall. Keitt & Jones (1926) showed that liberation of ascospores of the apple scab fungus *(Venturia inaequalis)* is correlated with rain. Hirst *et al.* (1955) trapped no ascospores of this fungus during dry weather in orchards, and during the first hour after the onset of rain they found only a few ascospores; yet dense concentrations occurred during the second and third hours. Rain at night led to lower concentrations than an equal amount falling by day. In general, most perithecial Ascomycetes must be wetted before ascospores are ejected. Spores of *Ophiobolus graminis*, the wheat take-all pathogen, do not occur in the air during dry weather, but they reached a concentration of 3700 per cubic metre in air over wheat stubble within 2 hours of the fall of 1·3 mm of rain (Figure 26): a few ascospores were liberated by as little as 0·25 mm of rain (Gregory & Stedman, 1958). Ascospores of some species are evidently discharged when the ground is wet with dew, and these types appear as part of the fine-weather night pattern air spora. Little is known about the species composing the damp-air spora, or about the concentrations in the air of spores liberated by rain-splash, and their study waits on improved technique (Plate 5A).

Low relative humidity can check release of basidiospores (Zoberi, 1964); by contrast low humidity can increase the release of aecidiospores (Pady *et al.*, 1969).

Hamilton (1957) studied correlations at two centres (London and Rothamsted) between spore concentration of twenty-eight visual types and the weather. Her main positive findings are as follows. Rainfall had no effect on the atmospheric concentrations of hyaline basidiospores (including those of *Nolanea, Lactarius, Tilletiopsis,* and possibly *Sporobolomyces*).

Concentrations of pollen and most types of fungus spore decreased with rain, but all ascospore types and *Helicomyces* increased with rain. In half of the types studied, concentrations were significantly increased by increases in temperature, dew point, or relative humidity. The only significant decreases were in grass pollen (and possibly *Ustilago* spores) with increased relative humidity, and in *Nolanea* with increased dew point. Sunshine had no significance, except for positive correlation with *Ustilago* and algal groups ('*Gloeocapsa*'). Increased wind speed significantly decreased the concentration of *Alternaria*, some basidiospores (including *Ganoderma*, *Tilletiopsis*, and *Sporobolomyces*), *Botrytis*, *Cladosporium*, *Entomophthora*, *Pullularia*, uredospores, insect fragments, and *Urtica* pollen. By contrast, plant hairs and algal groups ('*Gloeocapsa*') were increased by increasing wind speed – possibly because both are released by friction between plant parts. Gustiness was associated with increases in *Alternaria*, filiform ascospores, and *Ustilago*.

Hammett & Manners (1971) used a digital computer to calculate correlation coefficients, and showed that for *Erysiphe graminis* conidia, greater airborne concentration was positively correlated with a complex involving high wind speeds, dry leaf surface, high temperature and low relative humidity: rainfall acted independently of the complex. Onset of wind or rain was more important than continuation of either of them.

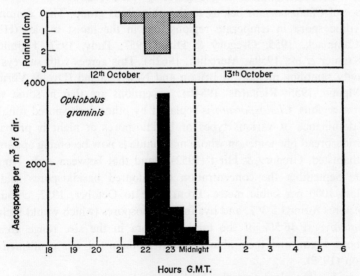

Figure 26. Liberation of ascospores of *Ophiobolus graminis* by rain. Numbers of ascospores in air at 45 cm above ground level in wheat stubble at Rothamsted Experimental Station, 26–27 October 1953. (From Gregory & Stedman, 1958; reproduced by permission of the *Transactions of the British Mycological Society*.)

SEASONAL CHANGES

Seasons affect the air spora profoundly. *Cladosporium* and *Alternaria* show pronounced seasonal periodicity in temperate regions, as of course do the pollens, and spores of mosses, pteridophytes and plant pathogenic fungi. By contrast *Penicillium* may show little seasonal change, or in cities it may even be more plentiful in winter than in summer (Maunsell, 1958; Hamilton, 1959). To the allergist winter is a period of allergen-free outdoor air. Spring brings the deciduous tree pollens, followed by those of conifers in early summer and, more important, by grass pollens characterizing the 'hay-fever' season. Late summer brings the moulds in dense concentration and what are known as the 'weed pollens' (including the notorious rag-weeds, *Ambrosia* spp. in North America), extending into early autumn. Late autumn, like winter is relatively free from allergens.

EFFECT OF LOCALITY

The air spora near the ground tends to be dominated by local sources, and these components of local origin are seen against a background of material from distant sources. Volumetric sampling shows that some spores are practically ubiquitous, whereas other are more or less confined to certain localities. Valuable surveys of airborne pollen are given by Hyde (1952, 1956, 1959a, b), and of fungus spores by Werff (1958).

Species of *Cladosporium* belong to the ubiquitous group. They dominate the daytime spora in temperate regions and in the moist tropics (Hirst, 1953; Cammack, 1955; Gregory & Hirst, 1957; Pady, 1957; Hamilton, 1959; Kramer et al., 1959a; Meredith, 1962b). This agrees with surveys by the gravity sampling method in Britain and New Zealand (Dye & Vernon, 1952; Menna, 1955; Richards, 1954b): exceptions are that in some very dry warm regions, *Cladosporium* is replaced by other dark spored fungi.

The dominance of various types of basidiospores at night is probably also a widespread phenomenon whose magnitude is now becoming apparent. At Rothamsted, Gregory & Hirst (1952) found that between early August and late September the concentration of coloured basidiospores seldom fell below 1000 per cubic metre. From June to October, 1952, coloured basidiospores formed 3·3%, and hyaline basidiospores (which would include *Sporobolomyces*) 46·5% of the total air spora in the size range greater than about 4 μm, and their abundance at Rothamsted was confirmed by Hamilton (1959).

It soon became clear that the hyaline basidiospores were mostly from colonies of *Sporobolomyces* occurring on leaves, and that their abundance varied greatly in different places. Last (1955), when sampling air within a stand of wheat, found large differences in concentration of *Sporo-*

bolomyces between manured and unmanured plots in the same field (Broad-balk field, Rothamsted). Concentrations of 1 million *Sporobolomyces* per cubic metre were found near Chichester Harbour, England (Gregory & Sreeramulu, 1958).

In any one place it is difficult to disentangle the respective contributions of local and distant sources. It might be expected that spore concentrations would be less dense in a city than in the nearby countryside. This was indeed found by Hamilton (1959), who compared continuous records from two Hirst traps, one at Rothamsted (2 metres above ground) and the other in London (16 metres above ground on a roof in South Kensington). The counts of total pollen were greater in London because of a large excess of *Platanus* (plane) pollen, but grass pollen grains were 50% more numerous in the country. Fungus spores outnumbered pollen grains by 75 to 1, and although counts in London were less than half those at Rotham-sted, they still averaged 6500 per cubic metre during the 1954 season. With some species, circadian changes in concentration tended to be less pro-nounced in London than at Rothamsted (Figure 27); this may perhaps have indicated that spores of the species trapped in London came mostly from distant sources.

BIOTIC FACTORS

Human activity also plays a part in affecting atmospheric spore concen-tration. Mowing and tedding grass can produce a great and immediate local increase in *Cladosporium* and *Epicoccum* spores, and (with an apparent delay of 2 hours) of grass pollen (Sreeramulu, 1958a). Pawsey (1964) found a great increase in spores of *Cladosporium* and *Pithomyces chartarum* during grass mowing up wind of a trapping site in Nottingham, England; and Brook (1959) demonstrated increase in *Pithomyces char-tarum* by mechanical disturbance of pastures in New Zealand.

Threshing of grain produced a local spore source (Heald & George, 1918). The role of overhead irrigation must not be overlooked, and spraying with insecticides and fungicides has been claimed to spread some fungal diseases.

MARINE AIR

The oceans, forming three-quarters of the Earth's surface, act as a vast source, putting a mainly bacterial population into the atmosphere. Com-pared with air over land, the microbial concentration in surface layers over the sea is usually very small. Processes by which marine organisms become airborne include: spray droplets from the breaking of waves on shore or at sea; foam blown off white-caps; and bursting of bubbles pro-duced by white-caps, rain or snow (Blanchard & Woodcock, 1957). These

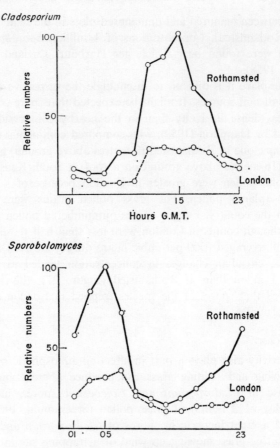

Figure 27. Circadian periodicity of airborne spores of *Cladosporium* and *Sporobolomyces* at London (South Kensington) and Harpenden (Rothamsted), based on Hirst trap data from May to September, 1954. (Reproduced by permission of Elizabeth D. Hamilton from *Acta Allergologica*, 1959.)

processes, however, also tend to remove suspended particles from sea air by the large liquid surface whose relatively constant temperature determines continued up-and-down movement in the lower layers of air.

Much of the older work is reviewed, and new data added, by ZoBell (1946). A critical appraisal of the whole subject of aerobiology comes from Jacobs (1951), who calculated, on the basis of salt concentration of the air and the bacterial concentration of sea water (which seldom exceeds 500 per cm^3), that the number of marine bacteria in air near the sea surface averages about 5 per cubic metre.

The microbial exploration of marine air was pioneered by Miquel (1885, 1886) with the help of a sea captain, Monsieur Moreau, during seven voyages. For visual examination of cryptogamic spores, an aspirator was worked by suction provided by an engine condenser – a device which sampled 700 litres per 24 hours. Bacteria were estimated by drawing air through glass-wool plugs in tubes at the rate of 1000 litres per 24 hours, washing the plugs, and inoculating aliquots into flasks of liquid beef extract. A total of 113 cubic metres sampled in the seven voyages averaged 1 bacterium per cubic metre, or 0·6 per cubic metre if samples taken within 100 km of land were excluded.

Visual counts over the ocean usually showed a few hundred cryptogamic spores and many pollen grains per cubic metre (1/10th of the usual number on land), but on one occasion a total of 3700 per cubic metre was encountered at a distance of 30 km from land off the coast of Senegambia; this comprised a spora very different from that found by Miquel in Paris. Near to continents the winds coming from land always brought impure air, but the sea rapidly purified it. In normal weather, bacteria from sea water were not abundant in the air, but in rough weather Miquel found that the sea contained a few marine bacteria.

The air in a ship's saloon always contained incomparably more microbes than sea air, but its purity increased rapidly in the early days of the voyage until it reached an equilibrium between purification by ventilation and contamination by vital activity on board, at a level of perhaps 1% of that of dwellings in Paris. Nevertheless, Miquel concluded, a ship travels in an atmosphere of self-contamination with bacteria, moulds and starch grains.

On a voyage to the Caribbean, B. Fischer (1886) found very few terrestrial microbes in ocean air, except near major land masses, where large numbers of bacteria appeared, apparently derived from the soil. Flemming (1908) sampled air on a voyage from Hamburg to Rio de Janeiro and Santos. Of the numerous 20-litre samples taken more than 200 km from land, two-thirds were sterile, but even at this distance he averaged 34 viable spores per cubic metre. These were mostly of moulds and yeasts, though bacteria increased in proportion nearer to land.

Although over the sea the lower air is extremely pure in comparison with air over land, most investigators on board ship have found bacteria, yeasts and mould spores wherever tests have been made. Bisby (1935) exposed petri dishes on a voyage from Montreal to England and isolated bacteria, *Botrytis cinerea*, and *Phoma hibernica*, all near the coast of Ireland. At higher altitudes over the sea, microbial concentrations are often greater than in the surface air (Chapter XII).

Microbes of marine air have been studied at the Scripps Institution of

Oceanography, California, by ZoBell & Mathews (1936) and ZoBell (1942). They claimed that less than 5% of bacteria in sea water will grow in freshwater nutrient media, and a still smaller percentage of freshwater bacteria will grow on seawater media. Petri dishes of nutrient media made up with distilled water (FW) or sea water (SW) were exposed horizontally at distances up to 1600 metres inland from a sea wall during a sea breeze of 2·6 m/s. The SW count decreased and the FW count increased with the distance inland, the SW/FW ratio decreasing steadily from 10 to 20 at the sea wall, to 1 at 400 metres, and 0·5 at 1600 metres inland. The number of mould spores usually increased with increasing distance from the sea.

In a land breeze, littoral spray puts into the air saltwater bacteria which can be detected at up to 8 km out to sea, beyond which the SW/FW ratio decreases to 1 owing to the predominance of terrestrial bacteria in the air for a distance of 160 km out to sea in fine weather. Exceptionally at 800 metres altitude on Mt Woodson, 32 km inland, plates exposed in a sea breeze following rain gave a ratio of SW/FW = 2·06, which was interpreted as indicating a predominance of marine bacteria in the air in a region where soil bacteria normally predominate. It has been calculated that an average of 12·7 cubic miles of sea water is put into the Earth's atmosphere each year in the form of splash droplets, and this would provide an average of only about one marine bacterium per square centimetre of the Earth's surface per year – a small quantity compared with the deposition rate from the land air spora (ZoBell, 1942).

Although all workers agree that marine air contains extremely few bacteria, ZoBell points out that the use of seawater media might be expected to increase the counts of earlier workers by a factor of 10 or 20 times. On these media gram-negative bacilli predominate, there are few cocci, no vibrio or spirilla forms, and fewer than half were spore formers. A pink yeast has been reported from ocean air by several workers. This spora contrasts strongly with the gram-positive rods, spore formers, cocci, Bacillaceae, and Micrococcaceae which, with mould spores, are normally abundant in air over land (ZoBell, 1942).

Rittenberg (1939) sampled in (presumably horizontal) petri dishes at 21 metres above the deck of the vessel *E. W. Scripps* off the Pacific Coast of California. Contrasts between seawater and freshwater media were not so clear as in previous tests, and the numbers varied widely at different stations; but, on average, moulds decreased in numbers with increasing distance from land (Table XXIV).

Detailed examination of 100 bacterial and yeast cultures taken at random by Rittenberg showed that 32% were yeasts, 30% cocci, 15% gram-negative rods, and the remaining 23% gram-positive spore-forming rods.

They included: *Bacillus subtilis, B. flavus, B. megatherium, B. mycoides, B. tumescens, B. cohaerens, B. laterosporus, Flavobacterium liquefaciens, Staphylococcus aureus, S. albus, S. citreus, Micrococcus flavus, M. candidus*, and *Sarcina flava* (but additional information from ZoBell (1942) indicates that these identifiable species were all from freshwater plates). One hundred mould cultures included: *Cladosporium (Hormodendrum)* 22%, *Penicillium* 18%, *Alternaria-Macrosporium-Stemphylium* 11%, *Cephalosporium* 7%. Others identified included: *Plenozythia, Catenularia,*

TABLE XXIV

NUMBERS OF MICROBES AND DISTANCE FROM LAND. SUMMARY OF 25 PETRI DISH EXPOSURES AT 21 METRES ABOVE DECK

(Rittenberg, 1939)

| Distance from land (km) | Average number of colonies per hour of exposure (on 4 petri dishes of each medium | | | |
| | Seawater medium | | Tapwater medium | |
	Bacteria and yeasts	Moulds	Bacteria and yeasts	Moulds
0–20	45	115	20	200
20–300	48	79	13	69
300–800	71	20	39	36

Spicaria or *Paecilomyces* and *Trichoderma*. To avoid the usual embarrassment of the microbiologist, all sampling in free air seems to have been done with sugar-free media; otherwise moulds would have been many times more abundant. Rittenberg points out that this marine airborne microbial flora is unlike that of sea water itself, but resembles the air spora over land.

Although the air spora over the sea is clearly largely of land origin, petri dish sedimentation tests are difficult to interpret and cultural and volumetric work is needed on the contribution from the ocean itself, both in the zone of littoral influence studied by ZoBell, and far from shore. The mode by which the ocean purifies the air flowing over it, and the fate of airborne spores trapped by the ocean, are still obscure.

Earlier workers, including Miquel and B. Fischer, found marine air almost free from pollen, but this freedom is now seen to be only relative. Erdtman (1937) operated a vacuum-cleaner filter trap at the masthead of the M.S. *Drottningholm* during a voyage from Gothenburg to New York extending from 29 May to 7 June, 1937. Compared with the average of 180 pollen grains per cubic metre recorded during spring at Västeros (110 km west of Stockholm), he found only 0·18 per cubic metre in the North Sea, and 0·007 per cubic metre in mid ocean, with an increase again on approaching North America. Temporarily greater concentrations ('pollen rains') occurred three times: of *Pinus* (0·13 per cubic metre) in the North

Sea; of *Alnus viridis* (0·045 per cubic metre) and Cyperaceae (0·006 per cubic metre) at a distance of 250 to 600 km off Newfoundland; and of combined grasses, *Plantago*, and *Rumex* (totalling 0·1 per cubic metre) at 220 to 300 km from Nova Scotia and Massachusetts. During strong western and north western winds, about mid way between Iceland and Ireland, Erdtman caught tree pollens (*Alnus, Betula, Corylus, Juniperus, Myrica, Picea, Pinus, Populus, Quercus, Salix, Tilia, Ulmus*) and herb pollens (Chenopodiaceae, Cruciferae, Cyperaceae, Ericaceae, Gramineae, *Plantago*, Umbelliferae, and *Urtica*), as well as spores of *Dryopteris* and *Lycopodium clavatum*.

Erdtman's volumetric sampling firmly establishes the occurrence of pollen in small but measurable concentration near the surface of the sea right across the Atlantic, and there is no reason to doubt that the land air spora extends to all parts of the globe. Confirmatory evidence comes from transatlantic sampling by Dyakowska (1948) and Polunin (cf. 1955a).

Bishop Rock Lighthouse stands on a low rock at the southwestern extremity of the Scilly Isles, which are a group of small, rocky islands with few trees (mainly *Ulmus* and *Pinus*). Gravity slide sampling on the lighthouse platform 38 metres above sea level (Hyde, 1956) showed mainly pollen of *Betula, Quercus*, and *Fraxinus*, with some *Pinus*. The total tree-pollen deposit was quite large (2800 per cm² per year, compared with 2200 at Aberdeen and Brecknock Beacons, and 10 000 at Cambridge). The proportion of tree pollen at Bishop Rock was 27%, and this is typical of country areas in Britain (in towns it may reach 50%). It is remarkable that the greatest deposition of grass pollen recorded for any centre in Britain during Hyde's gravity slide survey was 1679 grains per 5 cm² at Bishop Rock on 29 June 1953. Whether this resulted solely from high concentration, or whether it was aided by high efficiency of turbulent deposition in strong winds, is not yet clear. The observation becomes easier to understand in light of the concentrations encountered during high level sampling over the North Sea discussed in the following chapter.

Sreeramulu (1958b) used a Hirst trap at 20 metres above sea level on a voyage in the Mediterranean in October and November, 1956. At 5 to 50 miles from land he found an average of 56·4 fungus spores and 1·6 pollen grains per cubic metre. In Malta harbour the concentrations were 121 and 12 per cubic metre respectively. At sea, *Cladosporium* predominated with 16 spores per cubic metre, followed by smut spores at 5 per cubic metre, and coloured basidiospores at 7 per cubic metre. Also of interest was the occurrence of spores of *Helminthosporium, Alternaria, Torula herbarum, Nigrospora, Curvularia*, and *Epicoccum*, as well as hyphal fragments.

Pollen in marine air must come from land plants; the mould spora is more characteristic of above-ground sources than of the soil; the bacteria, however, may well come largely from sea water and the soil.

THE AIR OF POLAR REGIONS

The air of polar regions seems to be still purer than that over the sea. Levin (1899), who aspirated air through powdered-sugar filters, obtained only three bacterial colonies and a few moulds in a total of 20 cubic metres of air sampled at various points in Spitsbergen (Svalbard).

During 2 years on an island near Graham Land, Antarctica, Ekelöf (1907) exposed petri dishes at intervals; 40% of them grew bacteria, which he thought came from the soil. On the average, one colony arose per 2-hours exposure.

Pirie (1912) exposed petri dishes in the 'crow's nest' of the *Scotia* in the Weddell Sea, Antarctica, during the summer of 1903, for as long as 20 hours, and also on a glacier at Scotia Bay during winter; they all remained sterile. E. Hesse (1914) exposed petri dishes while at sea south of Spitsbergen and also found the air to be almost sterile.

Darling & Siple (1941) exposed jars and dishes of media in remote places in Marie Byrd Land, Antarctica, and from their isolations identified: *Achromobacter delicatulum*, *A. liquidum*, *Bacillus albolactis*, *B. fusiformis*, *B. mesentericus*, *B. subtilis*, and *B. tumescens*. They concluded that, although some bacteria had been brought to Antarctica by man and migrating animals, the vast majority must have come as atmospheric dust in subsiding air.

Recent work in the Arctic has demonstrated a fair range but sparse 'population' of microbes present in the air in summer near ground and sea level in various parts of these regions. Polunin (1954, 1955a) organized the exposure of sticky slides at several points ranging eastwards from Point Barrow, Alaska, to Spitsbergen, in 1950, and found a considerable variety of pollen grains and 'probable moss spores' at each station. Remarkably enough the pollen grains caught most plentifully through most of that summer in Spitsbergen were of *Pinus* – several hundreds of kilometres from their nearest possible source. In 1954 Polunin (1955b) was responsible for the exposure of sticky slides (cf. Polunin, 1960) off the north coast of Ellesmere Island and, in 1955, for the exposure of sticky slides (Barghoorn, 1960) and petri dishes of nutrient medium (Polunin, *et al.* 1960; Prince & Bakanauskas, 1960) on Ice-Island T-3 when it was floating in the North Polar Basin at about 83° N. Here again the air spora was very sparse compared with middle latitudes, but it included pollen grains which, in some instances, indicated long-range transport (cf. Barghoorn, 1960, p. 91, despite p. 88). From the T-3 exposures a few slow-growing fungal isolates (all identified as *Penicillium viridicatum*) and more of Actinomycetes were obtained, but no bacteria.

THE ORIGIN OF THE AIR SPORA

There is no reason to doubt the conclusion of Miquel and of Proctor (1935) that most airborne bacteria originate from the soil or from the oceans (ZoBell, 1946). But it is doubtful whether the soil makes a substantial contribution to the fungus-spore content of the atmosphere, as has been argued by some writers. It seems more likely that this air spora is derived predominantly either from moulds, plant pathogens and other fungi growing on vegetation, or from surface-growing fungi equipped with mechanisms which liberate their spores actively into the freely-moving turbulent air layer (Dransfield, 1966; and cf. Chapter IV). In the soil, bacteria, Penicillia, and Aspergilli predominate; but *Cladosporium* predominates in the air, seconded by basidiospores (true, agarics mostly grow on the ground, but so do trees and no one claims that pine pollen originates from the soil!). Similarity between the soil and air sporas results mainly from the soil being the ultimate 'sink' to which most of the air spora is destined, washed into the soil and awaiting extinction.

Much of the air spora comes from wild vegetation. Industry pollutes the atmosphere mainly with inorganic particles and gases. It is not generally appreciated that agricultural practices may pollute the air with plant pathogens and with respiratory allergens on a large scale.

For microbes to get into the air in the concentrations observed at peak seasons, a take-off mechanism is necessary. However, the most unlikely organisms occasionally get into the air. Siang (1949) isolated one colony of the aquatic phycomycete, *Hypochytrium catenoides*, from air on a roof at McGill University, Montreal, Canada. Probably almost every kind of microbe would be found if sampling were continued long enough 'but one swallow does not make a summer'.

From recent work on the air spora near the ground, we learn that its composition and concentration often fluctuate enormously, sometimes within quite short time-intervals. The significance of this for plant pathologists and plant breeders is obvious. Some constituents, such as the grass pollens, are important in respiratory allergy (Chapter XV). In the course of 24 hours we inhale perhaps 50 microgrammes of a mixture of microbes. Although some constituents of this mixture are harmful, this dose may also bring in useful quantities of organic compounds. Chauvin & Lavie (1956) found antibiotics in *Salix* and maize pollen, and their presence in fungus spores too, would not be surprising (*see* Whinfield, 1947). We may wonder whether the reputed beneficial effects of country air for human health are attributable not only to its freedom from smoke and fumes, but also to positive gains from the air spora.

Finally, to balance loss with gain, as the Earth's surface is the ultimate receptacle of almost all spores liberated, it can be estimated that the soil must receive a dose of fertilizer from the air spora equivalent to 5 kg of nitrogen per hectare per annum – an amount negligible on fertile land, but enough to aid plants colonizing barren places.

XI

Survival in the Atmosphere

While airborne, microbes are assumed to be resting or inactive, either because they are in the modified state as spores, or because of external limitations imposed by desiccation and starvation: only oxygen is adequate. But although the particle is at rest, it is undergoing changes that sooner or later lead to death unless it is replaced in a favourable environment. For this reason the airborne phase, while it persists, is at best regarded as a retarded decay process, and death of the cell is sometimes rapid.

There are definite spheres of interest where survival is irrelevant. An allergenic particle remains active even when dead, so long as chemical denaturation has not proceeded too far (Chapter XV, p.   ). In palynology also, in its restricted sense, viability is unimportant.

The degenerative changes may be graded in a series.

(1)   A pathogen may lose infectivity long before grosser changes can be detected.

(2)   A cell may lose susceptibility to bacteriophage replication at an early stage (Cox & Baldwin, 1964).

(3)   Mutations may occur affecting growth pattern when the microbe is returned to a favourable environment. These mutations commonly appear after a period of aerial suspension and one of the best known is the so-called 'petite colonies' type (Dimmick & Akers, 1969, pp. 75, 365).

(4)   Loss of ability to grow and reproduce (generally accepted as evidence of death) may occur with cells still able to respire and carry on some metabolic functions. It is not always easy to determine the status of a particular spore or cell. If it can grow it is clearly viable, but failure to grow may merely reflect failure of the investigator to provide suitable conditions.

(5)   Loss of allergenicity may be delayed long after growth death and

170

metabolic death, and seems to result from more drastic alterations to the protein molecule.

(6) A final grade of degradation that has been imagined by some authors, is disintegration and vanishing into thin air, but for this there seems no real evidence, even with the presence of chemical fumes or smog in the aerosol. Dissolution of wall and capsular layers seems unlikely. Pollen grain exines are highly resistant, as evidenced by persistence in peat. So too are the chitinous or cellulose walls of fungal and plant spores, and the muramic acid-containing walls of bacteria, which would need either flame, or conditions favourable to the activity of enzymes of autolysis, or action of other saprophytes to make them disappear.

DEGREES OF HARDINESS

Different species probably fall into an almost infinitely graded series in respect to powers of survival in the aerosol state. For convenience we can identify two broad categories, but the boundaries are shady, as distinctions often are in biology.

(i) *Hardy microbes.* This group includes a wide variety of organisms or parts of organisms that are obviously adapted in their evolution to wafting about the countryside. For example: pollens of anemophilous flowering plants; spores of ferns, mosses, many fungi, Actinomycetes, spore-forming bacteria, and encysted protozoa. For these organisms, which are normally disseminated extramurally, the question of survival hardly arises, until we come to consider dispersal to long distances (Chapter XVII).

(ii) *Tender pathogens.* This group, although regularly dispersed in the aerosol form, normally spreads among individuals in close proximity intramurally. Here are placed the common inhalent pathogens, conveyed by droplets expelled from infected individuals.

Here too must be included artificial aerosols of pathogens that are conveyed in nature by other means, but that can be spread by artificially generated aerosols. For example the pathogens of yellow fever and Q-fever, not well adapted for aerial dissemination, but becoming airborne when 'taken by surprise' as in deliberate atomization, or incidentally in industrial processes.

MEASUREMENT OF SURVIVAL

The death-rate of a microbial pure culture is usually taken ideally to proceed exponentially with time, the same fraction of surviving individuals dying in each successive time-interval – a process analogous to radioactive decay. But natural populations are often heterogeneous, and these mixtures of individuals with different histories, or of strains and species with different tenacity of life, may deviate from the ideal exponential die-away curve. Measurements of survival time are often expressed as the time taken for most or all of the organisms to die under a particular set of conditions. Yarwood & Sylvester (1959) point out that this limit is difficult to measure accurately, and that a more useful concept is the half-life of the populations (as used for decay of radioactivity). Apart from being easier to measure than the endpoint, the half-life is more logical than an arbitrarily selected value such as 90% or 99% of the population, because it is the time at which all the individuals in a population which were alive at the start have an equal chance of being alive or dead. As an example, Yarwood & Sylvester gave the half-life of basidiospores of *Cronartium ribicola* under the particular conditions tested as 5 hours.

The literature on aerobic microbial survival under various conditions of stress is immense, but most of it is irrelevant to the airborne state. This is because most tests were conducted on microbes held in liquid suspension or in films dried down, usually on glass, where at best only about half the microbial surface is exposed to the air, and even that half is deeply immersed in the still boundary layer and therefore in an environment remote from that of the naked organism in aerial suspension. Knowledge of survival in aerosol condition has increased in the last fifteen years with the development of improved techniques. This research, directed towards survival of the 'tender pathogens', is a specialized and rapidly developing subject, which has been authoritatively summarized by Anderson & Cox (1967), and by Dimmick & Akers (1969) in their *Introduction to Experimental Aerobiology*.

EXPERIMENTAL METHODS

To study the decay in percentage viability of a cloud of microbes suspended in air is technically difficult because the particles also sediment under gravity.

Earlier tests in wind-tunnels are reviewed by Dimmick (1969), and those in stirred sedimentation chambers by Dimmick & Hatch (1969).

To overcome the problem of sedimentation, Goldberg et al. (1958) introduced an ingenious 'dynamic aerosol toroid', a drum six feet in diameter rotating on its axis at 2–3 r.p.m. Samples of ageing aerosol suspension withdrawn through the axle showed that clouds of a bacterial suspension of particles 1–2 μm in equivalent diameter could be held for up to 2 days with only a 50% loss of particles by sedimentation to the walls of the drum (contrasting with a 90% loss in a stirred settling chamber in the same period). This apparatus is of little use for particles greater than 5–6 μm equivalent diameter because they settle too fast. The technical difficulties encountered with all these methods are emphasized by Anderson & Cox (1967) who summarize the position thus: 'Owing to the large number of variables which affect survival in the aerosol, both the results and their interpretation are markedly dependent upon the precise technique employed.' In artificially generated aerosols they note effects on survival due to age of culture, suspending fluid and method of generating the suspension, and of storing and sampling the aerosol.

Causes of death in the suspended state suggested include desiccation, particularly loss of bonded water from the microbial protein (Webb, 1965, 1967), or a toxic effect of oxygen (Anderson *et al.*, 1968; Cox, 1968; Cox & Baldwin, 1967). Webb shows that the presence of some bacteriostatic substances in the fluid sprayed with the artificial aerosol particle may increase survival (e.g. inositol).

How far tests in closed chambers are relevant to real conditions in the atmosphere is uncertain. In the rotating drum and the stirred sedimentation chamber the particle is held continuously in the same mass of air, whereas in nature it is free to fall from one layer to another. Further, the purified air used in laboratory tests may not have the same properties as external air. This was demonstrated by the 'micro-thread' technique (Druett & Packman, 1968; May & Druett, 1968) devised at Porton, which overcame some of these limitations. The method uses the ability of small, web-forming spiders to manufacture strong, inert silk threads, fine in gauge (about half a micrometer in diameter) which are moreover bacteriologically sterile. In the laboratory, threads are wound on a frame which is then placed in an air current bearing an artificially generated aerosol so that the threads become loaded with microbial particles. The threads are then exposed in a test environment for periodical assessment of viability during ageing.

When tested in the laboratory with *Escherichia coli* the microthread method gave survival curves substantially similar to the rotating drum method. However, unexpectedly, when *E. coli* was sprayed on microthreads and exposed, not to laboratory air, but to outdoor night air, on some occasions viability decreased much faster than in the laboratory. Further work at Porton has shown the occurrence of a previously unknown toxic 'open air factor', which was shown to be destroyed by sunlight, and is thought by Druett & May (1968) and May *et al.* (1969) to arise from some form of atmospheric pollution (*see* Cox & Penkett, 1971).

The microthread technique brings the anchored microbe into much more intimate contact with flowing air than do the earlier methods. The appropriate wind speed for relevant tests appears to be that of the Stokesian sedimentation speed of the particle, though a range of speeds on each side of this can be tested at will. The method also offers scope for work with larger particles that would sediment too rapidly for convenient handling in the rotating drum. But dry spores may blow off the threads unless fixed with an adhesive, and this adhesive could itself modify the survival of the organism. Testing viability in aerial suspension is still technically very difficult.

VIABILITY IN AIR SPORA SURVEYS

Some air spora surveys have systematically tested viability at different times of day and night.

R. R. Davies (1957) studied the percentage viability of *Cladosporium* spores in London. Using a slit sampler to measure viable spores and a cascade impactor for total count, Davies concluded that on average 91% of 'dispersion units' (clumps) germinated within 44 hours. On a few occasions the occurrence of soot and ash depressed germination. A similar study in Kansas (Kramer & Pady, 1968) gave little evidence of any circadian variation in spore viability, but large and consistent differences existed between different types. *Cladosporium* averaged 45%, *Alternaria* 80, *Cercospora* 53, fusiform ascospores 47, rust uredospores 32, basidiospores 6, and hyphal fragments 20% germination.

HAZARDS WHILE AIRBORNE

The environment of a microbe while airborne in the free atmosphere is peculiar. No matter whether the air suspending it is still or in rapid and turbulent motion, the spore moves through the air at a nearly constant speed: its characteristic terminal velocity of fall in still air (cf. Chapter II). This is important in relation to hydration of the particle. In still air water is lost by diffusion, at a rate depending on the absolute humidity of the air. For an anchored particle exposed to wind the relative humidity is the appropriate parameter, but a suspended particle is in a transitional region where diffusion (and therefore absolute humidity) become less important with increase in terminal velocity, which in effect, is the particle's characteristic wind speed.

The hazard of desiccation is greatest in daytime, and in air layers near the ground. At higher altitudes, and throughout the atmosphere at night, conditions are less favourable for evaporation, and spores may even be found germinating in the clouds: a phenomenon occasionally reported for the uredospores of rust fungi. We are still not clear how to relate meteorological observations to conditions for viability.

Radiation at wavelengths shorter than heat presents a much more serious hazard on which there is rapidly growing literature. The radiations most quickly lethal in the atmosphere are in the ultraviolet region, and these are largely absorbed by the air before they reach the ground. Ascent to the upper air, therefore, brings the risk of greatly increased dosage with ultra-violet radiation, except in the shelter of clouds. Bacteria carried on larger inorganic rafts, and pigmented spores, may also be shielded from radiation.

The shielding properties of pigments in the air spora have been noted by many workers: Marshall Ward, 1893; Tanner & Ryder, 1923; Fulton & Coblentz, 1928; Weston & Halnan, 1930; Weston, 1931, 1932; Whisler, 1940; Hollaender, 1942; English & Gerhardt, 1946; Gottlieb 1950; Stanier & Cohen-Bazire, 1957; Ashwood-Smith *et al.*, 1967. Stanier (1960) showed that photosynthetic bacteria were protected by the presence of carotenoids from damage by visible light. He also treated a variety of bacteria with photosensitizing dyes and found that organisms containing carotenoids suffered less damage than colourless species. Carlile (1970) also concluded that the presence of carotenoids in fungi protect against harmful illumination.

Receiving radiation by day and emitting it at night, one might expect that the temperature of an airborne microbe would fluctuate violently, but because of its small size and low heat capacity, its temperature will be almost instantly adjusted to that of the surrounding air mass (Druett, 1964). Radiation of shorter wavelength than heat, but longer than the ultraviolet may also be deleterious, and there is some evidence of lethal action of radiation in the visible region.

When a spore returns to ground level, the phenomenon of photo-reactivation by visible light of organisms that have been exposed to ultra-violet may perhaps decrease the damaging effects of radiation received at high altitudes. Photo-reactivation is defined by Jagger (1958) as: 'The reversal with near ultraviolet or visible light of ultraviolet radiation damage to a biological structure.' Visible light reverses lethal effects of ultraviolet radiation in many microbes, including bacteria (e.g. Webb, 1961), Actino-mycetes, fungi, yeasts, and protozoa, but the survivors of high altitude passage may be expected to show an increased mutation rate on their return to ground level.

The interactions of humidity, temperature, and radiation are not well known, and it may be that low temperature and desiccation protect a spore against radiation damage (Carlile, 1970).

Improvements to the technique of freeze drying show that damage to microbes is greatly affected by the temperature and speed at which they are desiccated. Repeated wetting and drying often lowers viability, but as experiments under controlled conditions show, many of the regular components of the air spora are resistant to desiccation. Less hardy organisms than these are probably better able to survive when they are high above the Earth's surface. Most micro-organisms will survive longer in a resting condition in the temperatures found in the upper air than they will at ground level temperatures. Temperatures in the upper atmosphere are preservative rather than lethal for most of the air spora, a conclusion reached by Meier (1936b).

The suitability of the atmosphere for the survival of 'aerial plankton' is summed up by Gislén (1948) as follows: 'While the lower cloudy air-strata – let us say under 3000 to 4000 metres – form a suitable medium for the transport of micro-organisms, the higher layers are very inhospitable to them, not so much because of the low temperature, drought and barometric pressure, as because of destructive radiation.'

XII

The Upper Air Spora

The air spora near the ground is dominated by fluctuations in its immediate local sources. In the upper air, however, the effects of local sources are smoothed, and attention can be focused on organisms undergoing long-distance transport. Concentrations in the upper air are sparse, and at great heights the necessity of keeping samples free from contamination is paramount: sterile technique for enumerating the microscopically small particles becomes exacting when they are exceedingly dilute.

VERTICAL DIFFUSION

Whereas spores and pollen grains are heavier than air, and tend to fall under the influence of gravity, atmospheric turbulence and convection tend to work in the opposite direction. As a result, the atmosphere is in a sense a spore suspension that generally decreases in concentration from ground level up to the base of the stratosphere. Eddy diffusion will bring spores to the top of the outer frictional turbulence layer: above this, convection will operate and, in the upper part of the troposphere, we would expect to find mostly components of the daytime air spora. When it first became possible to explore the air overhead, it was a matter of surprise to find how far up microbes could go. As methods have been developed for exploring greater and greater heights, we can begin to form a picture of the changes in concentration with height and of the circulation of spores of micro-organisms over the surface of the globe.

Evidence that concentration normally decreases with increasing height comes from two distinct sources of information which have often been confused: (1) observations at a standard height above local ground level at a chain of stations differing widely in altitude above sea level; and (2)

177

observations at widely differing altitudes above local ground level at a single station.

GROUND STATIONS AT DIFFERENT ALTITUDES ABOVE SEA LEVEL

Observations in this category are extremely fragmentary and have the flavour of holiday tasks on fine days in summer. Samples at various altitudes are taken at successive times as the climber reaches a suitable station – as in Pasteur's visit to the Mer de Glace, where the relative purity of mountain air was convincingly demonstrated (see p. 4). Using a volumetric method, Miquel (1884, p. 524) confirmed this conclusion. In July and August, while bacterial concentrations of 55 000 per cubic metre were current in the air of the Rue de Rivoli in Paris, and 7600 in the Parc Montsouris, Miquel found only 8 bacteria (and numerous moulds) per cubic metre at a height of 1 metre above the surface of a field at Lake Thun (570 m above sea level), and none at stations between 2000 metres altitude and the summit of the Eiger at nearly 4000 metres. Miquel attributed this purity to the effect of reduced atmospheric pressure doubling the volume of air and diluting its dust load, to the rarefied air less easily holding particles in suspension, and to the absence of local sources of contamination – especially in the regions of perpetual snow. In similar volumetric data from the Dauphine Alps recorded by Bonnier et al. (1911), bacteria decreased with height more rapidly than moulds.

No one has yet compared concentrations at different heights above ground level over plateaux with those over mountains, or over flat and convex surfaces at the same altitude.

The purity of the air in regions of perpetual snow is understandable, but it is surprising that air at one or two metres above ground level in mountain valleys should also contain so few microbes (R. R. Davies, 1969a, 1969b). Geiger (1965) envisages mountain slopes as covered with a skin of air having the usual characteristics of air near the ground but easily removed by wind and convection, except where protected by vegetation. Convex surfaces generally have a more extreme climate than flat surfaces, and concave surfaces are more equable.

THE ROLE OF TURBULENCE

The role of turbulence in diffusing spore clouds vertically was first emphasized by Schmidt (1918, 1925), although the theory was developed for heat transfer by Taylor (1915) with information derived from temperature records over the Great Banks of Newfoundland. Schmidt argued that, when a stable state of diffusion by eddies has been reached, the number of particles falling, under the influence of gravity, across any horizontal boundary is compensated for by the number of particles moved upwards by diffusion, and so the concentration of particles in the air should decrease exponentially with increasing height according to the equation:

$$\chi = \chi_0 \exp.\left[-\frac{v_s z}{A} \right],$$

where χ_0 = concentration at height $z = 0$,
$\quad v_s$ = terminal velocity of fall,
$\quad A$ = Schmidt's 'Austausch' or intermixing coefficient which is assumed to be invariable with height.

The total spore content of the column of air standing above 1 cm^2 would be $\lambda = \chi_0 A/v_s$. The logarithm of the concentration should give a straight line when plotted against the height.

This approach to the problem suffers from two defects in practice. The coefficient for diffusion (Schmidt's A or Taylor's K) is not invariable with height, and it is doubtful whether a steady state is ever reached because of the great diurnal changes occurring when the source consists of living organisms.

C. G. Johnson & Penman (1951) supposed that the vertical distribution of aphids at any one time is determined by the net effect of upward transport by turbulence and downward transport by the combined action of gravity and biological impulse – the mean clearance rate.

If χ is the concentration at height z, and ω is the 'mean clearance rate', they deduced that a graph of log χ against log z should yield a straight line.

Attempts have also been made to fit empirical curves to observational data on vertical gradients. Wolfenbarger (1946, 1959) used regression equations of the type: $Y = a + b \log x + c/x$. C. G. Johnson (1957) fitted records of insect-trap catches with: $f(z) = C(z + z_e)^{-\lambda}$, where $f(z)$ is concentration at height z, C is a scale factor depending on population size, λ is an index of the diffusion process and the profile, and z_e is a parameter whose significance probably depends on the rate of exchange of insects between the air and the ground.

Particles entering the air near ground level become mixed throughout the layer of frictional turbulence so long as the wind blows. Convection provides a local intermittent mechanism which distributes spores from the ground layer throughout the troposphere. Observed vertical concentration gradients sometimes fit theoretical lines quite adequately, and they may well describe long-term averages. But as demonstrated below, theoretical treatments of this problem are often unsatisfactory, especially in failing to predict concentrations at an instant of time in the first few hundred feet above ground.

EARLY STUDIES OF THE UPPER AIR

Measurements of microbial concentrations at heights above ground level were first attempted from towers and tall buildings by Miquel (1883), Carnelley *et al.* (1887), and, more recently, by Kelly and others (cf. p. 191). Miquel found that the bacterial contents of the air at the level of the Lanterne of the Panthéon in Paris was only 1/20th of that in the street below.

Probing upwards into the atmosphere for microscopic life started dramatically when the Manchester physician Blackley (1873) used two kites in tandem to lift sticky microscopic slides to a height of 300 metres and caught from 15 to 20 times as much pollen as on slides similarly orientated at 1·4 metres above the ground. Kites were also successfully used in India by Mehta (1952) to catch spores of the cereal rusts, *Puccinia* spp., and small balloons were used for the same purpose by Chatterjee (1931).

SAMPLING FROM BALLOONS

Cristiani (1893) obtained bacteria and a few moulds by volumetric sampling from a balloon at up to 1300 metres above Geneva (at a total of 1700 metres above sea level). He was obviously puzzled by his results which he regarded as inconclusive, attributing most of his catch to contamination from the surface of the balloon and its rigging, and remaining convinced that the upper air is extremely pure.

The credit for first demonstrating the existence of a *viable* microbial population in the upper air should probably go to the mycologist Harz (1904), who sampled during a balloon ascent over southern Bavaria on a sunny morning in March. At altitudes of between 1500 and 2300 metres he aspirated air through a Miquel-type filter of powdered sodium sulphate by suction obtained with a horse's stomach-pump; culturing the catch in nutrient gelatine, he found a few moulds, and bacterial concentrations ranging from 179 000 to 2 870 000 per cubic metre. At 1800 and 2000 metres there was a zone with 16 times the concentration at 1500 metres and 5 times that at 2300 metres. These phenomenally large bacterial concentrations were associated with a large temperature lapse and strong convection from hot dry soil. Moulds were identified as: *Penicillium glaucum*, *P. cinereum*, *P. atro-viride*, *Sporidesmium* sp., *Acremonium alternans*, *Mucor racemosus*, *M. mucedo*, *Oospora ochracea*, *O. ferruginea*, *Periconia atra*, *Hormodendron (Cladosporium) penicillioides*, *Arthrococcus lactis*, *Aspergillus niger*, and a sterile mycelium.

During ascents from Berlin with both captive and free balloons, Flemming (1908) used trapping methods similar to those of Harz. He found viable microbes up to 4000 metres, averaging 370 per cubic metre above 500 metres, and 12 900 per cubic metre lower down. Sterile samples were rare. Concentrations were not uniform but increased strikingly at the level of the cloud base. Species identified included: *Micrococcus radicatus*, *M. albus*, *M. nubilis*, *M. aerogenes*, *Bacillus ubiquitus*, *B. aurescens*, *B. aureo-flavus*, *B. terrestris*, *B. aerophilus*, *B. submesenteroides*, *B. mycoides*, and *Penicillium crustaceum* – all spore-formers. Above 2500 metres altitude he found *Bacillus terrestris*, *B. aerophilus*, and *Sarcina lutea*, and, at 4000 metres, *Micrococcus citreus*, *M. luteus* and *Penicillium crustaceum*. Flemming commented on the frequency of pigment-formers among bacteria and yeasts of the upper air.

From catches made during balloon flights over southern Germany, Hahn (1909) concluded that, on the average, bacterial and dust counts run parallel with each other and decrease with height because of sedimentation. He claimed that the air above a certain height was germ free, and that this limit was lower in winter than in summer.

SAMPLING FROM AEROPLANES

Exploration for microbes in the upper air from aeroplanes was started in 1921 when Stakman *et al.* (1923) exposed Vaseline-coated slides over the Mississippi Valley to trap cereal rust spores. Flights from Texas as far north as Minnesota and at altitudes up to 3300 metres yielded numerous pollen grains and fungus spores, among which *Alternaria* (often in chains) were most numerous, followed by *Puccinia, Helminthosporium, Cladosporium, Cephalothecium, Ustilago, Tilletia*, and *Scolecotrichum*. Among rust spores the uredo forms predominated, but some teleutospores and aecidiospores were also caught. Spores became relatively scarce at altitudes above 3000 metres. At 5400 metres (the highest altitude tested), two uredospores of *Puccinia triticina* were caught. *Alternaria* from altitudes of 1000 to 3000 metres germinated readily, as did uredospores from 2300 metres.

In the summer and autumn of 1923, Mischustin (1926) exposed petri dishes of nutrient agar during flights from Moscow. On rather slender evidence obtained from tests in a wind-tunnel, he concluded that he was sampling 20 litres of air per minute – probably a large underestimate. The plane was first flown *above* the layer to be sampled to free it from ground dust, and then lowered to the required height. At 500 metres the numbers of bacteria increased in windy weather to 7000 or 8000 per cubic metre, from Mischustin's normal of 2000 to 3000 at this altitude. *Micrococcus* and *Sarcina* decreased greatly in calm weather, but the numbers of bacterial rods and moulds increased. At 1000 to 2000 metres numbers were small. The proportion of spore-forming bacteria, of moulds, and of Actinomycetes, was greatest at the greater heights. The concentration of organisms over the city of Moscow at 2000 metres included an average of 650 bacteria per cubic metre, and was from four to five times as great as that over the surrounding countryside.

Pollen was found at up to 5800 metres over the Mississippi, with the greatest concentrations often at from 300 to 1000 metres (Scheppegrell, 1924, 1925).

In flights from Cambridge, England, Weston (1929) found that fungi and bacteria were abundant up to 3000 metres, but relatively scarce above this altitude. Air within clouds contained more bacteria and fungi than did air above or below clouds, a phenomenon noted by other workers, including Heise & Heise (1948).

Allergist and aeronaut, Heise & Heise (1948, 1949, 1950), flying their own plane over Wisconsin found the air in and near cumulus clouds (formed by vigorous convection of microbe-rich air near ground level) contained many times the amount of *Alternaria* and ragweed pollen that was present in air away from the cloud. The greatest pollen concentration was near 1000 metres by day, but was near the ground at night, being dependent on the temperature lapse rate. A negative lapse rate, and particularly a temperature inversion, kept a high concentration of ragweed pollen and *Alternaria* near 'breathing height' to the discomfort of allergic subjects.

On flights up to 2200 metres over the arid lands of southern Arizona, Browne (1930) isolated 'white and grey' bacterial colonies, *Aspergillus*, *Penicillium*, *Alternaria* and yeasts; but no spores of wheat rusts were observed on slide spore traps. Cotter (1931) studied dispersal of wheat rust in flights near Lake Michigan, trapping on oiled microscope slides. Uredospores were not more numerous in rain than in fine weather; fewer were caught over Lake Michigan than over nearby land, and more were caught over areas abounding in barberry (the alternate host of the pathogen). MacQuiddy (1935) exposed petri dishes and slides in flights up to 2000 metres over Omaha, Nebraska. Pollen was abundant up to 900 metres, and bacteria and mould spores began to decrease between 1200 and 1500 metres. MacLachlan (1935) made flights in early May over Massachusetts, to trace spores of the juniper rust *(Gymnosporangium biseptatum)*. Petri dishes exposed over the side of the plane for 1 minute gave viable spores up to 600 metres (the maximum height tested), although numbers and viability decreased steadily with increasing height.

Tasugi & Kurosawa (1938), flying over Tokyo, used sticky slide traps of the Stakman type and also petri dishes of nutrient medium; they also sampled air through a metal tube passing into tubes of medium. Fungi and spore-forming bacteria (e.g. *Bacillus subtilis*) were found at up to 2000 metres.

Klyuzko *et al.* (1960) studied bacteria in 530 samples from altitudes of 100 to 6000 metres near Lvov in relation to origin of air mass, time of day, proximity to the city and altitude.

Zhukova & Kondratiev (1964) investigated moulds at up to 225 metres over Moscow, and made the interesting observation that the air spora was not dominated by *Cladosporium*, as usual elsewhere, but were outnumbered by *Penicillium, Aspergillus, Botrytis* and *Mucor. Penicillium* numbers were at a maximum at 14.00–17.00 hours, and a minimum at 02.00 hours; whereas *Botrytis* was at a maximum at 23.00 hours and a minimum at 14.00 hours.

Hermansen *et al.* (1965) investigated rusts and mildews with a cascade impactor at altitudes up to 1500 over Denmark.

AEROBIOLOGICAL WORK OF MEIER

The pioneer work of Meier is known only from abstracts (*see* Chapter I, p. 14). Sticky slides were exposed by Colonel Charles A. Lindbergh, in collaboration with Meier (1935a, 1935b), in special containers at about 1000 metres altitude during a flight between Maine and Denmark. Material trapped over Davis Strait and East Greenland included algae, fragments of insects' wings, diatoms and possibly sponge spicules, volcanic ash and glass. Fungus spores were tentatively identified as belonging to: *Macrosporium, Cladosporium, Leptosphaeria, Mycosphaerella, Trichothecium, Helicosporium, Uromyces, Camarosporium,* and *Venturia.* Some of these were abundant over Maine and Labrador but diminished over Davis Strait, the ice cap of Greenland, and Denmark Strait.

Flights by Meier over the United States at from 150 metres to 5500 metres showed a varied spore 'population' which usually decreased in both numbers and variety above 2400 metres. Viable spores of *Pestalozzia* were caught over Washington, D.C. at 5500 metres on 22 March 1932. Other genera recognized included: *Acremoniella, Alternaria (Macrosporium), Aspergillus, Chaetomium, Cladosporium, Coniothyrium, Dematium, Epicoccum, Fumago, Fusarium, Helminthosporium, Penicillium, Sclerotinia, Stachybotrys, Stemphylium* and *Trichoderma* (Meier *et al.*, 1933).

After flights over the Caribbean Sea, Meier (1936a) came to feel that the trade winds might be important in disseminating microbes. Viable spores were found at 800 to 1200 km from land; but after rain squalls, petri dishes sometimes remained sterile after exposure at 60–240 m altitude a few kilometres to the leeward of islands – so demonstrating that showers remove spores from surface winds.

Sugar-beet pollen was trapped on agar plates during flights over a small sugar-beet seed-growing area in New Mexico. Viable beet pollen, mixed with pine pollen and fungus spores, occurred up to 1500 metres, which was the greatest height tested and the level of the dust horizon (Meier & Artschwager, 1938).

VERTICAL GRADIENTS

Hubert (1932) trapped spores during two flights at the time of an epidemic of yellow rust of wheat at Halle in Germany. During the second flight, for which data are more extensive, the numbers of spores trapped per square centimetre per minute of exposure at the various heights were: 30 metres and less: 1418 spores, 400 metres: 638 spores, 600 metres: 336 spores, and 800 metres: 82 spores.

During an epidemic of wheat rust in Manitoba in July and August 1930, Peturson (1931) trapped spores at different altitudes in eight aeroplane ascents. The average number of spores caught per 650 mm^2 of trap surface (presumably with comparable exposure times) were: 305 metres: 10 050 spores, 1520 metres: 1180 spores, 3050 metres: 28 spores, and 4260 metres: 11 spores.

Some of the records from the Canadian prairies are instructive in showing deviations from the ideal altitudinal profile. A selection of profiles from Craigie (1945) is illustrated in Figure 28, where curves A and D obtained during rust outbreaks approach the ideal pattern of a logarithmic decrease with height. Here the rust uredospores were presumably of local origin, carried aloft by thermal and frictional turbulence. By contrast curve B is interpreted as exemplifying the vertical profile when spores are being transported from a distant source into an area which itself is not producing spores but which instead is acting as a 'sink', removing spores from the lower part of the air mass. In this case the lower part of the spore cloud was being depleted by various processes (such as ground deposition and rain-wash), and the uredospore concentration was larger at 14 000 feet than it was at 1000 or 5000 feet. Profile C, during the height of a moderate rust outbreak in 1931, shows very uniform concentration from 1000 to 5000 feet, but very few spores at 7000 feet and higher. In Figure 28, assuming a plane air speed of about 150 miles per hour and a sampling efficiency of 100% with the apparatus used, the concentration at 1000 feet would be approximately 1000 uredospores per cubic metre for profile A, and 0·1 spores for profile B.

Rempe (1937) trapped tree pollen in a series of aeroplane flights by day and by night over German forests in spring. In general, with light to moderate windy weather and with cumulus clouds at about 2000 metres, the pollen concentration decreased only slightly up to 1000 metres, and the maximum number of grains might occur as high as 200 or even 500 metres. This was regarded as a sign of a complete inversion of the air masses. A similar distribution also occurred under high pressure conditions without

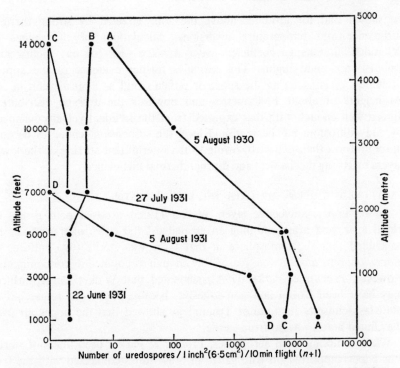

Figure 28. Vertical distribution of *Puccinia graminis* uredospores. Numbers of uredospores (plotted as: $\log_{10} n+1$) trapped per square centimetre of sticky slide trap per 10 minutes of flight at various altitudes over southern Manitoba during rust epidemics (Craigie, 1945).

clouds but with strong thermal turbulence. By way of contrast, conditions associated with a stratified cloud layer and high wind velocities showed a marked decrease of pollen with height.

In Rempe's night flights, the maximum number of grains was often reached at a height of about 200 metres, i.e. above the temperature inversion which often develops at night. At night the numbers trapped usually decreased with increasing height much more than by day. The total numbers trapped at all heights were also fewer by night than by day. From Rempe's data the mean numbers of pollen grains trapped per 1·275 square centimetres of trap surface per 20 minutes for all flights which extended up to 1500 metres were:

Altitude (m)	10–40	200	500	1000	1500
Day flights	904	849	852	581	267
Night flights	577	560	283	85	45

These records include pollen grains of several species; but taking $v_s = 3$ cm/s as a moderate value for terminal velocity of the grains, it can be shown that for altitudes above the zone affected by strong thermal turbulence and temperature inversions, calculations of the values of Schmidt's interchange coefficient gave $A = 2 \cdot 6 \times 10^5$ for day flights, and $1 \cdot 6 \times 10^5$ for night flights. This provides further evidence of the appropriateness of considering the spore or pollen cloud as a suspension in air. At heights of about 1000 metres and upwards the average distribution agreed well enough with that expected from the terminal velocity balanced by eddy diffusion, but near ground level the suspension tended to become more uniform than expected, owing to the intermittent stirring of the lower layers by strong mechanical and diurnal thermal turbulence.

SAMPLING THE UPPER AIR OVER THE UNITED STATES

In the upper convective layer, Walker (1935) exposed petri dishes of blood agar, and after sampling an estimated 2400 cubic metres of air, he concluded that the atmosphere in that layer was sterile (two cultures of *Staphylococcus aureus* were reasonably enough discounted as contaminants). However, Proctor & Parker (1942) suggested that Walker's agar surfaces may have been frozen and non-adhesive, because their own researches at the Massachusetts Institute of Technology showed that the upper air over the United States was far from sterile.

When sampling from aeroplanes, Proctor & Parker used filters of sterile lens paper supported on wire gauze in brass tubes connected with the free air and sampling about 28 litres per minute. The catches were examined both microscopically and by culturing. Bacteria averaged 12 per cubic metre over all flights, and 9 per cubic metre at 6100 metres or higher. There was sometimes evidence of a zone of greater concentration at a height of several thousand metres. Moulds were usually less numerous than bacteria, but the nutrient agar on which filter washings were plated was recognized as unfavourable to mould growth. Bacteria were mostly spore-formers, and those identified included species of *Bacillus*, *Achromobacter*, and *Micrococcus*. Among moulds, *Aspergillus* and *Penicillium* predominated, occurring with some other Fungi Imperfecti including *Cladosporium (Hormodendrum)* and *Fusarium*, as well as Mucoraceae, Actinomycetes and, occasionally, yeasts. Pollen was found on only three flights (Proctor, 1934, 1935).

The greatest mould count obtained in the M.I.T. studies occurred at an altitude of 200 to 300 metres in May over a wooded area, where 22 bacteria and 260 moulds per cubic metre were recorded. Particularly large counts of bacteria and moulds occurred during a dust storm which apparently came from Nebraska and South Dakota – the same dust storm during which Soule (1934) recorded massive invasion of his laboratories in Michigan by *Bacillus megatherium*. During this dust storm, at an altitude of 1500 to 3300 metres over the Boston area, bacteria, moulds, and dust particles totalled 140, 44 and 2800 per cubic metre, respectively. However, during the whole survey, dust particles were over 100 times as numerous as viable microbes, suggesting that much of the dust came from industry and combustion rather than from the soil (Proctor & Parker, 1938).

Petri dishes of nutrient agar were exposed during flights at from 300 to 3250 metres over Nashville, Tennessee, during winter by Wolf (1943). On this medium bacteria outnumbered moulds, the bacilli contributing 37·7%, non-spore-forming rods 24·6%, and cocci the remaining 37·7% of the total bacterial count. The bacteria were very similar to those found by Proctor (though with a smaller proportion of spore-formers) and further study supported the general conclusion that aerial bacteria are of types commonly found in soil and water, are generally unable to ferment common sugars with the production of gas, and are unable to produce indole.

From these flights by Wolf, *Actinomyces griseolus* was isolated twice, at 700 and 1400 metres, and *A. phaeochromogenus* once at 620 metres. A pink yeast was found at 1750 and 3050 metres. Fungi isolated, with their percentage frequencies, included: *Fusarium* 29, *Alternaria* 22, *Cladosporium (Hormodendrum)* 20, *Verticillium* 5, *Aspergillus* 3, *Penicillium* 1·6, and among others were: *Acladium*, *Brachysporium*, *Cephalothecium*, *Chaetomium*, *Helminthosporium*, *Macrosporium*, *Mucor*, *Oospora*, *Plenozythia*, and *Scopulariopsis*. The large proportion of *Fusarium* and the small amount of *Aspergillus* and *Penicillium* spores are remarkable. The average of all samples gave a concentration of 7·5 cultivable microbes per cubic metre, and varied from none at 850 metres altitude in December to a maximum of 42 per cubic metre at 460 metres in October. In general the concentration decreased with increasing height, but on 25 January there was a zone of greater concentration at 900 to 1200 metres altitude.

SPORES OF GREEN PLANTS IN THE LOWER TROPOSPHERE

Using a glass-wool filtration apparatus, Overeem (1936, 1937) sampled from aircraft over the Netherlands on six occasions extending from July to October at heights of 100, 500, 1000 and 2000 metres. Filter washings were inoculated into Pringsheim's culture solution for green plants and kept in the light. From a total of about 28 cubic metres of air she obtained the following cultures. Algae: *Chlorococcum* sp. 9, *Phormidium luridum* var. *nigrescens*, *Chlorella vulgaris*, *Pleurococcus vulgaris*, and *Stichococcus bacillaris* 3 each, *Aphanocapsa* sp. 2, *Actinastrum* sp., *Stichococcus minor*, and *Hormidium flaccidum*, 1 each. Moss: *Funaria hygrometrica* 2 (from 500 and 1000 metres). Fern: 1 (unidentified, from 500 metres). Total numbers at the various altitudes were in the ratios: 5 : 10 : 3 : 3 at 100,

H

500, 1000 and 2000 metres, respectively. This work is of particular interest as one of the few demonstrations that spores of green plants invade the troposphere in fair numbers and variety.

McGILL UNIVERSITY STUDIES

FLIGHTS OVER THE ARCTIC

Extensive exploration of the upper air by aeroplane was initiated by Polunin at McGill University, Montreal, in 1947, and continued until 1951 with the cooperation of the Royal Canadian Air Force and the United States Air Force. Flights during 1947–49 were primarily directed to the study of Arctic conditions.

In the summer of 1947 flights were over the Northwest Territories northwards to Cape Bathurst, then northeast from Cambridge Bay on Victoria Island to beyond the region of the North Magnetic Pole and back, and finally southwest from Cambridge Bay to Yellowknife and Edmonton, Alberta. Petri dishes with nutrient medium, and also sticky slides, were exposed from the plane by hand, mostly at about 1500 metres altitude (Polunin et al., 1947, 1948). There were small but measurable concentrations of fungus spores, and the composition of the air spora appeared to depend on the origin and sometimes on the trajectory of the air mass rather than upon the locality of sampling (Polunin, 1951a, 1951b, and cf. 1954). The bacteria were identified as: gram-positive rods, about 40% (two-thirds of which morphologically resembled Corynebacterium), Micrococcus 23%, Achromobacter or Flavobacterium 17%, spore-formers 4%, and Sarcina 3%. Fungi identified in culture included: Cladosporium over 40% of the total, Sporormia, Pullularia, Verticillium, Penicillium, yeasts, Phyllosticta, Leptosphaeria, Alternaria, Stemphylium, Chaetomium, Pestalozzia and Streptomyces (Pady et al., 1948; Pady, 1951; Polunin, 1951a, 1951b).

The sticky slides exposed over the Northwest Territories showed small concentrations of angiosperm and gymnosperm pollen, spores of pteridophytes and bryophytes, Alternaria, and Helminthosporium sativum, totalling about 1 per cubic metre. Uredospores of the cereal rusts, Puccinia graminis and P. glumarum, occurred in small numbers (except in the most northerly flight, though a few were found north of the Arctic Circle); their concentration rose to about 12 per cubic metre over northern Alberta, where there was also a small smut spore concentration of about 6 per cubic metre (Pady et al., 1950; Polunin, 1951b).

In further flights over the Arctic the McGill workers attempted more elaborate sampling methods to eliminate possible contamination from within the aircraft, which could have increased the counts in the first two or three exposures of the earlier flights. In September 1948, Polunin flew over the Geographical North Pole in a plane fitted with a breech loading tube to hold a petri dish projecting 30 cm forward of the nose. Before exposure the interior of the petri dish was coated with a silicon grease; and after returning to base, the dish was filled with molten agar and incubated.

Immediately over the North Pole in late summer at 920 metres neither bacteria nor fungi were caught. However, at greater heights over the Pole and at some other high latitudes, some petri dishes caught nothing, while others exposed at altitudes up to 6770 metres, grew a few colonies of bacteria or moulds; no Actinomycetes were found (Polunin, 1951a; Polunin & Kelly, 1952). Thus microbes appear to be present, though irregularly distributed, even over the poles.

During a flight over the Geographical North Pole under winter conditions in March 1949, the McGill workers used three kinds of sampler: (1) siliconed slides exposed in the tube forward of the nose of the plane; (2) an electrostatic sampler installed in a box through which a slow stream of air was passed; and (3) a filter tube packed with glass wool and lens paper. The electrostatic sampler and filters indicated a viable concentration of 26 bacteria plus yeasts and 1·6 fungi per cubic metre in some very high latitudes. The authors concluded that the air over the Pole and its environs is nearly sterile and that it is of very mixed origin. Here again there was evidence that at high altitudes the origin of the air mass was more important than the locality of sampling (Polunin, 1951a, 1951b, 1954; Polunin & Kelly, 1952).

The results of further arctic and sub-arctic flights are reported in detail by Pady & Kelly (1953) and Pady & Kapica (1953). On one flight from Winnipeg via Churchill to Baker Lake in the Northwest Territories at an altitude of about 1000 metres, using the G.E. Electrostatic sampler, cultures averaged: bacteria 10, and fungi 25 per cubic metre. Bacteria were predominantly cocci and spore-formers (though in a local flight over Churchill gram-positive pleomorphic rods predominated). Fungi were mainly *Cladosporium* and *Alternaria*, but included *Penicillium*, *Papularia* and *Stemphylium*.

In the summer of 1950, northern air was sampled daily by McGill workers for 21 days with the G.E. Electrostatic sampler and the slit sampler on a roof 17 metres above ground at Churchill near the tree line on Hudson Bay. This survey formed the standard of reference for two flights to Resolute Bay, Cornwallis Island, some 1600 km to the north, on 1–3 August, at an altitude of around 3000 metres. At Churchill the catch consisted of: gram-positive pleomorphic rods 46%, gram-negative rods 20%, spore-forming rods 18%, and cocci 15%. In the two flights to Resolute

Bay in the Arctic during this period, 51% of the bacteria were spore-forming rods (Pady & Kelly, 1953). Fungi were assessed both in culture, and visually on silicone-coated slides. The numbers per cubic metre (with the visual counts in parentheses) were: *Cladosporium* 17 (132), with its maximum in an air-mass of tropical origin, *Alternaria* 0·7 (2·1), *Stemphylium* 1·1 (1·8), smuts *(Ustilago)* (86), yeasts 3·5 (304), and *Penicillium* 2·1. Of the fungus cultures, 57% were non-sporulating. In addition, *Fusarium* was reported as common slides but rare in culture, and *Septoria* was sometimes abundant on slides. An interesting list of microbes which were caught only infrequently includes: *Pullularia*, Actinomycetes, *Botrytis*, *Aspergillus*, *Verticillium*, several Ascomycetes, and a single culture of *Cunninghamella* – one of the rare isolates of a mucoraceous fungus from the upper air. In addition there were numerous moss spores and pollen grains, which together averaged 20 per cubic metre (Pady & Kapica, 1953).

On the flights to Resolute Bay at 3000 metres, the fungi were essentially the same as at ground level at Churchill; but they were in much smaller concentrations averaging 12 per cubic metre (125 per cubic metre if determined visually), and were principally yeasts, *Cladosporium* and *Ustilago*. Pollen grains averaged 16 per cubic metre; and in warm air on the southern part of the flight moss spores averaged 47 per cubic metre. It was thought that the southern parts of these flights lay through an old continental tropical air mass, which had moved into the Arctic where most of the spores had died; north of this was cold polar air containing very few spores. The general conclusion reached was that the air spora over the Arctic comes mainly from the agricultural regions to the south (Pady & Kelly, 1953; Pady & Kapica, 1953). However, the large quantities caught at ground level at Churchill, and the numerous moss spores, suggests that the tundra also made an important contribution.

MICROBIOLOGY OF AIR MASSES OVER NORTHERN CANADA

The earlier McGill studies suggested that, in the upper air above the level of pronounced concentration gradients, microbial concentration depends on the history of the air mass. This was clearly shown in a series of ten flights over northern Canada (Kelly & Pady, 1953; Pady & Kapica, 1953) between September 1948 and August 1949. The history of air masses encountered during sampling and the positions of fronts were correlated with the results of sampling. The electrostatic sampler, loaded alternately with petri dishes and siliconed slides, gave the most consistent results. (However, as sampling was non-isokinetic, pollen and large spores may have been underestimated.)

On these flights bacteria varied in concentration less than fungi. At the end of December many samples were blank and the air was evidently almost sterile. Fungi were much more plentiful in June, July and August than in the rest of the year, but bacteria were most numerous in spring and autumn. Kelly and Pady suggest, reasonably enough, that the bacteria come mainly from soil which is exposed and cultivated in spring and autumn, giving the opportunity for wind erosion; but their suggestion that the fungi also come from the soil seems to be contradicted by the predominance of fungi in summer, and all the evidence points to the fungi coming mainly from vegetation and debris above ground level. All the bacteria isolated and examined in detail were regarded as typical soil forms; they were classified as: aerobic spore-formers 37·9%, gram-positive pleomorphic rods 23·8%, *Micrococcus* 18·8%, gram-negative rods (*Flavobacterium*, *Achromobacter*, or *Pseudomonas*) 4·8%, and *Sarcina* 4·6%.

Fungi obtained in culture were: *Cladosporium* 73%, *Alternaria* 7%, *Penicillium* 2·9%, *Streptomyces* 2·9%, *Stemphylium* 1·5%.

The percentages of fungi obtained in culture were: *Cladosporium* 73, *Alternaria* 7, *Penicillium* 2·9, *Streptomyces* 2·9, *Stemphylium* 1·5, *Aspergillus* 0·7, yeasts 0·7, and other fungi 11%. Many more fungi could be counted visually on silicone-coated slides than could be grown in culture – an effect which was exaggerated by the numerous smut spores that were obtained in one flight over the prairies in October 1948, which yielded: smut spores 52·4%, *Cladosporium* 32·4, *Alternaria* 3·3, *Helminthosporium* 0·3, and rusts 0·1%.

AIR MASSES OVER MONTREAL

In a further survey of microbes associated with different air masses, the McGill workers used the electrostatic sampler and the slit sampler between 10.00 and 13.00 hours on 113 days between September 1950 and December 1951, at the top of a high building in Montreal (Kelly & Pady, 1954; Pady & Kapica, 1956). Ten types of air mass were recognized, classified on the basis of exposure to agricultural land; but as most examples of any one type occurred at one time of the year, the effects of differences in origin of the air mass may have been confounded with seasonal effects. Further, at the altitude of 130 metres, sampling near midday is likely to be done in the frictional turbulence layer and to be dominated by local ground sources which are themselves affected by the temperature and humidity of the prevailing air mass.

This survey is therefore perhaps best regarded as a valuable contribution to knowledge of the local air spora near the ground; judging from knowledge about the upper air, the amount contributed by the air mass is likely to have been small. *Cladosporium* and yeasts were the chief constituents of all the air masses (even of fresh polar air), and on our interpretation the abundance of *Penicillium* is not surprising for samples taken in a large city.

Alternaria and *Fusarium* were more abundant in tropical air. Smut spores occurred in all air masses and at all seasons; basidiospores of agarics were suspected but not positively identified. Fungi were most numerous in July and August, when 625 cultivable (8610 visible) spores were recorded per cubic metre, and least numerous from December to February, when 36 (28) were recorded per cubic metre. Bacteria were present in greatest numbers in polar air during spring and autumn, increasing from fewer than 70 per cubic metre during March to 710 per cubic metre in June, then decreasing to the end of August and increasing to a second peak during November. In air of maritime origin the trend was irregular.

AIR MASSES OVER THE NORTH ATLANTIC OCEAN

In two flights from Montreal to London, England, at altitudes ranging between 2700 and 3000 metres, the McGill workers were able to study the relation between microbial concentration and air mass (Pady & Kelly, 1954; Pady & Kapica, 1955). Over the ocean, polar air had generally fewer bacteria and fungi than tropical air (Table XXV).

TABLE XXV

ANALYSIS OF PADY & KELLY'S (1954) DATA ON TWO RETURN FLIGHTS OVER THE NORTH ATLANTIC, SHOWING CONCENTRATIONS PER CUBIC METRE OVER LAND, AND MEAN CONCENTRATION OF BACTERIA AND FUNGI IN AIR MASSES OVER OCEAN

			Bacteria		Fungi	S	
Air mass	Month	Position	E	S	E	Culture	Visual
Polar	June	Over Quebec Province	—	—	—	—	240
Polar	August	Over Quebec Province	14·0	7·0	53·0	92·0	—
Tropical	August	Over Quebec Province	28·0	11·0	57·0	160·0	940
Polar	August	Over Quebec Province	6·4	3·9	18·0	32·0	12 700
Polar	August	Over Labrador	3·5	7·0	21·0	70·0	—
Polar	June	Over Ocean	5·9	7·5	3·9	19·0	26·0
	August	Over Ocean	8·2	4·6	2·8	16·0	56·0
Tropical	June	Over Ocean	6·8	6·9	35·0	140·0	208·0
	August	Over Ocean	5·2	15·0	23·0	194·0	67·0
Tropical	June	Over England	1·3	2·0	170·0	317·0	—
Tropical	August	Over England	13·7	53·0	52·0	215·0	580·0

E = electrostatic sampler   S = slit sampler   — = not investigated

Over Quebec Province, in one air mass which was classified by meteorologists as of polar origin, and which gave few bacteria or fungi in culture, very many fungus spores were caught on a silicone-coated slide in the slit sampler. The authors interpreted this as evidence of a load of non-viable organisms which could only have originated in the tropics, and suggested that the air had been carried into the Arctic, thence eastwards, and finally southwards, during which passage most of the suspended microbes had lost their viability. However, another explanation seems possible on further examination of the data. The visible total on the silicone-coated slides, amounting to 18 700 per cubic metre, was made up largely of yeasts (9900 per cubic metre) and yellow-brown spores (7500 per cubic metre). As 'about 50% of the latter had an apiculus and were considered to be basidiospores', the other 50% were probably also basidiospores lying in the alternative position in which the apiculus would be invisible. The flight may well have been through one or more thermals arising from coniferous forests of Labrador and Quebec Province, by which a polar air mass was becoming charged with the air spora of the ground layer.

The bacteria obtained on these flights were classified as

	June per cent	August per cent
Micrococcus and Sarcina	41·4	13·2
Gram-negative rods	4·3	20·7
Gram-positive pleomorphic rods	20·4	37·0
Aerobic spore-formers	33·2	29·0

The fungi identified occurred in the following percentages (mean of two flights): *Cladosporium* 82·3, *Alternaria* 2·6, *Pullularia* 2·3, yeasts 2·1, *Penicillium* 1·6, *Botrytis* 1·5, *Stemphylium* 1·1, non-sporulating colonies 3·2%. Of these, *Alternaria*, yeasts, *Botrytis* and *Penicillium* were noted as more abundant in tropical air, whereas *Stemphylium*, *Pullularia*, *Fusarium* and also *Papularia* were more abundant in polar air. *Sporormia* was found several times, always in polar air. Many other fungi occurred in small numbers.

Among the many interesting results that stand out clearly from these flights is the discovery that viable bacteria and fungi occur at an altitude of 3000 metres in air masses all the way across the North Atlantic, though the bacteria were so few that some samples of about 2 cubic metres appeared to be entirely devoid of them. There was, however, no gradual diminution with distance from land. *Cladosporium* is the dominant fungus over the oceans, as it is over land, but it probably tends to lose viability as the air mass travels.

THE STRATOSPHERE

We would expect the stratosphere to be almost devoid of organic particles, because of the apparent inadequacy of mechanisms able to carry them regularly above the top of the troposphere. At great heights the intensity of radiation would be less favourable to survival than lower down. However, well-documented evidence is worth more than any theory and, for the present we must admit that we know little about the possibilities of life in the stratosphere (*see* Junge, 1964). Apparently the only attempts made to sample microbes in the stratosphere have been made with free balloons. Rogers & Meier (1936a, 1936b) devised a sampler to be opened and closed by an aneroid between 21 000 and 11 000 metres during the descent of the balloon *Explorer II*. They obtained five bacterial cultures, all of which were species of *Bacillus*, and five fungi (*Rhizopus* sp., *Aspergillus niger*, *A. fumigatus*, *Penicillium cyclopium* and *Macrosporium tenuis*); in all, these were equivalent to approximately 0·14 viable organisms per cubic metre (Meier, 1936b).

The United States National Aeronautics and Space Administration sponsored unmanned balloon flights to sample microbes in the stratosphere (Bruch, 1967). A 'direct flow sampler' used a fan to draw large volumes of air (of the order of 10^4 cubic metres) through polyurethane foam filters (p. 134). *Micrococci*, *Alternaria* and *Cladosporium* were cultivated after sampling at 20–30 km altitude, and Bruch considers that 1 organism per approximately 300 cubic metres is the best estimate of the upper limit of organisms that could be present, in view of the fact that background contamination was observed on filters of control samplers (flown but not exposed).

ISOKINETIC SAMPLING FROM AIRCRAFT

Workers at the United States Air Force School of Aerospace Medicine designed and tested an isokinetic sampler. A sampling pipe projects 1·5 metres forward of the plane, and through this air is drawn isokinetically by pumps, passing through soluble gelatine foam filters (Timmons *et al.*, 1966). The apparatus was then used in extensive studies of upper air 'micropopulation', a term used for the cultivable microbes (bacteria, Actinomycetes and fungi). Over the city of San Antonio, Texas, a marine-influenced air mass averaged one-third the micropopulation and had less variability in concentration than the air land mass. Persistent temperature inversions prevented upward movement of microbes from the surface air layers (Fulton & Mitchell, 1966).

Variation of 'micropopulation' with altitude was studied by Fulton (1966a) with three aircraft flying at 690, 1600 and 3127 metres above the mean ground level. Noteworthy in this series is the unexpectedly small population at the middle altitude, associated with a persistent low-level inversion. Even in the upper air a circadian periodicity was apparent, with maximum concentrations between 12.00 and 24.00 hours, corresponding to convection from the ground. Three-quarters of the fungi cultured were *Cladosporium*, *Aspergillus* and *Alternaria*, and three-quarters of the bacteria were *Bacillus* and *Micrococcus*.

The air at a cold front, without rain, was characterized by vigorous surface turbulence and by a microbial content many times greater than either the preceding warm air, or the cold air which followed the front (Fulton, 1966c). By contrast, air at a wet cold front had few microbes.

In all these tests in Texas, great variability occurred between samples from a single air mass, and a single sample could be unrepresentative.

UPPER-AIR MICROBIOLOGY

The ideal situation, of a logarithmic decrease with height, is seldom realized, because conditions change too rapidly for a stable state to be attained. This is illustrated in Figure 28 by the rust spore profiles from Craigie (1945), and by the vertical changes in 'micropopulation' associated with temperature inversions by Fulton (1966a). Wind velocity increases with height, and layers of air at different heights in a vertical column at any time will have been over different places at different times as a result of 'wind shear'. The thickness of the turbulent layer of air fluctuates. Biological factors in a circadian cycle put vastly differing numbers of organisms into the air at different times, and the vertical concentration is continually building up or decaying. Temperature inversions will affect vertical diffusion. According to Jacobs (1951): 'The presence of a stable layer at the surface will prevent or retard the introduction of surface organisms into the upper atmosphere but will, at the same time, maintain higher concentrations of organisms in the surface layers; the presence of a discontinuity surface in the upper air will limit vertical transport in either direction, resulting in the concentration of organisms above or below such a surface.' Concentrations will often start to decrease from the active surface of a crop and not from true ground level.

Knowledge of upper-air microbiology is based on occasional samples and is affected by place, season, weather, air mass, and so on. There are no continuous records; but there are some hints that a 'biological zone' occurs at middle height, which can probably be explained in terms of temperature inversions, air masses, and precipitation. The evidence for greater microbial concentrations above a persistent inversion was shown clearly by Fulton (1966a, c) sampling over Texas.

Molisch (1920) introduced the concept of *aeroplankton* to denote the microbial complex referred to in this book as the air spora. It has been argued that the word 'plankton' suggests organisms based on the air during at least a vigorous phase of growth, whereas the air spora is only temporarily airborne, even though adapted to wind transport as a means of dissemination. Clearly this argument is valid for pollens and plant spores; but is there, in addition, a vegetative air-inhabiting plankton? We cannot yet give this apparently improbable hypothesis a decisively negative answer. Evidence in favour of it has been stated by R. C. McLean (1935, 1943), who wrote: ' "Dust to dust" seems to be the only cycle envisaged. Yet experiments of Trillat and others show at least the possibility that the air may be a vegetative habitat and the large proportion of non-spore-formers present . . . needs more than a conventional explanation.' Proctor & Parker (1942) noted that one-third of the bacteria collected from the upper air could grow at 0° C, and survive 48 hours exposure at −26° C.

If there is a truly indigenous aeroplankton, its habitat must be exacting in the extreme, and tolerable only to specialized bacteria, yeasts, or Actinomycetes. Frequent drying must reduce such a population to inactivity, though metabolism could be resumed in a cloud of water droplets when gaseous nitrogen and carbon compounds could be absorbed and used. In constant danger of being removed from the air by rain or snow, or by contact with the ground, the risk of removal would be increased by any attempt to parasitize organic particles brought up by convection from below. However, radioactive dust can persist for several weeks in the troposphere, and this is a long period on a microbial time scale. The aerial environment is not obviously beyond the range of exploitation by microbes, for the rate of loss by death or deposition might be no greater than for bacteria in the sea, and there would be freedom from predators (*see* Bruch, 1967). If anywhere, such an aeroplankton might be expected to ride clouds on the ascending side of a tropical convection 'cell' over the equator.

Although the origin of the upper-air spora from the soil has been assumed by most investigators, the circumstantial evidence suggests a wider range of sources. The bacteria are probably mainly soil forms with a small proportion from sea water. But hyphal fragments, especially conidiophores of *Alternaria* and *Cladosporium*, which are commonly reported from the upper air (Pady & Kapica, 1953, p. 321), evidently come from the ground vegetation layer rather than from the soil. The numerous yeasts, coloured basidiospores of toadstools, and smut spores, evidently originate above the soil surface. It is hard to believe that wind could burrow into soil, picking out the few spores of *Cladosporium* and *Alternaria*, yet leaving behind the far more numerous spores of *Penicillium*, *Trichoderma*, *Aspergillus* and Mucoraceae – not to mention clay particles!

The number of spores in suspension overhead is enormous. For a conservative but realistic estimate we may turn to the records of 24 days in summer 1952 at Rothamsted, during which air was sampled continuously at two heights (Gregory & Hirst, 1957). The mean concentration of a dozen selected spore types, ranging in size from grass pollens down to hyaline basidiospores (largely *Sporobolomyces*) at 24 metres height was 81·5% of that at 2 metres. Assuming a linear relation between log concentration and log height (Johnson & Penman, 1951), and assuming the mean spore concentration of 12 500 spores per cubic metre at 2 metres height as found for the 5 summer months at Rothamsted, a graphical summation indicates approximately 1000 spores in suspension over 1 square centimetre of ground surface up to the height of 1 km (taken as the top of the outer frictional turbulence layer). The number would reach 6000 if convection is assumed to distribute the spore cloud to the top of the troposphere at 10 km.

XIII

Deposition in Rain, Snow, and Hail

Airborne microbes may be deposited direct or they may be washed out of the air in raindrops, snowflakes or hailstones. Trillat & Fouassier (1914), from their laboratory experiments with artificial fogs condensing on a suspension of pathogenic bacteria in small vessels, thought that airborne microbes act as condensation nuclei. Most condensation nuclei are now thought to be sub-micron size hygroscopic particles, and it seems probable that already formed droplets collect spores by impaction (as described in Chapter VII for impaction on small spheres). McCully *et al.* (1956) estimate that, over all land areas of the globe, from 35 to 50% of the total atmospheric dust load is washed out each day. Here we will consider the results of rain-wash in nature, and the spore content of precipitation water.

Over the last three hundred years a score or more of people are known to have sought microbes in precipitation water – mostly in ignorance of each other's work. Collecting the sample has some pitfalls, however; the collecting vessel must obviously be clean, but the danger of contamination by rain-splashed soil has not always been anticipated, although with current knowledge of the magnitude of splash and its part in soil erosion, the danger is now clear (Laws, 1940). Much of the early work summarized below, however, is clearly trustworthy.

Animalcules in rain water delighted Leeuwenhoek (1676, *in* Dobell, 1932). Rain was collected in a clean porcelain dish set on a wooden tub to avoid earth being splashed by rain. Minute organisms were searched for in vain until after the rain water had stood for some days, by which time it could also have been contaminated by dry deposition, so we do not know whether or not Leeuwenhoek himself found microbes in precipitation water.

RAIN

The only systematic study of precipitation microbiology comes from Miquel 1884, p. 597; 1886, p. 530) at the Parc Montsouris, Paris. Miquel caught his rain in a metal funnel fixed at 1·7 metres above ground level on a pillar, well away from trees and buildings. Rain falling into the funnel was collected in a platinum crucible with a cover, both funnel and crucible having been heated to redness just before sampling started. The sample was then sown, drop by drop, in 50 to 100 flasks of beef broth. Miquel also designed apparatus which placed raindrops on a moving band of nutritive paper. After 6 days incubation the paper was dried and kept as a record of bacterial and mould colonies. The largest catches of bacteria occurred in the warmer months, when numbers varied from 0·0008 to 8·3 per cm^3 with a general mean of 4·3 per cm^3, but these figures excluded the first rain after several days when 200 bacteria per cm^3 might be recorded.

During prolonged rainfall the numbers fluctuated instead of continuing to diminish, suggesting to Miquel that the rainclouds themselves had a characteristic bacterial content, with the percentage composition of: *Micrococcus* 60, *Bacillus* 25, and *Bacterium* 15. Moulds fluctuated in the same manner as bacteria and averaged 4 per cm^3. Miquel estimated the annual precipitation of bacteria and moulds at Montsouris at over 4 million per square metre – a figure that was obviously too low as he still excluded the contribution of the first rain after dry days.

The pharmaceutical uses of rain water induced Lindner (1899), in Germany, to collect 28 samples of rain in a clean porcelain dish on a bleaching ground near his house. Samples were then added to sterile hay infusions, albumen, milk or blood serum. His liquid cultures gave a regular succession of bacteria, flagellates, and monads in the first day or two and, later on, stalked *Vorticella*-like ciliates, *Paramecium*, *Stylonychia*, and *Volvox*. Once he got two amoeboid forms, but never gregarines or coccidians. Lack of precautions against splash-contamination appears to leave the interpretation of his data in doubt.

In this century various workers have cultured microbes from rain. Minervini (1900) collected numerous rain samples on board ship in the North Atlantic Ocean. Bacteria were abundant, half the samples yielded pink yeasts, and a quarter of them *Penicillium*. He also obtained *Aspergillus glaucus*, *A. niger*, *Monilia candida*, and many other moulds. Busse (1926) recorded pine pollen in rain.

Rain water collected over the ocean at considerable distances off shore by ZoBell (1946, p. 179) averaged 1 to 10 bacteria per cm^3, with few or no mould fungi. Rain water collected on land at the Scripps Institution of

Oceanography, California, contained from 10 to 150 microbes per cm^3. As usual, the highest counts were obtained during the first rain and were associated with a predominance of mould spores.

Protozoa in rain were studied at Heidelberg by Puschkarew (1913), who collected ten samples of rain water in a sterile funnel, and then added nutrient solutions. At the start of rain he found many fungi and bacteria, and the numerous protozoa included a new species, *Amoeba polyphagus*, with species of *Bodo, Monas, Calpoda*, and other genera.

Twice in the month of November, rain was collected in sterile flasks on a roof at Leiden by Overeem (1937) and inoculated into flasks of nutrient medium favourable to growth of green plants in light. In a total of 221 cm^3 of rain water she obtained the following cultures. Algae: *Stichococcus minor* (8), *S. bacillaris* (5), *Chlorococcum* sp. (7), *Pleurococcus vulgaris* (4), *Chlorella vulgaris* (2), *Hormidium flaccidum* (2), and *Navicula minuscula* (1). Moss: *Brachythecium rutabulum* (1).

Another worker who made a rewarding study of autotrophic plants in rain water was Pettersson (1940), working at the Zoological Station at Tvärminne, Finland, in the summer of 1936. Glass funnels (176 cm^2 in area) were lined with filter paper, sterilized and taken, covered, to the trapping site. After exposure, the filter paper was sprayed with a nutrient solution and the funnel was covered with a glass lid and left to stand in a light place for a few days. Developing organisms were picked off and transferred to new culture vessels to continue their growth.

The originality of the method lies in the medium being unfavourable for the development of bacteria and fungi because the cellulose of the filter paper was the only carbon source provided. Pettersson, like Overeem, was therefore able to explore a novel part of the air spora. Snow traps were also used, consisting of shallow glass dishes 15–20 cm in diameter, with a thick bed of blotting paper and an upper layer of filter paper. These two methods gave an exceptionally rich harvest. A sample of snow taken at Pikis (Piikio) from the start of snowfall on 1 March 1936, gave thirty-six lichen thallus fragments and a moss gemma. On the next day, 2 hours after the start of another snowfall, a sample corresponding to 625 cm^3 of water yielded nineteen lichens and two *Chlorococcum* colonies. A third sample of 805 cm^3 of water, taken 4$\frac{1}{2}$ hours later, yielded six lichens, and three mosses which were identified after 6 months' growth as *Bracythecium velutinum, Hypnum cupressiforme*, and *Pylaisia polyantha* (*see also* Pettersson, 1936).

Pettersson's rain trap yielded a wealth of information from the fourteen samples investigated, for details of which the original paper must be consulted. The interest was taxonomic and qualitative rather than quantitative. For the early samples the funnel was placed on a low rock, 2·5 metres above sea level, on open grassy soil. Some of the organisms caught may possibly have come by splash from the ground, but not many can have done so because the largest catch of mosses belonged to a genus hitherto unrecorded in Finland (*see* Chapter XVII). The precaution of raising the funnel on a wooden base 1 metre high was adopted in later tests.

In a total of 1373 cm³ of rain collected, Pettersson obtained 1200 conifer pollen grains, 300 liverwort spores (all of *Marchantia polymorpha* except for one of *Metzgeria*), Myxomycete spores (*Stemonitis fusca* three times, and *Arcyria denudata*), and numerous algae. Blue-green algae were scarce, being represented only by *Nostoc commune* in separate samples. Green algae were abundant in almost every sample, those identified including: *Chlamydomonas nivalis, Chlorella vulgaris, Chlorococcum humicolum, Cystococcus pseudostichococcus, Prasiola stipitata, Roya* sp., and *Tatraedron punctulatum*. Some of these, Pettersson suggests, may have originated from lichen soredia. Lichen spores and soredia were not the main source of lichens in the traps, for the lichens mostly originated as thallus fragments and were evidently of fairly local origin.

The 2000 moss plants cultured from the spores caught in Pettersson's funnels included specimens of: *Aloina brevirostris, A. rigida, Amblystegium serpens, Bracythecium velutinum, Bryum* spp., *B. argenteum, B. pallens, Ceratodon purpureus, Funaria hygrometrica, Leptobryum pyriforme, Mniobryum carneum, Pohlia cruda, P. nutans,* and *Pylaisia polyantha*.

Observations on micro-organisms in rain were made at Rothamsted Experimental Station in 1951 by Gregory, Hirst & Last (*see* Hirst, 1959) while they were comparing various spore-trapping techniques. Two conical glass funnels, 20 cm in diameter, were exposed on a wooden structure at a height of 2 metres above ground level. One funnel was open to rain (rain trap), while the other (dry trap) was protected by a flat asbestos–cement disc held 25 cm above the mouth of the funnel – to keep off rain but still to allow dry deposition. Washings from both funnels were collected daily and the fungus spores separated by sedimentation on to a glass coverslip. On dry days the fully exposed rain trap consistently caught fewer microbes than the dry trap; but this was reversed after rain, especially the first rain after dry weather (Table XXVI).

Rain falling during one thunderstorm was studied in detail (Gregory, 1952; Hirst, 1959), and a detailed account of changes in the air spora during this period, observed with the aid of a Hirst automatic volumetric spore trap, has already been published (Hirst, 1953, pp. 382–385). A 7-day spell of warm, dry weather ended in a thunderstorm at 13.25 hours on 22 July 1951. The rain trap was cleaned immediately before the rain started, and the first 1 mm of rain which fell in the first half-hour of the storm was

TABLE XXVI

GEOMETRIC MEANS OF RATIOS OF CATCHES BY RAIN TRAP TO DRY TRAP,
2 METRES ABOVE GROUND, ROTHAMSTED, JUNE–SEPTEMBER 1951
(Hirst, 1959)

	Ratio for dry days	Ratio for all rainy days	Ratio of single rainy days to the first of a succession
Smuts (mainly *Ustilago*)	0·6	3·8	6·0
Cladosporium	0·8	1·3	1·8
Alternaria	0·8	3·7	6·9
Pollens < 20 μm	0·8	1·4	2·3
Pollens > 20 μm	0·8	1·5	2·4

collected separately from the succeeding 3·75 mm, which contained many
fewer spores (Table XXVII).

As Hirst (1959) remarks in discussing this series of observations: 'Spores
released during rain are presumably removed from the air as readily as
spores already there when rain starts to fall, so that concentrations of airborne spores measured during rain represent, not the total released, but the

TABLE XXVII

SPORES BROUGHT DOWN BY THUNDER RAIN TERMINATING 7-DAY DRY SPELL,
ROTHAMSTED, 22 JULY 1951
(Gregory, Hirst & Last, *unpublished*)

	Number of spores per cm³ of rain	
	in 1st 0·95 mm of rain falling 13·35–13·55 hours	in succeeding 3·75 mm of rain falling 13·55–08·25 hours (23 July)
Smuts (mainly *Ustilago*)	455	55
Cladosporium	1770	205
Alternaria	370	20
Erysiphe	280	10
Small pollen grains	270	10
Large pollen grains (over 20 μm diameter)	120	5

excess of those released over those removed. Rain-scrubbing seems an ideal
method of deposition for air-dispersed soil fungi. For foliage pathogens
its biological significance is far from clear. Many spores may be lost in
"run-off" unless they can attach themselves to the leaf surface or penetrate
into crevices they would be unlikely to reach when deposited from dry air.'

Maguire (1963) cultivated numerous algae and protozoa from small samples of rain water collected in Texas in the month of July, but in four samples from Colorado, where the atmosphere was washed by daily rains, only 3 moss protonemata and one unicellular alga were obtained.

During a severe outbreak of rice blast, Suzuki (1965b) found 8 times as many spores of the pathogen, *Piricularia oryzae*, in 1·5 mm of rain as had been deposited in the same time on a similar area exposed to wind but screened from rain.

SNOW

Janowsky (1888), Pettersson (1940; *see* p. 268) and others have found a few organisms in falling snow. Only Gazert (1912) gave a negative report from the Antarctic on the microbial content of freshly fallen snow in Kaiser Wilhelm II Land. A. L. McLean (1918), on the contrary reported numerous organisms in snow and ice in Adelie Land; but it is not certain whether they were brought down with the snow, or deposited dry in fine weather from the atmospheric dust which settles over the Antarctic. However, on three occasions McLean caught falling snow in a sterile basin: '. . . elaborate precautions having been taken to prevent contamination, the thawed-out samples showed under a coverslip cocci, motile bacilli, and, invariably, zoogloea masses of bacteria in moderate numbers. Diplococci, and occasionally cocci, were observed to be invested by a pale capsule . . . A glucose agar slope culture of falling snow showed a few small greyish colonies.'

Atkinson isolated a motile bacterium believed to have been carried to the Antarctic by upper-air currents and brought down by the snow (Scott, 1913). Most of the arctic and antarctic snow samples were taken from fallen snow, and organisms could therefore possibly have reached the snow by dry deposition (e.g. Salimovskaja-Rodina, 1936; Darling & Siple, 1941) as considered in Chapter 8.

HAIL

Large numbers of microbes were recorded by Bujwid (1888), who collected hailstones in the month of May, washed them in sterile water, and, on plating out the melt water, found 21 000 bacteria per cm^3. They included *Bacillus fluorescens liquefaciens*, *B. f. putridus*, and *B. janthinus*. From these numbers Bujwid concluded that surface waters must have been carried aloft and frozen.

During a hailstorm in St Petersburg, windows were broken by hailstones the size of walnuts. Foutin (1889) washed some of these and, on plating out the melt water, obtained 628–729 bacteria per cm^3, but no fungi or yeasts. Abbott (1890), recorded 40–300 microbes per cm^3 of hail melt water.

In July storms at Guelph, Ontario, Harrison (1898) collected hailstones, washed them in 1 in 500 mercuric chloride solution and, after rinsing, plated out the melt water. One storm gave 955 colonies per stone, of mixed bacteria and moulds, including *'Penicillium glaucum'*, *Mucor* sp., *Aspergillus* sp., *Bacillus fluorescens liquefaciens*, *B. f. non-liquefaciens*, and *Proteus vulgaris*. A later storm averaged 1125 colonies per cm³ but these included fewer moulds than the first. Harrison concluded that the bacteria must have come from surface water, but that the moulds were picked up from the air. Belli (1901) obtained 140 organisms per cm³ of hail melt water, of which eight were *Aspergillus* or *Penicillium* and the remainder bacteria. Hail was also sampled by Dubois (1918).

SIGNIFICANCE OF PRECIPITATED MICROBES

Spores in raindrops appear to play a part in some processes of plant infection. Dry wind-blown spores of barley loose-smut *(Ustilago nuda)* rarely affect the ears of susceptible varieties but, when drops containing spores in suspension fall on the flowers, the spores are brought into direct contact with the ovary, and infection follows (Malik & Batts, 1960).

Asai (1960) introduced the useful method of filtering rain through membrane filters under reduced pressure to extract the microbial load. He failed to obtain uredospores of *Puccinia graminis* by this method, although they were known to be in suspension in the air at the time of the tests. However, Roelfs *et al.* (1970) found them regularly by using a similar method for routine monitoring of rain for detection of cereal rust spores in Minnesota. A 15·25 cm diameter funnel headed a cylindrical reservoir, fitted at the bottom with a cellulose acetate filter disc (8 μm pore size) through which rain filtered under gravity, leaving its spore load on the filter for microscopic examination.

With this device, Rowell & Romig (1966) were able to detect immigrant spores in the overhead air mass by collecting spring rain, much earlier than the inefficient wind-vane slide traps then in use for sampling dry air near ground level. This was no doubt because, in falling through the lower part of the troposphere, the raindrops efficiently sampled a very large volume of air. (Wind-vane traps tended to detect only later, when the first local infections were sporulating.) Moreover, it appears that the uredospores, scavenged by rain from the higher altitudes, were the spores actually initiating the spring outbreak, and the rain provided the water film over the wheat leaf in which the spores could germinate. The rain trap therefore gives relevant information on an important discovery about how rust epidemics are initiated on spring wheat.

Among other possible effects of microbial material brought down by rain must be mentioned the discovery by Parker (1968) that rain brings down to the soil appreciable amounts of vitamin B12. While on the debit side, L. P. Smith & Hugh-Jones (1969) discuss the possibility that pick-up of airborne foot-and-mouth disease virus by raindrops may be significant in contaminating fodder plants and establishing foci of disease many kilometres from the source.

The organisms in precipitation water remain almost unstudied. The little we know from existing records is tantalizing. Precipitation water is non-sterile, whether on land, over the oceans, or about the poles. A wide variety of organisms has been recovered from such waters, including bacteria, fungi (moulds, yeasts and plant pathogens), algae, liverworts, mosses, pollens and protozoa. Microbes are found in rain, hail, and snow, when collected as they fall, before the possibility of ground contamination.

The largest counts are recorded from hail and, at present, these are perhaps the most reliable records because hailstones can be surface sterilized. The first rain after a dry spell is heavily contaminated and, even during prolonged wet weather, the spore numbers in rain remain substantial.

Conceivably, spores may undergo reconcentration within a cloud. Rising convection bubbles may bring new spores to the top of the cloud, where they can be collected and washed down by raindrops to the base of the cloud. Here the drops might evaporate, allowing the spores to be carried up again, perhaps eventually to be brought down to earth in hailstones. The abundance of microbes in hail, and the reports of a 'biological zone' at several thousand metres altitude, supports the suggestion that convective clouds may be spore concentrators.

Exploration of organisms in precipitation water needs an experimental study of methods of sampling from ground, ships and aircraft. Systematic sampling could then be attempted with some prospect of learning what part such precipitation plays in terrestrial microbial circulation. Microbial sampling of precipitation is just passing out of the naïve stage. Methods have not been tested, and we still do not know how a collecting vessel should be placed to avoid contamination from soil and vegetation.

A spore liberated near ground level has a high probability of being deposited dry; but wash-out by rain, hail or snow probably most often terminates the journey reaching the tail end of the dispersal gradient (*see* Chapter XVI). Dry deposition (dependent on p or v_g) depletes only the layer of air in contact with the ground and its vegetation. But rain depletes the whole thickness of the layer through which it falls.

XIV

The Air Spora of Enclosed Spaces

A small but important fraction of the atmosphere is walled-in and provides microbes with an environment different from the outdoor world. Indoor air hygiene is an aspect of medical science with a voluminous literature which can be approached through such works as: *Aerobiology* (Moulton, 1942), *Studies in Air Hygiene* (Bourdillon *et al.*, 1948a), *Airborne Contagion and Air Hygiene* (Wells, 1955), *Mould Fungi and Bronchial Asthma* (Werff, 1958), and *Aerobiology* (Silver, 1970). The brief treatment given here of 'intramural aerobiology' presents an ecological instead of the normal medical viewpoint.

Outdoor air moves as wind flowing over surfaces, and a point near the ground is immersed in a continually flowing stream of fresh air. Rooms, by contrast, are *ventilated*, a process by which fresh air is assumed to mix thoroughly with the existing air instead of displacing it bodily. By one 'air change' (ventilation turnover) is meant the introduction of a volume of fresh air equal to the volume of the room; and an equal volume of mixed stale and fresh air is displaced during the process, leaving a mixture of stale and fresh air in the room. Unless continually renewed, any microbial concentration in the air of an enclosed space will tend to diminish with time as a result of ventilation and deposition. Concentration of viable organisms will also decrease with time, following the natural death-rate or because of any disinfection that may have been applied.

Wind is a powerful agent in ventilation. Walls and roofs facing the wind are under positive pressure which forces air into openings. Lateral and lee walls sustain suction. Air buoyancy is another agent, assuming importance in calm weather and with air temperature indoors differing from that of the outside air (Daws, 1967, p. 35). Direction of flow across a room, or other enclosed space, may fluctuate with wind direction minute by minute. Artificial ventilation with centrifugal fans, by contrast, stabilizes the direction of flow. This is important if internally-generated microbial contamination is to be controlled: this is best done by using two fans, one on the inlet and one on the exhaust opening.

DIE-AWAY OF CONCENTRATION

Die-away of concentration is a phenomenon seen most clearly indoors, because outdoors a concentration is carried away bodily by the wind.

Indoors, decrease of concentration with time is the result of a combination of factors: (1) exchange with outside air (i.e. ventilation); (2) deposition on walls, ceiling, floor and other surfaces by various processes including sedimentation; and (3) reduction in the viable count through death. Ventilation does not immediately sweep away the whole microbial load, but progressively dilutes it exponentially. Thus, n air changes will reduce concentration in the ratio: $1/e^n$. Decreases in concentration due to deposition, death or disinfection, may also follow a logarithmic law, and these factors can then be expressed in units of equivalent ventilation turnovers for ease of comparison.

Ways of expressing rates of removal or death of bacteria are discussed by Bourdillon *et al.* (1948a), and are based on the constant K, in the equation: $N = N_0e^{-KT}$, where N_0 is the number present at time $T = 0$, and $e =$ the base of Naperian logarithms. K is the die-away, the rate of removal of bacteria by all processes during the period, and may be subdivided. Thus K_D is the death-rate; K_R is the rate of removal by ventilation only and is identical with the ventilation rate in air changes per hour; and K_S is the rate of removal by sedimentation. In an example of die-away rates of bacteria from all causes in a bedroom with open windows during fine weather at midsummer, K was equivalent to 6·1 air changes per hour after the occupants settled down to sleep at 23.00 hours; $K = 4·9$ after they went down to breakfast at 07.55 hours; and $K = 6·8$ after the making of beds at 08.50 hours (Lidwell, 1948, p. 253). Another example is shown in Figure 29. The case of die-away with stirred settlement has been discussed by C. N. Davies (1947), and Dimmick (1969) [In Dimmick & Akers, 1969, pp. 127–163], and *see also* Chapter II.

MICROBIAL MOVEMENT IN CONVECTION CURRENTS

The characteristics of air movement indoors differ considerably from outdoor wind (where pressure is exerted over a wide area). Indoors, pressure from wind or buoyancy from heaters, is applied at a few points or lines, and not over an area, except in the special case of laminar flow rooms (p. 213). The 'feel' of the subject of air movement in buildings comes

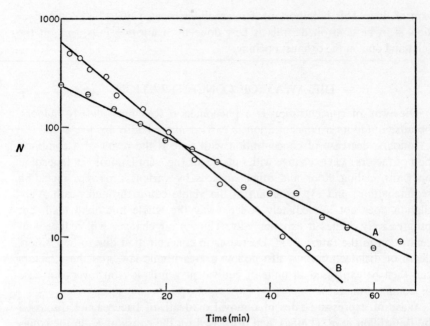

Figure 29. Exponential form of the die-away of bacteria-carrying particles from the air of a room. Line A: In an observation military canteen after the occupants had left suddenly. Line B: Observations on die-away following a group of sneezes in a small room. (From Lidwell (1948); reproduced from *M.R.C. Special Report* No. 262, *Studies in Air Hygiene*, by permission of the Controller of H.M. Stationery Office.)

across from the paper by Daws (1967), with its photographs showing air movement visualized in an experimental transparent room by light shining on to paraformaldehyde particles.

By definition of an enclosed space, the ventilation ports, whether doors, windows, chimneys or cracks, are small in relation to the area of the walls. Even indoors air movement tends to be turbulent. Ventilation produces currents across a room. Heat sources (including occupants) produce vertical convection currents within the room.

Speed of air flow in the body of the room is given approximately by dividing the volume of air supplied in unit time by the cross-section area of the room. Only very close to the outlet the air converges from all directions and accelerates towards the vent. For this reason, to extract contaminants the outlet must be as near the source as possible. Provision of a hood helps to prevent escape of contaminants from the exit stream into the turbulent circulation within the room. (When dangerous microbes are being extracted the exhaust may pass to an incinerator.) When air temperature is the same outside and inside the room, air enters 'isothermally', entraining room air as it enters, slows down and becomes part of the general flow across the room. If an inlet stream persists until it reaches a room surface (wall, ceiling, etc) it clings to that surface for a time by reason of the Coanda effect (Daws, 1967, Plate 4).

Air entering a room because of external pressure tends to slow down as it mixes with room air. By contrast a convection current arising within the room may accelerate as it rises, entraining surrounding air and becoming turbulent till it reaches the ceiling and spreads out, again showing the Coanda effect. If cooled at the ceiling or walls, air may flow down the walls to floor level, whereas air retaining its heat descends a short distance only and then tends to accumulate in the upper parts of the room, leading to pronounced thermal stratification (Daws, 1967, Plates 8 and 9).

Transparent room measurements showed convected air rising at 30 cm/s above the head of an occupant. The action of walking across the floor propelled dust particles into the air at up to 90 cm/s, but when the disturbance ceased, normal speeds of air movement were re-established in 15 seconds (Daws, 1967, Plates 12 and 13). The convection current over an occupant was also demonstrated by Lewis *et al.* (1969) using Schlieren cinematography, and with ambient temperature at 15° C they measured ascending convection currents of 30 to 50 cm/s in the region of the head.

Evidently air movement within an occupied room is normally stirred to a degree adequate to maintain microbial circulation, and renders the die-away equation given above applicable.

Convection currents alone, in an enclosed space without access of outside air, are often sufficiently active to diffuse fungus spores evenly through the whole volume of air. With fruit-bodies of Basidiomycetes enclosed in chambers, Falck (1904) found that vertical tiers of horizontal paper shelves became covered with spore deposit in a remarkably uniform manner. By contrast, suspending the pileus of an agaric in a small glass vessel often resulted in 'curious and fantastically' irregular spore deposits on a piece of paper placed underneath. These were interpreted by Buller (1909) as due to convection currents in the vessel being of a velocity comparable with the terminal velocity of the spores. The heat from a lamp was sufficient to alter a previously established convection system.

Even without any ventilation, air *circulates* in a room because of thermal convection. Heating of air by rock surfaces in a mine may result in a flow along an adit and up a shaft. Heating of glasshouses in sunlight leads to strong convection currents.

Within a building the temperature of the air may be less changeable than outside, and this may lead to characteristic air-movement patterns. Warmer walls will generate an up-draught, colder walls a down-draught – each being balanced by opposite currents in the centre of the room (Figure

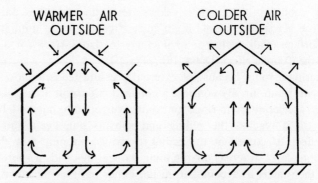

Figure 30. Diagram showing changes in circulation in a room
according to relative temperature of walls and of inside air.

30) and often moving fast enough to counteract particle sedimentation under gravity. Circulation of air between rooms of a house is complex, but there is evidence of a fairly rapid exchange of air and of its suspended spore load throughout a house. C. M. Christensen (1950) experimented with spores of *Hormodendrum resinae*, a mould that is peculiar for its ability to grow on a coal-tar creosote medium, and therefore suitable for use as a marker spore in dispersal experiments. Spores were liberated in a room on the lowest floor of a house while all doors to the central hall were left open. Within a few minutes spores were found deposited on petri dishes in rooms communicating with the hallway, but situated one, two and three storeys higher.

INTRAMURAL SOURCES

Microbes in the outdoor air may come from the outdoor air spora by ventilation, or they may originate within the enclosure, in which case they are usually limited in variety but may occur in high concentration.

Defective timber attacked by fungi may be an important source of spores in dwelling houses; so also may moulds growing on damp walls. A. W. Frankland & Hay (1951) showed that some asthmatics are sensitive to the spores of the dry rot fungus *(Merulius (Serpula) lacrymans)*, and concentrations ranging from 1630 to 360 000 spores per cubic metre have been recorded in buildings with active fructifications of this fungus (Gregory *et al.*, 1953). Timber in mines is particularly liable to fungal decay and may also bear superficial mould growths. Extensive growth of *Sporotrichum beurmanni* (the pathogen of human sporotrichosis) was found on fresh timber of mines in Transvaal by Brown *et al.* (1947). The fungus was isolated from the air, and ventilating currents of 1 m/s could detach spores from wood provided its moisture content was less than 80%.

In coughing and sneezing, large numbers of droplets of mucus and saliva are propelled with explosive violence into the air. Jennison (1942) gave photographic evidence of 20 000 droplets being put into the air from a single sneeze. The largest number observed was 40 000, and a weak, stifled sneeze gave only 4600 droplets. A cough produced a few hundred droplets, and the enunciation of consonants was also productive. Sneeze droplets, ranging in diameter from a lower limit of 5–10 μm, and with 20–40% smaller than 50 μm, could evaporate instantaneously to 'droplet nuclei'. (*See* Chapter XV, p. 22.)

The concept of 'droplet nuclei', developed by Wells (1955), has proved fruitful. 'Droplet nuclei' are the particles formed from the smallest droplets which evaporate before falling to the ground and so remain suspended in air. They consist of the solid residue of the evaporated droplet, together with any bacteria or virus particles, and may be coated with semi-dried up mucus which tends to preserve activity and viability. Few droplets are actually propelled more than a metre; but when evaporated, the resulting droplet nuclei, with any bacterial cells or virus particles, would remain in suspension for a long time. The droplet nuclei have no independent trajectory but move with the slightest air currents, and are emitted in large numbers. Thus although most airborne bacteria seem to be carried on rafts of dust particles which settle rapidly, they appear to be relatively innocuous saprophytes; the pathogens are present only in special environments, being carried in the much smaller and more insidious droplet nuclei which are small enough to be capable of entering, and being retained by the alveoli of the lungs.

THE AIR OF DIFFERENT ENVIRONMENTS

DWELLING HOUSES

In spite of ventilation, *Penicillium* dominates the air inside most houses, in contrast to *Cladosporium* outside, and bacteria tend to be more abundant indoors in winter than in summer.

Microbial concentration indoors varies greatly with the amount of mechanical and human activity. Carnelley *et al.* (1887), using Hesse's tubes in schools and mills in Dundee, Scotland, observed that, in densely-populated rooms, stirring up dust increased the total air load and increased the ratio of bacteria to moulds. When air in rooms is left undisturbed the bacteria (or particles to which they are attached, such as human epidermal scales) settle out rapidly, but the moulds do so much more slowly. Clearly sedimentation soon purifies the air of undisturbed rooms. Adams & Hyde (1965) offer the practical suggestion to hay-fever subjects that 98% of outdoor pollen and fungus spores can be excluded by keeping doors and windows shut. Dingle (1957) found that ragweed pollen penetrated normal 'crackage' areas of a test house at wind speeds greater than 1 mile (1·6 km) per hour.

Maunsell (1954a, 1954b) used the slit sampler in bedrooms and found that shaking beds, brushing carpets, and any building repair work, increased the mould-spore content of the air up to 17 times, but that it rapidly returned to normal when activity ceased. Other studies of the air spora of houses are summarized by Werff (1958).

Tests with a portable spore trap show that the airborne dust in inhabited rooms is commonly dominated by what appear to be fragments of human skin in the form of minute flattened scales from the stratum corneum of the epidermis. Concentrations of several thousand of these potential bacterial 'rafts' per cubic metre are common indoors, and 390 000 per cubic metre have been noted after bed-making (Gregory, *unpublished*). These epidermal scales which have an average equivalent diameter of 8 μm (terminal Velocity $\backsimeq$ 2 mm/s), probably carry a large proportion of the airborne bacteria of indoor air (Davies & Noble, 1962). (Plate 5B).

HOSPITALS

In studies of hospital air over a period of 15 months, Miquel (1883) found a mean value of 11 100 bacteria per cubic metre in the crowded wards of the Hôpital La Pitie, Paris, the counts varying from 5100 in June to 23 100 in December. The general improvement in hospital hygiene since that time is illustrated for example by Colebrook & Cawston (1948) for a Birmingham hospital, where they found from 210 bacteria and moulds under quiet conditions, to 2800 per cubic metre with bed-making in progress. (One very high count of 22 000 per cubic metre, including many moulds, was obtained with the ward windows closed.)

Recommendations for the maximum tolerable number of particles carrying bacteria in operating theatres are 700 per cubic metre for minor operations, and down to 70 or even 15 per cubic metre for dressing burns and for operations on the central nervous system (Bourdillon & Colebrook, 1946; Bourdillon *et al.*, 1948b). Spread of strains of bacteria resistant to antibiotics has again raised interest in the possible role of air in hospital cross infection. (*See* Fincher (1969) in Dimmick and Akers (1969).)

UNDERGROUND ENCLOSURES

London's underground railways were investigated by Andrewes (1902) and by Forbes (1924), and the New York Subway by Soper (1908). Studies in caves are few, but include those of Lurie & Way (1957) and by Mason-Williams & Benson-Evans (1958).

Miquel (1880), in Paris, gave special attention to the sewer in the Rue de Rivoli near its junction with the large collector of the Boulevard Sebastopol. He found a steady load of from 800 to 900 bacteria per cubic metre. Pollens were absent, and cryptogamic spores were only one-third to one-quarter as numerous as in outdoor air at the same time. The contamination of the air in the nearby Rue de Rivoli was lower in winter but higher in summer than it was in the sewer. Comparable results were reported from London in sewers under the Palace of Westminster (Carnelley & Haldane, 1887).

FARM BUILDINGS AND GLASSHOUSES

High microbial concentrations often occur in farm buildings, such as cowsheds, where hay is being fed to animals, or in barns where threshing or cleaning is in progress (Baruah, 1961). Inhalation of the spore cloud may produce Farmer's lung (p. 228) or various diseases of farm animals. Moist storage of grain in open-topped silos presents another hazard (Lacey, 1972).

The air spora of glasshouses and mushroom sheds has been little studied, despite the fact that workers and crops may be exposed to dense concentrations of microbes. Glasshouses may act as important spore emitters by means of convection through open ventilators (Hirst, 1959).

SHIPS, AIRCRAFT AND SPACE VEHICLES

Air in holds and living quarters on board ship was studied early by Miquel (1886). A study of the microbial content of air in ships, including submarines which are relatively clean, is reported by Ellis & Raymond (1948). The incidence of upper respiratory infections in relation to air contamination of Polaris submarines is reported by Watkins et al. (1970 (in Silver)).

Similar problems of confined and more carefully controlled environments arise in space vehicles where numerous studies are reviewed by Gordon (1970 (in Silver)), who gives an extensive reference list to official reports. In fact very few changes from normal pattern were observed, although by analogy with animal experiments some simplification of the microbial flora might be expected when a group of people pass a long time in a crowded environment.

LAMINAR-FLOW AREAS

The turbulence inherent in normal room ventilation acts to scatter any contaminants liberated internally, as well as any introduced from outside. This may be dangerous in some hospital and industrial requirements, and the concept of laminary-flow chambers was therefore developed in order to minimize the risk of microbial dissemination by sweeping away contaminants with a minimum of lateral diffusion. Such an installation has more in common with a wind-tunnel (p. 92) than with either outdoor wind or indoor ventilation. Bourdillon & Colebrook (1946) developed a piston-like movement of air in rooms for dressing burns or wounds. Clean, warm air, introduced at the ceiling under positive pressure, formed a stable descending layer, pushing contaminated air down below it. In industry horizontal flow systems are often preferred (Sykes, 1970 (in Silver, pp. 46–56)). Both methods, with air entering the chamber through a filter occupying the whole area of wall or ceiling, give near laminar flow except in the wake of bluff objects, but flow must be fast enough to overshadow convection from lights, operators and other equipment, involving flow rates of near 50 cm/s. Lidwell & Towers (1970 (in Silver, p. 109)) studied the behaviour of a horizontal flow system in a room used to nurse several persons with a high degree of inter-patient isolation, and found that at velocities as low as 18 cm/s particles containing spores of *Bacillus subtilis* var *niger* could not be detected moving against or across the general direction of air flow, except when persons were moving about the room, and even then transfer across the flow was only one-tenth that of a turbulently ventilated room.

Laminar-flow hoods and chambers used in laboratory contamination control are discussed by Chatigny & Clinger (1969 (in Dimmick & Akers, Chapter X, pp. 194–263)).

XV

Inhaled Microbes

The air spora merits study both in its own right, as a natural phenomenon, and because it is a component of the air we breathe.

Popular belief supposes that inhaled particles are promptly exhaled again, but Lister (1868) showed that this idea was erroneous. Ordinary air set up putrefaction when bubbled through blood but air exhaled from the lung did not. Experiments show that exhaled air is substantially cleared of particles in its passage through the respiratory tract, where most of the inhaled material is deposited by various processes during the cycle of inhaling and exhaling. Exhaled air is normally microbe free, unless contaminated by droplets from the upper respiratory tract emitted with sneeze, cough or speech.

This filtering action is a disadvantage that goes with the lung's main function of exchanging carbon dioxide for oxygen, because a fraction of the retained particles are potentially harmful to human or animal recipients by acting as irritants, pathogens, allergens or antigens.

THE RESPIRATORY TRACT AS AN AIR SAMPLER

Considered in its accidental role as a 'vacuum cleaner' the respiratory tract is seen as a complicated instrument in which particles of different sizes are roughly sorted by a variety of deposition mechanisms acting in succession. The site where spores are deposited depends on the anatomy of the respiratory tract and on the terminal velocity of the particles (i.e. on their aerodynamic size), in a manner illustrated from theoretical calculations that have been substantiated in general by experiment, as shown in Figure 31 (after Lidwell, 1970; and *see* Hatch & Gross, 1964; Druett, 1967).

215

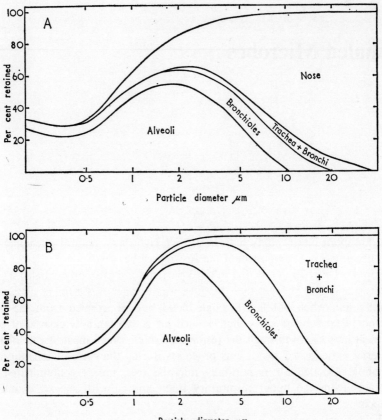

Figure 31. Retention of inhaled particles in the respiratory tract, showing relation between particle size and site of deposition following normal nose breathing, and mouth breathing. A = curves for nose breathing; 8 = curves for mouth breathing. (After Lidwell, 1970, by permission of O. M. Lidwell, and of Niels Jonassen, editor of *Termisk og Atmosfaerisk Indeklima*, Lyngby, 1970.)

Fractionating air samplers, such as the cascade impactor, the Andersen sampler and the multistage impinger (Chapter IX) show analogies with the respiratory tract, as indeed they were designed to do (Table XXVIII).

Austwick (1969) draws an analogy between the respiratory tract and 'an extremely thick membrane filter with a single pore leading to a system of bifurcating tubes of ever decreasing diameter'.

During normal breathing the volume of air inhaled in one breath is customarily referred to as the *tidal air*, to distinguish it from the *residual air*, i.e. the volume that is not normally expelled from the lungs. In the

TABLE XXVIII

STAGES OF FRACTIONATING AIR SAMPLERS APPROXIMATELY CORRESPONDING
TO DIFFERENT PARTS OF THE RESPIRATORY TRACT (FOR PARTICLES OF SAME
DENSITY AS WATER)

Sampler	Through-put (litres/minute)	Nose	Bronchi and bronchioles	Alveoli
			Stages	
Cascade impactor	17·5	1+2	2+3	3+4
Andersen sampler	25	1+2	3+4	5+6
Multistage impinger	55	1	2	3

normal pattern of breathing for man (*see* Druett, 1967) the air involved in each breath is made up as follows:

600 cm³ of tidal air (= The mean volume of air inhaled or exhaled at one breath. Of this about 140 cm³ is 'dead space' = the air filling the trachea, bronchi and upper bronchioles, never reaching the alveoli.)

1300 cm³ of reserve or supplemental air (= The additional volume of air that can be forcibly expelled by effort if needed.)

1600 cm³ of residual air (= The volume of air that cannot be expelled.)

2000 cm³ of complemental air (= An additional volume that can be drawn in by forcible inhalation.)

The complexities of the respiratory tract are described by Hatch & Gross (1964) and Druett (1967). The fate of inhaled spores in terms of deposition and removal processes can be described under anatomical headings as follows.

UPPER RESPIRATORY TRACT

The nose is divided internally into two parallel chambers. Behind the external opening are stout hairs, the vibrissae, serving mainly to exclude the largest particles. Inhaled air then passes over a complicated series of plate-like folds of tissue covered with mucus-producing cells and ciliated cells. The moving cilia continually sweep the layer of mucus with its entrapped particles proximally towards the pharynx. Incoming particles larger than 20 μm are deposited on this sticky labyrinth with nearly 100% efficiency. Some smaller particles are also deposited here but their efficiency of deposition in the nose decreases rapidly as their diameter decreases.

Pollens of anemophilous flowering plants, and the larger fungus spores will be retained in the nose (Figure 31A) until swept into the back of the mouth and swallowed. Habitual mouth breathers and rhinitis sufferers, however, will by-pass the nose filter: then Figure 31B applies, instead of Figure 31A, and it is clear that in mouth breathing all the larger material normally filtered off in the nose and removed to the digestive tract, is now deposited lower down the respiratory tract in the trachea and bronchi.

The trachea and bronchi. Leaving the nose, and flowing past the throat, the inhaled air next passes through the straight tube of the trachea, at the bottom of which some particles may impact where the tube forks into branches forming the left and right main bronchus. With repeated forking into numerous smaller bronchi, the air enters the lobes of the lung.

With normal breathing comparatively little material is deposited in the trachea and bronchi.

During inhalation flow rates through the respiratory tract from the nose up to the larger bronchi are approximately 100 cm/s, but with the branching into hundreds of secondary and tertiary bronchi the effective cross-section of the duct system increases and the flow-rate consequently decreases to the order of tens of centimetres per second.

Bronchioles. With still further branching the stage is reached where the air passes through bronchioles less than 1 mm in diameter, in which flow-rate decreases to about 1 cm/s.

Up to this point all these ducts are lined with mucus-producing cells and ciliated epithelial cells which continually propel the mucus blanket towards the trachea. Particles reach the mucus wall by impaction and sedimentation and are quickly cleared in the mucus blanket.

LOWER RESPIRATORY TRACT

Terminal respiratory bronchioles and alveoli. The bronchioles branch successively into the terminal respiratory bronchioles, then into the alveolar ducts and finally end in the alveolar sacs – the latter numbering perhaps 10^7 or more in man, each closely associated with blood vessels where gaseous interchange occurs, making up a total respiratory surface of 30 m².

Each alveolar chamber has a diameter of about 0·5 mm, and here 1–4 μm diameter particles are deposited mainly by sedimentation during the pause of a second at each inhalation. But for particles in the 0·5 μm range the duration of each inhalation is too short to allow all the particles to sediment; consequently many are swept out again, thus accounting for the low retention efficiency shown by the 0·5 μm particles in Figure 31A, B.

Still smaller particles will be subjected to Brownian diffusion and jostled into contact with the walls by irregular bombardment by gas molecules. Hence the increased efficiency postulated at equivalent diameters less than 0·4 μm – a matter of interest in the deposition of some inorganic dusts.

Particles deposited here are removed comparatively slowly because in this part of the respiratory tract the wall is not ciliated. Some foreign particles are taken up by phagocytic cells which transport them to the moving mucus blanket higher up; others penetrate to the lymphatic system; still others remain, for a long time at least, in the lung tissue.

In spite of some analogy with the air sampling devices referred to earlier, the respiratory tract differs fundamentally from any of the instruments in general use because there is no continuous air flow through it, neither is the flow-rate through it constant. Instead, the residual air oscillates from alveoli to bronchioles and bronchi where it meets and mixes with the tidal air and its entrained microbes. Unlike the fractionating sampling devices mentioned in Table XXVIII (p. 217), the largest particles are deposited in the region of maximal flow-rate (flow-rate in nose and trachea, over 100 cm/s), whereas the smallest are deposited in the respiratory bronchioles and alveoli (flow-rate less than 1 cm/s). Throughout the respiratory tract the speed and direction of air flow change rhythmically with each breath. This complexity results in a less sharp fractionation of particles by size than in mechanical sampling instruments. Further, individuals differ in breathing habits, and the same individual inhales different volumes at different speeds according to his demand for oxygen, depending on whether he is at rest or in strenuous exercise.

Respiratory disease will still further alter the pattern of deposition. The most obvious alteration being the effects of rhinitis in hay-fever leading to mouth breathing, thus diverting spores and pollens of over 10 μm diameter into the trachea and bronchi (Figure 31A, B).

Most pollens, spores and droplet nuclei are hygroscopic and may increase considerably in diameter as the air carrying them into the body becomes saturated – a process completed rapidly and mostly in the nose (Proctor, 1966). This alters their aerodynamic size and deposition site. A further complication, still awaiting investigation, is the effect, if any, of the electrostatic charge carried by the airborne spore (*see* Proetz, 1953).

WHAT IS INHALED?

We inhale the ambient air spora, but this statement needs qualifying by comparing the results of breathing in still air and in a wind.

Breathing from still air. Indoors, at ambient temperatures below 37° C, the heat of the body generates an upward convection current (Daws, 1967). Lewis *et al.* (1969) demonstrated this current by schlieren cinematography, and found, at 15° C ambient temperature, ascending currents of 30 to 50

I

cm/s in the region of the head. As a consequence we may suppose that the air inhaled comes from close to the ground, and air from this level commonly is more contaminated than air at head level. Moreover, in a vertical current intake into the nose may even approximate to isokinetic sampling. Under temperate conditions and in nearly still air the human body will draw up towards the nose the particle-rich ground layer of the air – a disadvantage not shared by mammals which lack man's upright posture. In hot environments, with little difference in temperature between air and body, this convection current will disappear.

Breathing from wind. As wind speed increases, the effect of any body-generated convection disappears. The nose then receives the oncoming air spora appropriate to its height above ground and samples it non-isokinetically. Possibly when facing the wind the nose acts on the principle of the 'stagnation point' sampler described by May (1967).

INFECTION BY INHALATION

In relation to airborne infection the intramural and extramural environments must be considered separately.

INTRAMURAL ENVIRONMENT

Inhalation anthrax (wool-sorter's disease) and pulmonary tuberculosis are well established as diseases acquired by inhalation (Wells, 1955; Brachman et al., 1966). So too are many rhinovirus and adenovirus infections; also influenza, measles, pneumonic plague, pneumococcal pneumonia, and streptococcal infections of the respiratory tract. But not all respiratory infections are acquired by inhalation, and conversely some infections so acquired affect other parts of the body.

Some tumour viruses are airborne. McKissick et al. (1970) show that in animal houses the Rauscher murine leukaemia and Yaba viruses are readily transmitted to susceptible hosts, probably by inhalation.

Inhalation of pathogenic organisms forms one possible route of infection, along with other possible routes including contact with infected animals or fomites, and by ingestion.

Processes by which infectious diseases are transmitted through the air have been matters of vigorous controversy, and the answers given have influenced social habits and prophylactic measures. Cornet (1889) held that pulmonary tuberculosis is normally acquired by inhaling dust of dried, infected sputum, but Flügge (1897) believed that infection was from germs expelled from the mouth and nose when coughing. Riley et al. (1962) used guinea-pigs as traps to monitor the air of tuberculosis wards, and from the number of infections obtained they estimated the concentration of infective particles (probably droplet nuclei) as 1 particle per 310 to 570 m³ of extracted air. However, Wilson & Miles (1964) conclude that present information is insufficient to evaluate the relative importance of dust, droplets and droplet nuclei in pulmonary tuberculosis.

With many diseases the quantitative role of the different possible routes of entry is controversial. Probably, under a given set of conditions, one route only is predominant, but may not be the same in a different environment. The role of airborne infection by bacteria is discussed by Williams (1967), that of viruses by Tyrrell (1967), and both are reviewed by Pappagianis (1970).

Droplets from sneeze, cough and speech show a wide range of diameters. The larger droplets, carrying more inoculum, settle rapidly. Smaller droplets tend to dry down to 'droplet nuclei' (Wells, 1955) and are then easily inhaled. The smaller the initial droplet (and therefore the smaller the dose of pathogen) the greater its surface to volume ratio and the more rapidly it loses viability while circulating about a room. Droplet size may profoundly alter infectivity, for example the number of cells of *Franciscella tularensis* required to kill 50% (LD 50) of exposed rhesus monkeys varied from 17 with 1 μm droplets, to 3000 with 22 μm droplets (United Nations, 1964, p. 61).

EXTRAMURAL ENVIRONMENT

Outdoors, inoculum is diluted faster than indoors, and fewer outdoor inhalant diseases are considered important. The viruses of both fowl pest (C. V. Smith, 1964), and of foot-and-mouth disease are undoubtedly airborne, but the route of infection, whether by inhalation or ingestion on food, is still controversial (Norris & Harper, 1970; Chamberlain, 1970).

The pathogen of Q-fever *(Rickettsia burnetii)* is known to be airborne. Welsh *et al.* (1958) inoculated pregnant sheep and sampled the air in the pen by drawing samples through a cotton plug at 10 litres/minute. Air was free from the pathogen from the 9th to 14th day, but became contaminated at parturition. The virus is concentrated in the placenta, and it was concluded that the act of parturition is responsible in some manner for generating an aerosol of the *Rickettsia*. In a notorious outbreak, Wellock (1960) reported on 75 cases of Q-fever occurring in a triangular pattern extending several miles down wind of an animal-fat rendering plant in San Francisco Bay.

Few bacterial diseases, it seems, arise from inhaled particles outdoors, but there is an important group of fungal infections so acquired. These were first reported from warmer climates as rare systemic and sometimes fatal infections; it now seems likely that infection is common but transient in normal persons who develop immunity. Fuller references to relevant literature are given in Emmons *et al.* (1970).

Nocardia asteroides, a thermotolerant Actinomycete causing 'farcy' in man, and sometimes in dogs and cattle (*N. farcinica* is commoner in bovines), is here grouped with the fungi for convenience. Primary pulmonary lesions are assumed to follow inhalation of fragmented mycelium (arthrospores).

Cryptococcus neoformans has long been known as an invader of the central nervous system. The initial lesion is now considered to be in the lung and derived from inhalation of airborne cells, to which many people must be exposed. The fungus is abundant in many parts of the world in accumulations of old pigeon droppings (Emmons, 1955). The discovery that *C. neoformans* is the imperfect state of a basidiomycetous fungus (Shadomy, 1970) suggests that the fungus may become airborne by the usual basidiospore mechanism, as well as by mechanical disturbance of bird excreta.

Coccidioides immitis, first recognized as a rare disseminated fatal mycosis in man, is now known to cause a common transient fever in the south western United States, where Emmons (1943) showed that it infected the lungs of many species of rodents; it is known to grow freely in desert soil, and at least 25 000 people are estimated to be infected annually in the United States. Exposing monkeys and dogs to desert air, Converse & Reed (1966) considered that inhalation of from 10 to 100 arthrospores would set up infection in dogs, and only 10 or fewer were required in *Macaca mulatta*. The arthrospores are well adapted to aerial dispersal and their small size is compatible with deep penetration of the lung (Levine (1969) in Dimmick & Akers, 1969).

Histoplasma capsulatum in America, and *H. duboisii* in Africa cause histoplasmosis, usually originating as a pulmonary infection from inhaled spores, occasionally followed by severe or chronic invasion of other organs, but normally remaining mild or even asymptomatic. *Histoplasma* has been isolated frequently from soil enriched by excreta and the fungus abounds in caves and other places inhabited by bats. In some areas of the south eastern and central United States from 60 to 90% of the human residents react positively to intradermal injection of histoplasmin, indicating that they have become sensitized – probably by inhalation of spores. Take-off mechanism for this organism is unknown, but it is assumed to be in wind-blown dust.

Blastomyces dermatitidis causes blastomycosis in warm regions in America and Africa. Infection is believed to arise from inhaled spores, giving usually a mild respiratory disease, but occasionally becoming disseminated or affecting the skin. The source of the inhaled spores is still obscure: they are assumed to be aleuriospores of the type found in culture, but with the demonstration of the existence of an Ascomycete state of the fungus, *Ajellomyces dermatitidis* by McDonough & Lewis (1968), this question needs re-examining.

Paracoccidioides brasiliensis, causing conspicuous oral and cutaneous lesions in South and Central America, is also considered to start with pulmonary lesions, perhaps often mild and healing spontaneously, derived from inhaled spores whose origin is obscure.

Adiaspiromycosis (haplomycosis) is a peculiar disease caused by moulds of the genus *Emmonsia*. *E. parvum* is common in desert rodents in the southwestern United States where it is often associated with *Coccidioides immitis* infection. *E. crescens* is widespread among rodents of moist environments in temperate regions (Jellison, 1969). Their life-cycles are obscure. In culture at room temperature their mycelium produces aleuriospores, 3–4 μm in diameter, which are adhesive and easily spread by contact. At 37° C, either in culture or in the body, the aleuriospores swell enormously, reaching 14 μm in *E. parvum* and 400 μm in *E. crescens*; the wall thickens and the cell (the 'spherule') remains dormant. If returned to room temperature the spherule sprouts numerous hyphae thus reproducing the sporogenous mycelium. Injected spores produce spherules in various organs, but naturally infected rodents have the spherules concentrated in the alveolar region of the lungs, indicating infection by inhalation. The disease occurs in many species of mammals including man (*see* Emmons *et al.*, 1970). For a sticky aleuriospore to become airborne is somewhat unexpected, and the saprophytic life-cycle, habitat and spore-launching mechanism all need further study outdoors.

Evidence suggests that we have here a heterogenous group of mycotic infections of man and other mammals, not transmissible from one individual to another, all primarily arising as lung infections from inhaled spores, and all (except *Emmonsia*) in rare cases becoming disseminated to other organs. The saprophytic phases of *Histoplasma* and *Coccidioides* have been investigated extensively, and knowledge about *Cryptococcus neoformans* is growing, but the life stories of other members of this heterogenous group are little known. Possibly all abound saprophytically in specific ecological niches, launching their spores into the air by mechanisms at present unknown. Detection of the airborne phase is technically difficult, and most success has followed the exposure of 'sentinel' animals (Emmons, 1943; Converse & Reed, 1966), or inoculation of animals with air filtrates (Ibach *et al.*, 1954; Rooks, 1954; Ajello *et al.*, 1965). The pathogenicity of this group may be accidental, so far as their life-cycle is concerned, but all are able to grow at body temperature or even higher. It is possible that inhalation by small mammals may be a normal adaptation. Examining corpses of small mammals, and also the old excreta of carnivores and snakes which feed on small mammals, in areas where these mycoses are endemic might prove rewarding in tracing possible sources of the airborne spore form.

ALLERGY TO INHALED MICROBES

An allergic person differs from normal by having become sensitized (altered in reaction) by exposure to one or more specific substances which are referred to as allergens. Exposure may take place by various routes including contact, injection or infection, but only inhalation is considered here. Inhalation exposure may be either intensive, or slight but frequent. People may be grouped into two broad categories (atopic or non-atopic) according to their reaction to allergens (Table XXIX; and *see* Pepys, 1969).

The atopic group, comprising about 10% of the population, have a predisposition to becoming sensitized, often beginning with infantile eczema, followed by rhinitis or asthma. Sensitization develops by ordinary exposure to normal environmental concentrations of allergen, often leading to multiple sensitization to a variety of substances. Among the inhaled allergen-bearing particles are grass pollen, house-dust mites, insect hairs, animal hairs, and spores of fungi and Actinomycetes. Atopic subjects develop non-precipitating antibodies (reagins, immunoglobulin E or IgE) that sensitize the skin and the respiratory tract down as far as the larger bronchioles. Characteristically in the atopic group, symptoms of asthma or rhinitis develop rapidly after an already sensitized person inhales a dose of the allergen; likewise an urticarial weal develops within 10 to 15 minutes after an intradermal skin test with an extract of the allergen (Type I allergy, *see* Table XXIX).

Non-atopic subjects, comprising the majority of the population, are less predisposed to allergy but all may be sensitized by sufficient exposure; consequently their sensitivity is commonly to a single specific allergen. When sensitized they produce precipitating antibodies (IgG or IgM). On subsequent re-exposure to the allergen, symptoms commonly of alveolitis, less often asthma, appear after a delay of 5 to 7 hours (Type III allergy).

Some people show a mixture of Types I and III allergy. The presence of a positive skin test (due to reagins and/or precipitins) to a potential allergen is evidence of exposure to the allergen, but is not necessarily evidence that the substance in question is the cause of the patient's symptoms – the clinical relevance of a positive skin test may have to be evaluated by inhalation tests.

TABLE XXIX

TYPICAL REACTIONS TO AEROALLERGENS

Subjects affected:	Atopic persons (± 10% of population)	Non-atopic persons (majority of population)
Sensitization:	By normal exposure to inhalant, often multiple	By massive exposure, not typically multiple
Antibodies:	Reaginic (non-precipitating, transferable) IgE	Precipitating (non-transferable), IgG and IgM
Reaction:	Immediate (Type I) as a rule (but a night attack may result from daytime exposure) No complement needed for reaction	Late (Type III) after 5 to 7 hours Reaction needs complement
Skin test:	Urticarial weal in 10–15 min, gone in 2 hours	Often negative (or extensive oedema in 3 to 4 hours, gone in 24 hours)
Examples of Aeroallergens:	LARGER SPORES Alternaria Cladosporium Basidiospores Ascospores	Aspergillus fumigatus (is also pathogenic)
	POLLENS Grasses (Gramineae) Ragweed (Ambrosia) Nettle (Urticaria, and Parietaria)	SMALLER SPORES Cryptostroma corticale Aspergillus clavatus Micropolyspora faeni Thermoactinomyces vulgaris Thermoactinomyces sacchari
Site of reaction:	UPPER RESPIRATORY TRACT (trachea, bronchi and bronchioles)	LOWER RESPIRATORY TRACT (respiratory bronchioles, alveolar ducts, alveoli)
Condition:	Rhinitis Typically asthma	Typically alveolitis

The reaction of the individual is only one factor in inhalant allergy. Size of the particles inhaled is also important. By contrast with inorganic dusts, which have a broad size-spectrum, a microbial species usually varies within a restricted size range. Pollen grains or fungus spores of a single species tend to be 'like as peas in a pod', a statement implying a degree either of uniformity or variability, depending on one's point of view. Species and size thus determine approximately where in the respiratory tract an allergen-containing particle is likely to be deposited, and this in turn affects its action (Table XXIX).

THE AEROALLERGENS

For convenience we speak of inhaled allergens, but more accurately we should refer to 'allergen-bearing particles', in contradistinction to purified allergens.

Components of the air spora must fulfil certain requirements in order to qualify as allergens, although all spores must be considered potentially allergenic. According to Wodehouse (1945) the requirements are: 'buoyancy, abundance, and allergenic toxicity'.

A sufficient concentration of microbial particles in the size range 1–50 μm must occur regularly or occasionally either indoors or outdoors. The concentration can be smaller in respect to atopic subjects, but needs to be large in order to sensitize and provoke attacks in ordinary (non-atopic) persons: severity of symptoms is related to the quantity of pollen inhaled (Maunsell, 1958). Abundance in air implies both an abundant source and an efficient launching mechanism.

The species must be allergenically potent. Species differ in potency, either by concentration of the allergen or by its specific activity. Pollens from grasses and ragweed (*Ambrosia*) are highly potent, sensitizing a large proportion of atopic subjects exposed to inhalation. One grass pollen grain in an extremely sensitive subject may induce rhinitis. By contrast, conifer pollen, to which large populations in the North Temperate regions are exposed annually in heavy concentration, rarely induces allergy. *Streptomyces* species abound in mouldy hay, but only *Micropolyspora faeni* and *Thermoactinomyces vulgaris* are involved in Farmer's lung disease, and of these two species the former is more potent and usually less abundant.

The existence of subjects sensitized to the species is also an essential requirement. The non-atopic individual may become sensitized to inhaled allergens in massive doses, but the principles on which an atopic subject picks out from the environment his private group of sensitivities are at present obscure.

Emphatically, the viability of a microbe is *not* required of an aero-allergen. An allergenic microbe is still active when dead, so long as it is not too far denatured chemically. The glycoprotein allergen in house dust remains active after the sugars are split off the molecule, but activity ceases when acid hydrolysis is carried to the splitting off of amino acids (Rimington *et al.*, 1947).

The aeroallergens investigated are based on protein, or less often on polysaccharides. Targow (1966) found two types of fungal allergen: 'AP' in *Alternaria* and *Phoma*, and 'F' in all other genera of fungi he tested. But Mandell (1967) considers that there is a distinct allergen for each fungal genus. The allergens of grass pollen were studied by Wodehouse (1965), and purified by Augustine & Hayward (1962). Goldfarb (1968) found several allergens in *Ambrosia* pollen (*see* Newmark, 1970).

COMMON AEROALLERGENS

Some components of the air spora cause allergy in many people, but rare allergic subjects can be found who are sensitive to one of a vast range of substances, towards which most people are indifferent. For instance, patients have been found sensitive to the spores of puff-balls (Strand *et al.*, 1967), or of the Myxomycetes (Peterson & Cohrs, 1966). McGovern *et al.* report sensitivity to airborne algae (*Hormidium, Bracteococcus,* and *Tetracystis*).

Hay fever (pollen rhinitis) is caused by inhaling pollen (often also deposited in the eye, causing conjunctivitis). In Europe, atopic subjects are most commonly sensitive to grass pollen, but in North America ragweed (*Ambrosia* spp.) pollen is also an important allergen. Hay fever plants in various parts of the world are reviewed by Newmark (1968, 1970). Pollen from trees is less often involved than grasses and ragweed, but pollen from other herbaceous dicotyledons (grouped by the allergist as 'weed pollen') is often important, especially the nettle *(Urtica)* and in southern Europe the related *Parietaria*. Other significant 'weeds' are plantain *(Plantago)*, dock *(Rumex)* and the cultivated sugar beet and its allies *(Beta vulgaris)*. Furthermore, some typically insect-pollinated flowers shed a significant amount of pollen into the air and are reported to be allergens.

Asthma. Pollen asthma can follow rhinitis, or it may occur without rhinitis.

The commonest fungal spore allergies are towards *Alternaria, Cladosporium* spp. (including *C. fulvum* on tomatoes under glass) and *Aspergillus* species, especially *A. fumigatus* which is described in more detail below. Also important are the dry rot fungus in buildings *(Serpula (Merulius) lacrymans)*, smuts *(Ustilago* spp.) and *Chaetomium* spp., but spores of the rust fungi (Uredinales) are seldom allergens, although Cadham (1924) observed a few cases in Manitoba.

Very many asthmatics are sensitive to house-dust allergen produced by a mite, *Dermatophagoides pteronyssinus*, which eats desquamated human epidermal scales and flourishes on the surfaces of mattresses (Maunsell, *et al.*, 1968; and *see* Bronswijk & Sinha, 1971).

Inhalation of enzyme preparations from *Bacillus subtilis* has led to asthma in factory workers making enzyme detergent powders (Flindt, 1969; Pepys *et al.*, 1969), but housewives using the more dilute commercial preparations are rarely affected (Belin *et al.*, 1970).

Allergic alveolitis (pneumonitis). The allergens involved in this group of diseases are carried by small, deeply-penetrating particles. The now classic example is Farmer's lung disease, caused by the inhalation of massive doses of spores of the Actinomycetes: *Micropolyspora faeni* and *Thermoactinomyces vulgaris* (Pepys *et al.*, 1963). The same organisms are implicated in one type of 'fog-fever' in cattle (Jenkins & Pepys, 1965), and perhaps in mushroom worker's lung (Sakula, 1967). The same or similar thermophilic Actinomycetes have also been found growing in domestic air-conditioning equipment and causing a pneumonitis in residents (Banaszak *et al.*, 1970; and *see* Dworin, 1966 and Goodman *et al.*, 1966). Bagassosis, a disease of workers handling mouldy sugar cane residue, is caused by a related thermophile, *Thermoactinomyces sacchari* (Lacey, 1971).

Other kinds of alveolitis are caused by true fungi including *Aspergillus fumigatus*. *Aspergillus clavatus* causes alveolitis when inhaled by malt workers (Channell *et al.*, 1969), and maple bark strippers' disease is caused by inhalation of spores of *Cryptostroma corticale* (*see* Wenzel & Emanuel, 1967). An inhalation disease caused by spores of *Papularia* sp., known as 'mal de canes de Provence', may belong to this group, but shows some affinities with an immediate, Type I, allergy.

Knowledge of the biology of aeroallergens can be a useful adjunct to treatment. The more a patient knows about the natural history, seasonal and diurnal periodicity and local occurrence of his personal allergens the better he will be able to take avoiding action.

ASPERGILLUS FUMIGATUS

This common mould deserves special note for the variety of its pathological manifestations in animals. It is thermotolerant, able to grow over a temperature range from below 20° C up to 50° C, and growing well at over 40° C. It abounds in vegetable matter decomposing in warm environments, such as self-heating hay and composts. It occurs in outdoor air, and also indoors, even in hospitals (Noble & Clayton, 1963). It has also been found growing in air-conditioning systems (Wolf, 1969). Paradoxically in temperate regions its concentration in outdoor air is greater in winter than summer, possibly because in winter farm animals are being fed with partly moulded fodder.

Pepys (1969) described infection of the lung by *A. fumigatus* leading to a variety of disease patterns. Atopic subjects may show asthma, with reagins; asthma and pulmonary eosinophilia occurs in patients with bronchopulmonary aspergillosis, accompanied by precipitins and reagins. In non-atopic subjects, damaged lung tissue may carry a saprophytic growth of *A. fumigatus* in the form of 'fungus ball' (aspergilloma), while inhalation of massive doses of the spores may lead to alveolitis. Invasive aspergillosis may occur in patients with reticulo-endothelial disturbances such as induced by massive X-radiation.

Ainsworth & Austwick (1959) summarize reports of aspergillosis (mainly *A. fumigatus*) in many species of birds and mammals, both domesticated and wild, infected primarily by inhalation and starting as a pulmonary disease but sometimes involving other organs. *A. fumigatus* is also one of several moulds implicated in bovine mycotic abortion, probably initiated by inhalation, but with alimentary infection from spores ingested in mouldy fodder as another possible route (Hugh-Jones & Austwick, 1967).

ROUTINE MONITORING OF OUTDOOR ALLERGENS

In allergy, knowledge of the air spora is needed for a variety of reasons: for clues to the possible identity of unknown allergens; for avoiding known allergens by precise information on the time of day, season and locality of occurrence; and for precision in choice of material for treatment by hyposensitization.

For airborne pollens and spores of sizes down to about 5 μm diameter, the Hirst trap (p. 137) has given good service during 15 years almost continuous monitoring at the Wright-Fleming Institute, London, and for shorter periods at St David's Hospital, Cardiff, and in many other centres in other parts of the world, except in North America where the gravity slide method has been preferred (despite its gross inaccuracies which justify its description as a 'logarithmic lie' (Gregory, 1952; and *see* Gregory & Stedman, 1953)). Personal experience in scanning both gravity slides and continuously recording volumetric trap slides show that the latter are infinitely more rewarding, enabling changes to be followed hour by hour in terms of approximate airborne spore and pollen concentration. Potential allergens which are smaller than pollen are under-recorded and pass unnoticed on the gravity slide, but can be followed easily with the volumetric traps.

With pollens, gravity slide surveys over a period of years give substantially the same indications as the volumetric method (Hyde, 1959), but daily records can be widely in error, and hourly records are impracticable.

IDENTIFICATION OF UNKNOWN ALLERGENS

Knowledge of the air spora gained by continuous sampling has led to a widening of the range of allergens known from the restricted group of pollens and moulds formerly recognized. The abundance of basidiospores found in the air immediately suggested that they should be examined as potential allergens (Gregory & Hirst, 1952). Confirmation came slowly, but Evans (1965) reported allergy to *Sporobolomyces*. Later, Herxheimer *et al.* (1969, and *see* Adams *et al.*, 1968) showed that at Cardiff 16% of summer asthmatics gave positive skin reactions and/or inhalation tests to extracts of basidiospores from one or more species of mushroom or toadstool, or of the mirror yeast, *Sporobolomyces roseus*.

Again, on the basis of routine spore trapping in London from 1954 to 1957, combined with clinical records, the possible implication of airborne ascospores in late summer asthma was suggested by Maunsell (1958). This was confirmed by Ganderton (1968) who showed that ascospores of species of *Leptosphaeria* were among the allergens associated with late summer asthma. Imperfect states of some Ascomycetes, classified in the imperfect genus, *Phoma*, were shown to be allergens twenty years earlier, following their isolation from outdoor air by exposing petri dishes of culture medium (e.g. Richards, 1954a), but the ascospore state becomes airborne more easily than the conidial state, and may well be more important as the allergens.

Incrimination of a species as an allergen may involve following a complicated series of clues, as illustrated in the search for the cause of Farmer's lung disease. Farmer's lung was known to result from inhaling airborne dust from mouldy hay. We therefore ignored the total microbial population of the hay and concentrated attention on dry dust blown from samples of hay that had been associated with patients diagnosed as having Farmer's lung. Judging from the organisms found, the hay had evidently heated spontaneously at some stage in preparation. The dust contained surprisingly few particles derived from the herbage itself, and attention was therefore focused on thermotolerant moulds and thermophilic Actinomycetes. Extracts of pure cultures were tested by double diffusion in agar gel and by immuno-electrophoresis against precipitins in patients' sera. Because symptoms appeared in the peripheral or alveolar region of the lung, the smallest particles, in the 1–5 μm range, were extensively explored. Eventually the Farmer's lung hay antigens were located in two species of thermophilic Actinomycete, now known as *Micropolyspora faeni* and *Thermo-*

actinomyces vulgaris, and the causal link was confirmed by producing symptoms in Farmer's lung patients by inhalation tests (Gregory & Lacey, 1963; Pepys *et al.*, 1963; Kobyashi *et al.*, 1963).

INHALATION TESTS

In experimental inhalation to produce infections or provoke an allergic response it is technically difficult to reproduce natural conditions, in which the inhalant usually consists of monodispersed, dry, airborne spores. The usual practice of exposing the test respiratory tract to inhalation cf an aerosolized aqueous suspension of spores or an extract is far from ideal. Cells harvested in water from a sporulating culture have several abnormal features: (1) addition of leachates from the culture medium (Krishnan & Damle, 1965); (2) removal of material by leaching from the spores themselves (Lingappa & Lockwood, 1964); and (3) alteration of the aerodynamic particle size by the carrier drops of the aerosol, affecting the site of deposition. Ideally spores that are naturally released and inhaled dry should not be wetted in handling.

ATMOSPHERIC HUMIDITY: RELATIVE AND ABSOLUTE

The effects of atmospheric humidity are complex. Increased retention of hygroscopic organic particles, swollen in nearly saturated air in the nose, has already been noticed (p. 219). Effects of *relative* humidity on the survival of spores or of microbes in droplet nuclei have been discussed in Chapter XI. Little attention has been given to the effects of *absolute* humidity on the respiratory tract in relation to infection. At first sight absolute humidity readings may seem paradoxical.

Absolute humidity is the mass of water vapour contained in unit volume of air, and can be expressed in grammes per cubic metre of air, but it is more usual to express it as vapour pressure in millibars. The maximum amount of water vapour the air can hold (its saturation) depends on its temperature (Figure 32). For example, at $0°$ C saturated air has a water vapour pressure of 6·1 millibars, and contains 4·8 g/m³ of water. At 20° C the values are 22 mb and 17·4 g/m³; and at blood heat of 37° C, 62 mb and 43·9 g/m³. Thus saturation sets an upper limit to the absolute humidity of the air, which is only slightly exceeded in mist or fog.

Relative humidity is the ratio of the actual amount of water vapour contained in unit volume of air to the amount that would be present if the air were saturated at the same temperature, expressed as a percentage. More simply, relative humidity is the percentage saturation. Relative humidity is commonly used as a measure of how damp or how dry the

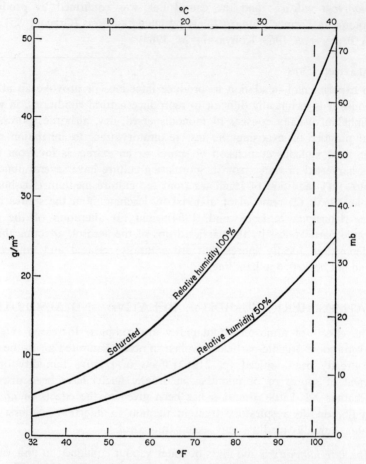

Figure 32. Relation between absolute and relative humidity, and temperature. Left-hand scale: Water-vapour content: g/m³ on left-hand scale; millibars vapour pressure on right-hand scale. Temperature: degrees Celsius top scale; degrees Fahrenheit on bottom scale. Curves shown for saturation (100% relative humidity), and for 50% relative humidity.

air *feels*. It affects such phenomena as drying the laundry, swelling of colloids (hence the hair hygrometer), spread of potato blight, and is useful in many biological contexts.

The relation between relative humidity, absolute humidity and air temperature can be obtained from tables (e.g. Deutschen Wetterdienst, 1963).

At different temperatures the same relative humidity can be arrived at with widely differing absolute humidities. For example, examination of the wet and dry bulb thermometer readings at Rothamsted Experimental Station for the whole year 1963 showed that, to take extremes, air with 100% relative humidity, which would feel very moist, was experienced with vapour pressures ranging from 2·1 mb (very cold dry air) to 16·5 mb (very moist warm air); this means that 100% relative humidity was experienced over almost the whole annual range of absolute humidity. Conversely vapour pressures around 10 mb, common in summer, produced relative humidities ranging from 55% to 100%. The significance of these facts in the present context is that a phenomenon bearing no obvious relation to relative humidity could nevertheless be strongly correlated with absolute humidity.

Unlike air temperatures indoors, which we now try to regulate within a narrow comfort range, the absolute humidity of the air in a building is still often close to that of the ambient outside air, although the relative humidity will normally be lower in a heated building than outdoors. Draughts around vents normally provide several changes of air each hour, even in a nominally closed room; when windows are open this exchange will increase greatly. But while we heat our dwellings in cold weather we seldom greatly modify the absolute humidity, which tends to be substantially that of the outdoor air mass. The absolute humidity inside ordinary dwellings is thus well documented by meteorological records, except where special conditions obtain, such as drying the washing or deliberate humidification.

Normally in passing through the respiratory tract inhaled air is warmed and more or less saturated at 37° C, whatever its initial temperature and humidity (Proctor, 1966); this implies that it is raised to approximately 62 mb vapour pressure or 43 g/m^3 water content (Figure 32).

Whereas we use *relative humidity* for most biological purposes, it seems reasonable to examine absolute humidity in relation to stresses on the respiratory tract, because stresses depend on how much water has to be given up by the moist layers to make the inhaled air up to 62 mb. Relative humidity would be a completely unreliable guide to such stress, because the respiratory tract saturates the incoming air at constant temperature. Antweiler (1958) and Ewart (1965) indicate that excessive dryness interferes with the functioning of the muco-ciliary system, slowing ciliary movement and making the mucus blanket thicker and less mobile. Waddy (1957) found the incidence of cerebrospinal meningitis in Africa correlated with absolute humidity. In New York Greenburg *et al.* (1967) noted an increase in asthma at the onset of cold spells which necessitated room heating in September and October. The increase was unrelated to sulphur dioxide, smoke, carbon monoxide, pollen or fungus spore concentrations, but correlated with extremely low levels of relative humidity, and hence at the low outdoor temperature correlated with low absolute humidity (alternative explanations were not excluded however).

Absolute humidity may well prove to be the significant factor stressing the respiratory tract and affecting speed of removal of inhaled pathogens and allergens.

AEROSOL IMMUNIZATION

The demonstration of antibodies in the individual's serum after inhalation of naturally occurring antigens, has been followed by research aimed at using the phenomenon as an alternative to existing methods of immunizing individuals against infectious diseases, which use injection or oral administration of vaccines. Immunization by inhalation has been explored, for example, against tularemia in man (Hornick & Eigelsbach, 1966). Hottle (1969), in reviewing progress with aerosol immunization, stresses the importance of particle size in controlling the region of deposition. Most progress with the method has been made with active immunization by inhaling living vaccines against respiratory diseases, especially Newcastle disease virus, and avian bronchitis virus in poultry.

XVI

Deposition Gradients and Isolation

Chapter VI described efforts to formulate changes in concentration of the microbial cloud while it diffuses and travels down wind. These concentrations are relevant to air pollutants inhaled into the respiratory tract. We must now discuss the more complex phenomena of deposition gradients: the decrease in number of microbes deposited per unit area of ground surface at increasing distances from the source. When deposition gradients give rise to infection gradients they are relevant to crop protection and plant breeding. An infection gradient can also arise as an *environmental gradient*, due to variation across the field of such factors as ecoclimate or soil type; their occurrence is then compatible with either uniform or non-uniform distribution of inoculum over the area. Environmental infection gradients are not considered here. The discussion in this chapter is based on crop disease, but its application to human and animal pathology, and to plant breeding will be obvious.

The infection of a plant by an airborne spore is a complex process of which physical transport is an important part. Infection may fail at any one of a chain of stages; and, for a deeper understanding of the whole process, it is necessary to understand the parts. Spore diffusion and deposition are stages selected for special attention in this chapter.

It is impossible to predict, from knowledge of spore deposition gradients, how many infections will be acquired by a plant at a given distance from a source of known strength, because conditions may be unsuitable for infection; but there is a possibility of being able to predict an upper limit, and to use this knowledge in choosing safe isolation distances. This chapter deals with gradients measured up to distances ranging from a few metres to a few kilometres from the source; long distance dispersal is discussed in Chapter XVII. The discussion assumes the simplified conditions dis-

235

cussed on p. 74; but even so, deposition and infection gradients have complications.

FACTORS COMPLICATING INFECTION GRADIENTS

(i) *Deposition coefficient.* With sources at or near ground level, the diffusing cloud of spores, unlike a gas or smoke, is robbed by heavy deposition close to the source. Concentration and area dose at a point down wind therefore depend on two factors – the diffusion history and the deposition history which the relevant part of the cloud has experienced.

In view of the evidence given in Chapter VIII, velocity of deposition may be taken as proportional to the terminal velocity, and the importance of deposition will increase with spore size.

(ii) *Viability.* It is assumed, in the absence of experimental evidence to the contrary, that viability is not affecting plant disease gradients over the short distances discussed here, though the future may show this to be an over-simplification.

(iii) *Available sites.* Deposition gradients do not necessarily give rise to observable infection gradients. For instance, a fruit-body of *Ganoderma applanatum* will emit spores continuously and copiously for months, and the spores will be diffused and deposited; yet the occurrence of *Ganoderma* fruit-bodies remains erratic throughout a forest, their occurrence being limited by what may be expressed in the ignorance-blanketing phrase 'availability of sites'. An infection gradient can only develop when sites such as trap surfaces, nutrient medium, susceptible host plants, ripe stigmas or burnt soil, are freely available. The deposition gradient is a regular phenomenon: an infection gradient follows when the deposition gradient is superimposed on unoccupied sites. Ecologically an infection gradient is a stage in succession, not a characteristic of a balanced state.

(iv) *Multiple infections.* As long as the number of available sites is large, the slope of the infection gradient will be parallel to that of the deposition gradient. But as soon as available sites begin to be used up, the infection gradient will be flattened; the flattening, beginning nearest the source, and the relation between the two gradients under simple conditions is given by the multiple-infection transformation (Gregory, 1948).

When disease incidence is recorded as the number or percentage of plants attacked irrespective of whether the plant has one or many lesions, aphid punctures, etc, the percentage will have to be suitably transformed before the formula for deposition can be applied.

The need for the transformation may be illustrated by considering a hypothetical example of 100 potato plants, uniformly susceptible and exposed to infection by *Phytophthora infestans* from a distant source. The first spore that causes an infection must obviously infect 1% of the plants. A second infecting spore, so long as it falls *at random* among the 100 plants will have one chance in 100 of alighting on the one plant already infected, instead of infecting a second plant. As the percentage of infected plants increases, the probability that each additional infection falls on a plant already infected (thus producing no increase in the percentage of plants infected) increases greatly. When 99% of the plants are infected, another infection will have only one chance in 100 of falling on the single remaining healthy plant. Different parts of the percentage range, therefore, correspond to very different spore densities per unit area, and the transformation can be neglected only in the lower-percentage categories.

Thompson (1924) applied the Poisson distribution to the problem of multiple infection and showed that if N = number of hosts available, and y = average number of hosts infected after the deposition at random among the hosts of x parasites, then $y = N \, (1 - e^{-x/N})$. Table XXX gives x calculated for values of y varying from 1 to 99·9%; it shows that whereas only one infection is required to bring about an increase from 1 to 2%, the increase from 98 to 99% requires sixty-nine infections (that is, 460 minus 391 according to the table).

If the distribution of infections is at random among the hosts, a straight line will be obtained when the logarithm of the percentage that remains uninfected (100 minus the percentage infected) is plotted against the number of infections. The slope of the line is given by $b = -0 \cdot 00434$. Blackman (1942), in ecological studies of flowering-plant communities, found that the occurrence of plants on quadrats may depart from the expected random distribution and yet give a reasonably straight line when plotted as before. The slope of the observed line, however, differs from that of the expected line. Blackman's 'correction factor', the ratio of the slope of the expected line to the slope of the observed line ($K = b_{expected}/b_{observed}$), then gives a useful measure of the departure from the random arrangement. Various other mathematically plausible formulations of this deviation have been attempted (*see* Fracker & Brischle, 1944; Greig-Smith, 1964).

Non-random distribution may result from various factors such as repulsion between individuals ('under-dispersion' as understood by ecologists) and K will then be less than unity. Aggregation may be due to such causes as local spread of infection, progeny remaining near parent, or to local differences in susceptibility. Plank (1946) gives a useful test for detecting aggregation in the field.

(v) *Infection efficiency.* Gradients may be observed either by directly counting the numbers of spores deposited on equal areas at different

TABLE XXX

MULTIPLE INFECTION TRANSFORMATION: PERCENTAGES TO INFECTIONS
(CALCULATED FROM 5-FIGURE LOG TABLES)

y%	x	y%	x	y%	x	y%	x	y%	x	y%	x
1	1·00	23	26·14	44	57·98	65	105·0	83	177·2	93·5	273·3
2	2·02	24	27·44	45	59·78	66	107·9	83·5	180·2	94	281·3
3	3·05	25	28·77	46	61·62	67	110·9	84	183·3	94·5	290·0
4	4·08	26	30·11	47	63·49	68	113·9	84·5	186·4	95	299·6
5	5·13	27	31·47	48	65·39	69	117·1	85	189·7	95·5	310·1
6	6·19	28	32·85	49	67·33	70	120·4	85·5	193·1	96	321·9
7	7·26	29	34·25	50	69·31	71	123·8	86	196·6	96·5	335·2
8	8·34	30	35·67	51	71·33	72	127·3	86·5	200·2	97	350·7
9	9·43	31	37·11	52	73·40	73	130·9	87	204·0	97·5	368·9
10	10·54	32	38·57	53	75·50	74	134·7	87·5	207·9	98	391·2
11	11·65	33	40·05	54	77·65	75	138·6	88	212·0	98·5	420·0
12	12·78	34	41·55	55	79·85	76	142·7	88·5	216·3	99	460·5
13	13·93	35	43·08	56	82·10	77	147·0	89	220·7	99·1	471·0
14	15·08	36	44·63	57	84·40	78	151·4	89·5	225·4	99·2	482·8
15	16·25	37	46·20	58	86·75	79	156·1	90	230·3	99·3	496·2
16	17·44	38	47·80	59	89·16	80	160·9	90·5	235·4	99·4	511·6
17	18·63	39	49·43	60	91·63	80·5	163·5	91	240·8	99·5	529·8
18	19·85	40	51·08	61	94·16	81	166·1	91·5	246·5	99·6	552·1
19	21·07	41	52·76	62	96·76	81·5	168·7	92	252·6	99·7	580·9
20	22·31	42	54·47	63	99·43	82	171·5	92·5	259·0	99·8	621·5
21	23·57	43	56·21	64	102·2	82·5	174·3	93	265·9	99·9	690·8
22	24·85										

With acknowledgements to Drs S. B. Fracker and H. A. Brischle of the United States Departmen of Agriculture.

distances, or by counting some consequent effect such as colonies, leaf-spots or diseased plants. Usually these gradients will be the same when numbers are plotted against distance, but the infection gradient will be much lower than the deposition gradient. Even with a nearly 100%-viable spore suspension, the general experience in inoculation tests with plant pathogens is that only a small proportion of the spores deposited will give rise to a lesion – even when conditions for infection are as favourable as possible; in unfavourable conditions the formation of lesions falls to zero. The proportion of spores successfully infecting is termed 'infection efficiency' by Gäumann (1950, p. 157), and values recorded by various workers under highly favourable experimental conditions include: *Phytophthora infestans* 6·5%, *Alternaria solani* 1·7%, and *Septoria lycopersici* 0·2% (all on tomato leaves by McCallan & Wellman, 1943); *Botrytis* sp. on *Vicia faba* 5% (F. T. Last, personal communication; and *see* Wastie, 1967); *Peronospora tabacina* approximately 1% (Waggoner & Taylor, 1958); *Phytophthora infestans* 1–4% (Knutson & Eide, 1961).

Rust fungi show relatively higher efficiencies. Thus Petersen (1959) observed penetration by 30% of uredospores of *Puccinia graminis tritici* on wheat leaves; but, at the dense concentrations tested, over 100 uredospores were required to produce one sporulating uredosorus (*see* Durrell & Parker, 1920). Petersen's results are examined in detail by Plank (1963). With the same fungus, Rowell & Olien (1957) obtained as many as eleven uredosori per 100 spores applied. McCallan (1944) evidently obtained about 10% efficiency from uredospores of *Puccinia antirrhini*. R. H. Cammack (personal communication) obtained 15 to 23% efficiency when inoculating *Puccinia polysora* to susceptible maize.

Considering its importance in plant pathology, it is remarkable how little attention was given to infection efficiency until Garrett (1960), Dimond & Horsfall (1965), and Wood (1967) studied the theory of inoculum. Infection efficiency must vary with dispersal conditions, but commonly the height of the infection-gradient curve above the *x*-axis will be only about one-hundredth that of the deposition curve, though the slopes of the two curves would be similar.

(vi) *Secondary spread*. From turbulence theory we can predict only primary dispersal gradients. As Waggoner (1952) points out: 'Because proximity of source is relatively more important than strength of the source, spatial distribution of diseased plants becomes more uniform as secondary infection progresses.' The infection gradient will therefore be flattened if observed long enough after deposition for secondary spread to occur around the primary lesions. Examples of this are found in Pape & Rademacher (1934), Zogg (1949, Figures 10 and 11), Waggoner (1952, Figure 2), and Cammack (1958a).

(vii) *Sampling*. At low levels of infection, the size of samples must be increased, as otherwise the sampling error becomes large (*see* Finney, 1947).

HISTORY OF EMPIRICAL METHODS

As noted in Chapter VI, gradients can be represented by either an empirical or a theoretical model. In the empirical method we make the curve fit the data, but in the theoretical method we test the fit of the data to the curve.

Frampton *et al.* (1942) concluded that the incidence of some insect-transmitted virus diseases decreased logarithmically with distance. Zentmyer *et al.* (1944) studied the spread of the Dutch elm disease pathogen, *Ceratostomella ulmi*, which is transmitted from tree to tree by the elm-bark beetle, *Scolytus multistriatus*. Their data, to distances of about 84 metres, indicated that the probability of infection decreased with the logarithm of the distance. Most subsequent curve-fitters agree that such decrease is logarithmic.

Wolfenbarger (1946, 1959; *see also* Wadley & Wolfenbarger, 1944), in valuable surveys of literature on the dispersal of bacteria, spores, seeds, pollen, and insects, concluded that the observed data could be fitted by one of the two following equations:

$$E = a + b(\log x), \quad \text{or}$$
$$E = a + b(\log x) + c(1/x),$$

where E = the expected value, x = distance from source, and a, b, and c are parameters derived from the observed data. Values of the parameters a, b, and c, obtained by Wolfenbarger showed enormous variation between the numerous sets of gradient measurements published in the literature, and it seems impossible to make any kind of generalization using this method, or to use the parameters, as given to predict gradients.

E. E. Wilson & Baker (1946a, 1946b) made field observations in California on the dispersion pattern of apricot brown-rot (*Sclerotinia laxa* on *Prunus armeniaca*), and they also liberated *Lycopodium* spores experimentally. They fitted the following equations to their data:

(1) $y = 100/x^2$ for the gradient of aerial spore concentration (concentration at distance x_2 and at subsequent distances being expressed as a percentage of the concentration at x_1);

(2) $y = 100(1+a)^2 / (x+a)^2$ for the gradient of infection by airborne spores, where a is a parameter of the experiment.

Dispersal by insect vectors is usually fitted by a logarithmic expression. Bateman (1947c) found that the proportion (F) of contamination of seed crops by insect cross-pollination at distance (D) was fitted by the expressions:

$$F = ye^{-kD^{\frac{1}{2}}},$$
$$\text{or} \quad F = \frac{ye^{-kD}}{D},$$

where y = contamination at zero distance, and k expresses the rate of decrease of contamination with distance.

Gregory & Read (1949) concluded that data for insect-borne viruses could be fitted well by the empirical expression: $\log I = a + bx$, where I = number of infective punctures at a distance x from the source after a given time, and a and b are constants for any one given set of field conditions.

With the empirical method an equation can usually be obtained, containing at most three parameters, which gives a good fit to any particular set of data. However, it is not easy to compare the results obtained by different workers. The empirical method is difficult to use because point, strip and area sources have not in the past been distinguished, the multiple-infection transformation has not been applied, even when appropriate, and the parameters as calculated from the data correspond to no obvious natural phenomena; they may even conceal different units of measurement varying from centimetres to nautical miles! Progress with the empirical method requires attention to these matters and, especially, the adoption of a standard metric unit of distance.

DIFFUSION AND DEPOSITION THEORIES

The more difficult, but potentially more useful theoretical approach is derived from the diffusion phenomena described in Chapter VI.

W. SCHMIDT'S THEORY

Schmidt (1918, 1919) used his diffusion theory to calculate Q_x/Q_0, the fraction of the eddy-diffusing spore cloud which remained in the air at distance x. To do this he assumed that any part of the cloud whose diffusion path would have brought it down to ground level, would have been removed from the cloud by deposition. He represented the terminal velocity of the particles as equivalent in effect to tilting the ground, and he gives a table from which the 'dispersal limit' under average conditions of wind speed and turbulence could be read for particles with different terminal velocities. Dispersal limit (V) was defined as the distance exceeded by only 1% of the particles liberated.

Schmidt's theory was developed further by Rombakis (1947), who brought the fall velocity of the particles into the differential equation and replaced the arbitrary 99% of Schmidt's dispersal limit by the concept of 'probable line of flight'. Rombakis postulated that a point P at height z and time t will be a point on the probable line of flight when it is statistically equally probable for a particle to occur above or below P. At the 'probable flight range' (a distance of about one-tenth of Schmidt's V), 50% of the particles liberated will have been deposited; 'probable flight height', and 'flight duration', are similarly defined. Rombakis also reached the interesting conclusion that the 'probable final velocity' of a particle is half its terminal velocity in still air. These concepts were applied to the dispersal of fungus spores by Schrödter (1954), who used Falck's (1927) calculations of terminal velocities to predict probable flight ranges, altitudes and durations for various spore sizes, wind speeds, and values of the turbulence coefficient. In his later review of the topic Schrödter (1960) uses Rombakis's method for estimating distance of dispersal, but uses Sutton's equations for the concentration of the spore cloud.

DEVELOPMENT OF SUTTON'S THEORY

Sutton's equations predict concentrations when there is no loss by deposition. To adapt them for particles which deposit appreciably during travel, it is necessary to calculate Q_x, the number of spores remaining in suspension after the cloud has travelled a distance x, from the equation (given on p. 112):

$$Q_x = Q_0 \exp\left[\frac{-2px^{(1-\frac{1}{2}m)}}{\sqrt{(\pi)}\ C\ (1-\frac{1}{2}m)}\right].$$

Expected depositions at various distances from point, line, strip and area sources can then be calculated (Gregory, 1945, and unpublished), from the basic formulae (p. 77) as follows in Table XXXI.

TABLE XXXI

NUMBER OF SPORES DEPOSITED AT DISTANCE x FROM SOURCE

Point source

(1) total on annulus 1 cm wide at distance x from source

$$D = \frac{p\ 2\ Q_x}{\sqrt{(\pi)}\ C\ x^{\frac{1}{2}m}}.$$

(2) mean per cm² on annulus 1 cm wide at distance x from source

$$d = \frac{p\ 2\ Q_x}{2\ \pi^{3/2}\ Cx^{\frac{1}{2}\ (m+2)}}.$$

(3) per cm² down wind of point source

$$d_w = \frac{p\ 2\ Q_x}{\pi\ C^2 x^m}.$$

Line source

(4) per cm² down wind of line source (N.B. $d_{lw} = D$)

$$d_{lw} = \frac{p\ 2\ Q_x}{\sqrt{(\pi)}\ C\ x^{\frac{1}{2}m}}.$$

Strip or area source

(5) per cm² down wind of strip or area of width w.

$$d_{aw} = \frac{p\ 4\ Q_x}{\sqrt{(\pi)}\ C(2-m)}\ x^{1-m/2}\left[\left(1+\frac{w}{x}\right)^{-m/2}-1\right].$$

Gregory (1945, p. 69: Appendix by Mrs M. F. Gregory) gave a set of calculated values for D, d, d_w, and d_{lw}, assuming $Q_0 = 10^{10}$, $p = 0.05$, $C = 0.6$ (cm)$^{1/8}$, $m = 1.75$ and 1.24. With further knowledge of these parameters, values from revised calculations are now given (pp. 245–46).

In a series of papers on pollen contamination of seed crops, Bateman (1947a, 1947b, 1947c) gave many examples of dispersal gradients, adopting and simplifying the Gregory formulae and using a regression method for testing the adequacy of the diffusion theories of W. Schmidt and of

Bosanquet & Pearson (1936). Bateman's method also made it possible to estimate p/C and Q_0, provided all measurements were expressed in the same units. Regression tests showed that pollen dispersal was best fitted by the 1945 formulae of Gregory, but the data were inadequate for choosing between the values 1·76 and 1·24 for the parameter m of Sutton's equations.

Combining field inoculation studies with eddy-diffusion models, Waggoner (1952) made an important contribution by adapting the formulae: (1) to allow for non-isotropic turbulence, and (2) to incorporate the ratio of spores deposited to percentage of leaves or leaflets infected. Waggoner used the findings of E. E. Wilson & Baker (1946b) and of Scrase (1930), to put the variance of concentration of vertical distribution $(\sigma_x{}^2)$ as equal to $4/9$ of the variances in the x and y planes. This led to the formula (in our symbols):

$$\chi = \frac{3\ Q_x}{2\ \pi^{\frac{1}{2}}\ x^{3m/2}}\ \exp\left[-x^{-m}\ (r^2 + 9z^2/4)\right],$$

where $r^2 = x^2 + y^2$.

From observed gradients of late-blight *(Phytophthora infestans)* around artificially inoculated potato plants in the field, Waggoner took k as the ratio of spores deposited per square centimetre to the proportion of leaflets diseased, and estimated the parameters p (deposition rate) and Q_0/k from the equation:

$$D = \frac{0\cdot135p(Q_0/k)}{x^{15/8}}\exp\ (-6\cdot78px^{1/8}),$$

where D = proportion of leaflets diseased. (In Waggoner's tests D was comparatively small and did not need the multiple-infection transformation.)

In his experimental potato plots in Iowa in 1949 and 1950, respectively, Waggoner found $p = 0\cdot12$ and $0\cdot15$, and $Q_0/k = 18 \times 10^4$ and $7\cdot8 \times 10^4$. Subsequently for *Peronospora tabacina*, Q_0 was estimated at approximately 10^3 spores per cm^2 of lesion per day (Waggoner & Taylor, 1958).

For assessing concentrations of radioactive clouds, Chamberlain (1956, pp. 20–27) at Harwell developed expressions combining the eddy-diffusion equations of Sutton (1947) and the allowance for loss by deposition (Gregory, 1945). Chamberlain also modified the equations to allow for elevation of the source above ground level, illustrating the fact that elevating the source greatly reduces the loss by deposition.

RECALCULATION OF THE DEPOSITION GRADIENT

With new information and further development of the statistical theory, deposition gradients can now be calculated for a wider range of conditions than was possible earlier. The values worked out here are offered as giving a useful indication of trends, but much further work still remains to be done.

Chamberlain's modifications have been adopted, in preference to Waggoner's, because the latter uses the parameter k, which can vary over a wide range down to zero – depending on leaflet size in a particular crop, and on how favourable conditions were for infection.

Two sets of graphs are provided. Figure 33 gives Q_x for $m = 1.75$,

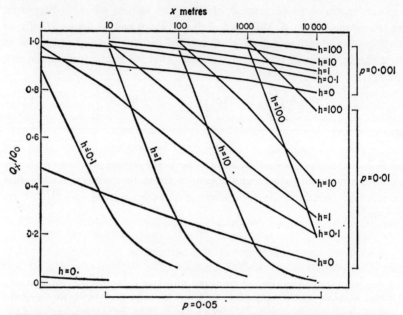

Figure 33. Fraction of spore cloud remaining airborne (allowing for loss of spores from spore cloud by deposition to ground), expressed as Q_x/Q_0. (Calculated from Chamberlain (1956): for $m = 1.75$; $C_z = 0.12$ (metres)$^{1/8}$; height of source above ground, $h = 0.0$, 0.1, 1, 10 and 100 metres; distance from source, $x = 1$ metre to 10 km.

$C = 0.21$ (metres)$^{1/8}$, and a variety of liberation heights, values of the deposition coefficient, p, and distance from source. (For distances within 100 metres of source the values for Q_x given by Tyldesley (1967, p. 27) could be used with advantage.)

Another set of graphs, Figures 34 to 36, gives values of D, d, d_w, d_{lw}, and d_{aw} on a logarithmic scale for sources of various geometrical form, assuming no loss from the cloud by deposition. To predict a gradient, the deposition values are first read off for various distances, using the appropriate line on Figures 34 to 36 and allowance can then be made for loss from the cloud at each distance by reading off Q_x as a fraction of Q_0 in

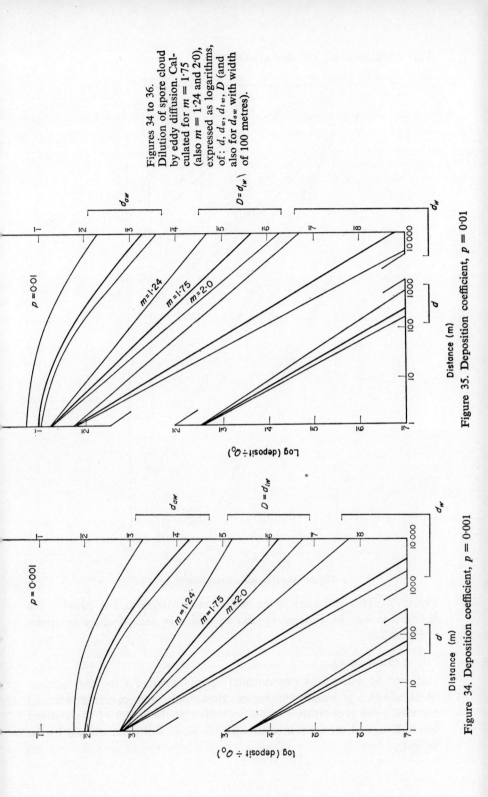

Figures 34 to 36. Dilution of spore cloud by eddy diffusion. Calculated for $m = 1.75$ (also $m = 1.24$ and 2.0), expressed as logarithms, of: d, d_w, d_{iw}, D (and also for d_{aw} with width of 100 metres).

Figure 34. Deposition coefficient, $p = 0.001$

Figure 35. Deposition coefficient, $p = 0.01$

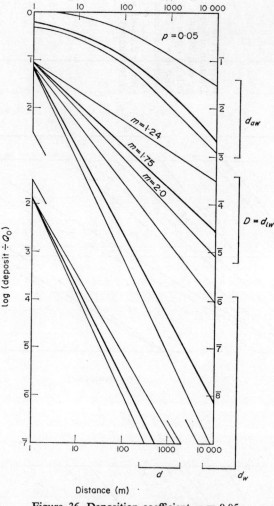

Figure 36. Deposition coefficient, $p = 0.05$

Figure 33. (For calculations on which these diagrams are based I am indebted to Mrs M. F. Gregory who was aided by the 'Mercury' computer at Harwell.)

CALCULATION OF Q_x

Except in some field experimental spore-liberation tests, the quantity liberated, Q_0, is usually unknown. However, cloud concentration and deposition are proportional to Q_0, so, although the *height* of the gradient line will depend on the source-strength, its *slope* will be unaffected by the value Q_0.

Elevation of source has been allowed for by using the equation number 52 of Chamberlain (1956), to calculate value of Q_x for heights $h = 0$, 0.1, 1, 10 and 100 metres.

Figure 33 shows that elevating the source decreases deposition on ground near the source, and in using Figures 34 to 36 for elevated sources it is important to neglect those parts of the curve before deposition starts. This decreased deposition near an elevated source was confirmed experimentally by Colwell (1951), who liberated ^{32}P radioactive *Pinus* pollen at 3.5 metres above ground level into a wind averaging 8.1 metres per second. Colwell sampled simultaneously with vacuum cleaners and petri dishes on the ground, and estimated the number of pollen grains with the help of a Geiger–Müller counter. Maximum deposition in this experiment was obtained at 5.8 metres horizontal distance from the source.

For Sutton's parameters, the values adopted here are: $C_x = C_y = 0.21$ (metres)$^{1/8}$, $C_z = 0.12$ (metres)$^{1/8}$, and $m = 1.75$.

As a measure of deposition, p has been chosen in preference to Chamberlain's v_g/u, because the wind speed in which deposition occurred is usually not known in field observations. However, the curves, which are calculated for $p = 0.05$, 0.01 and 0.001 in Figures 34 to 36 respectively, can be used with Chamberlain's velocity of deposition if known, because provisionally it is taken that $p = v_g/u$.

Some pathogens have large, readily-impacting spores. When liberated at, or near, ground level, although a small proportion may travel vast distances, most of these spores probably do not move very far from the source. Figure 34 suggests that for 'impactor' spores with a deposition coefficient of $p = 0.05$, and liberated at 10 cm above ground level, 94% would be deposited within 100 metres of the source.* But 'penetrators', with $p = 0.001$, would be deposited only to the extent of 6% within 100 metres under similar conditions.

The claim (Schrödter, 1960; Ingold, 1971) that Rombakis's treatment of diffusion is more realistic because it included an explicit term for terminal velocity is based on a misunderstanding. Far from being an oversight, the neglect of a term for terminal velocity (Gregory, 1945) was a deliberate choice because sedimentation under gravity is only one of many deposition mechanisms. The deposition coefficient, p, was intended to express the thickness of the spore cloud cleared by all mechanisms, including terminal velocity (*see* Raynor *et al.*, 1970, p. 893). Schrödter's treatment leads to longer probable flights than those estimated by our method (Gregory, 1966).

---

* My mistake (Gregory, 1952: *see* correction in Gregory, 1958), in stating that 99.9% would be deposited within 100 metres, was due to misreading ' $Q_0 = 10^{12}$ ' as 10^{10} in my own table (Gregory, 1945, Table XXI) and not to the misunderstanding suggested by Schrödter (1960, p. 178).

Chamberlain (1966a) considers present evidence is in favour of 'median ranges of order 1 km for grass pollen and other pollen liberated at knee height, and tens of kilometres for pollen liberated at tree-top height'. By contrast, within a forest canopy, Andersen (1967) after examining moss humus concluded that the majority of pollen was dispersed less than 20–30 metres. And Raynor *et al.* (1970) estimate that about 1% of ragweed pollen grains released from experimental sources remained airborne at a distance of 1 km.

COMPARISON OF GRADIENTS BY THE REGRESSION METHOD

Plotted on a linear scale, with distance on the x-axis and infections on the y-axis (the boundary between source and receptors being taken as $x = 0$), dispersal gradients characteristically show a hollow curve (Figure 37), decreasing rapidly at first near the source, then flattening and even-

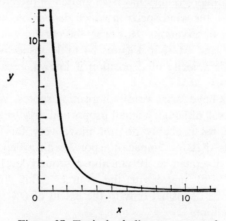

Figure 37. Typical hollow curve of gradient from a point source plotted with linear scales of x and y.

tually becoming nearly parallel with the x-axis. Such curves are difficult to compare with one another, and the point of inflexion of the curve depends very largely on the scales chosen for the axes. Further, the slopes of the gradient change from point to point along the hollow curve. (The amount of disease at the source itself is, of course, excluded in tracing the gradient.)

Much more useful is a graph on logarithmic scales, either using double logarithmic paper or plotting log of infection against log of distance (Figure 38). This has the effect of substantially straightening the line. A straight line fitted to these points by eye represents the general slope, which is con-

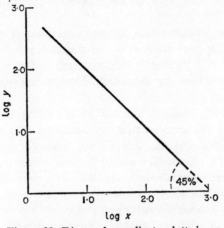

Figure 38. Dispersal gradient plotted on double logarithmic scales. The regression coefficient, $b = \tan \Theta^\circ$.

veniently characterized by the tangent of the angle it makes with the x-axis. For more precise comparison the linear regression of $\log y$ on $\log x$ can be calculated, using the regression equation: $\log y = \log a + b \log x$. The coefficient, b, is a measure of the gradient, and is numerically equivalent to the tangent of its angle with the x-axis ($b = \tan \Theta$), and it is usually negative. The constant, a, depends on the units adopted and measures the average disease incidence; it is usually only interesting when gradients within one experiment are being compared.

This double log regression is merely a convenient way of presenting and comparing gradients. Emphatically it is a descriptive and not an interpretative statistic. A more realistic analysis would be a weighted linear regression, giving less weight to low counts, which are usually unreliable and confounded with background contamination. Alternatively values for y could be transformed to angles and plotted against $\log x$, but in spite of statistical advantages, the angular transformation seems inconvenient with the wide range of values for y sometimes encountered in a single gradient.

Dimond & Horsfall (1960) use b to *define* the gradient. Here b is taken as the measure of the gradient, but the gradient itself is regarded as a real phenomenon occurring in fields and seen on a linear rather than a logarithmic scale.

STATEMENT OF BASIC PRINCIPLES

(1) A dispersal gradient implies a local source of inoculum. Background inoculum coming into a crop from afar will not show a dispersal

gradient, though it may show an environmental gradient such as an edge effect.

(2) A dispersal gradient also requires a population of susceptible host tissue. A spore dispersal gradient is not necessarily followed by an infection gradient – a fact repeatedly demonstrated in fungicide spraying trials by the clean cut-off at the boundary between sprayed and unsprayed plots.

(3) The primary gradient, defined as consisting of infections all of which come from the source under consideration (but not necessarily all on one spread occasion), is steeper than gradients in which secondary spread has occurred. Secondary spread flattens a primary gradient.

(4) Background contamination also flattens a primary gradient.

(5) The multiple infection transformation (p. 238) must be applied before computing b, when it seems appropriate.

(6) 'We shall find the real epidemic muddy and uncomfortable.' (Waggoner, 1962.)

Turning a blind eye to Waggoner's principle of the muddiness of epidemics, and assuming that dispersal is going on over an ideal and topographically uniform surface, a few simple geometrical concepts are essential for interpreting gradients by the regression method.

The shape of the source affects the gradient and its coefficient, b. (The strength of a source affects the amount of disease, altering the constant, a.

The concept of a point (or near point) source of propagules is essential as a basis for discussion. Other sources can be regarded as made up of a close assemblage of point sources, which are not additively effective because they receive (and therefore 'waste') some of each others disseminules.

Gradients from a point source can be measured in three principal ways, which yield different slopes. The amount of disease at each distance can be:

(1) *Averaged* per unit area in all directions from the point source at a particular distance, $= d_p$ (where d stands for deposition or disease); d_p is therefore the mean amount of disease per unit area on an annulus of unit width around the point source.

(2) *Summed* per annulus of unit width at a particular distance around the source, $= D_p$.

(3) *Measured* per unit area at points in a single direction (e.g. down wind of a point source), written d_{pw} (or at right angles to a line or strip source, written $d_{l,\,90}$).

The values for d_p and D_p are related, for $D_p = 2\pi.xd_p$. Also, deposition per unit area at a point down wind of a line source, $d_{l,\,90} = D_p$, assuming the emission per unit length of the line is equivalent to emission from one point source. The gradient coefficient, b, for $d_{l,\,90}$ will be the same as for

D_p, and 1 more than the value of b for d_p; b for Dp or $d_{l,\ 90} = b$ for d_p+1.

Simple geometry gives no information about the gradient down wind of a point source, d_w, but values can be drawn from spore dispersal tests.

Other things being equal, gradients from a point source are steeper than those from a line or area source. Close to a strip or area source the gradient is often flattened, and near a large area source the flattening may persist for a long distance. Clearly the geometry of the source must be considered in interpreting the gradient.

It is useful to select a few of the possible rates of decrease in the amount of infection with increasing distance from a point source, and work out the values of the regression coefficient, b (Figures 39 to 41).

Possibility 1. Suppose that the gradient approaches infinity: $b \to \infty$. This is the case, for example, in successful spraying experiments where there are no available infection sites (Principle 2) because the plants are protected by fungicide.

Possibility 2. Suppose that d_p varies inversely as the square of the distance from the source: $d_p = k_1.x^{-2}$. (This is interpreted as three-dimensional spread by Dimond & Horsfall, 1960.) The regression equation then becomes:

$$\log d_p = \log k_1 - 2 \log x, \text{ therefore } b = 2;$$

and for D_p or $d_{l,\ 90}$, the equation becomes:

$$\log D_p = \log d_{l,\ 90} = \log 2 + \log \pi + \log k_2 - \log x, \text{ therefore } b = -1.$$

Possibility 3. Suppose that d_p varies inversely as the first power of the distance (interpreted as two-dimensional spread by Dimond & Horsfall): $d_p = k_1.x^{-1}$. Then

$$\log d_p = \log k_1 - \log x, \text{ therefore } b = -1; \text{ and}$$
$$\log d_{l,\ 90} = \log D_p = \log 2 + \log \pi + \log k_2, \text{ therefore } b = 0.$$

Possibility 4. Suppose that d_p varies as in Sutton's eddy diffusion theory, taken over a short distance and without modification to allow for losses from the spore cloud by deposition:

$$d_p = k_1 - \tfrac{1}{2}(m+2), \text{ and if } m = 1.75, b = 1.875; \text{ and}$$
$$\log d_{l,\ 90} = \log D_p = \log 2 + \log \pi + \log k_2 - m/2(\log x),$$
therefore $b = m/2$, and if $m = 1.75, b = 0.875$.

Possibility 5. Suppose that d_p is invariant with distance from the source, then $d_p = k_1.x^{-0}$. This possibility is incredible in relation to an effective

K

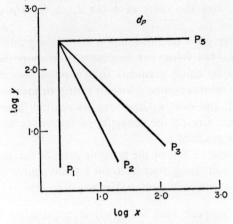

Figure 39. Some of many possible gradients showing effect on b (calculated for d_p). (P1, P2, P3 and P5 = possibilities 1, 2, 3 and 5.)

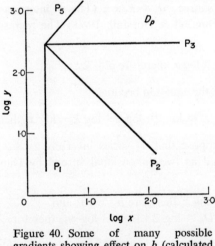

Figure 40. Some of many possible gradients showing effect on b (calculated for D_p). (P1, P2, P3 and P5 = possibilities 1, 2, 3 and 5.)

point source, as will be obvious when D_p is considered. Such an apparent relationship would be a clear indication that the supposed point source is nearly or quite inactive, and that the infections observed came from background contamination. The source is thus a 'pseudo-source', and for 'd_p', $b = \pm 0$, whereas for 'D_p', $b = +1$, because ever widening concentric circles take in more and more randomly-arriving background infections (Figure 41).

In practice a field may show a mixture of infections from a local source plus infections from numerous distant sources (background). Danger to a field from local and distant infections has been compared by Plank (1960), Waggoner (1962), Chamberlain (1967b), and Raynor *et al.* (1968).

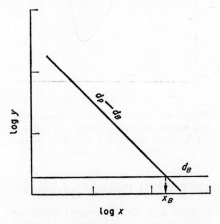

Figure 41. Changes in d_p with distance in source, contrasted with average background contamination, d_B. ('Background point' $= x_B$.)

In computing b, Figure 41 suggests that it is inadvisable to use points more remote from the source than the 'background point', x_B, defined as the distance where the gradient equals background. The calculation should be stopped at $y =$ twice background concentration, because including further points will meaninglessly flatten the gradient, with the amount of flattening depending on how many background points are included.

Possibility 6. The observed gradient may be U-shaped (Figure 42). A curve falling to a definite minimum and then rising steeply with increasing distance is usually evidence of approach to another localized source, as noted by Kerling (1949) with *Mycosphaerella pinodes*.

An inverted U-curve at the start may indicate 'skip-distance'. Where the source is elevated above the suscepts, a spore or insect cloud may have to travel some distance down wind before it has diffused to ground level at maximum concentration. This effect is well shown by Sreeramulu & Ramalingam (1961) who report readings nearest the source often smaller than the next farthest.

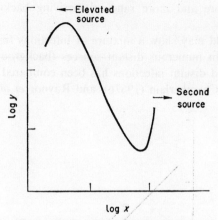

Figure 42. Inverted U-shape near elevated source (skip-distance effect), followed by U-shape indicating proximity of another infective source.

EXPERIMENT WITH *Puccinia polysora* IN NIGERIA

Most of the principles can be illustrated from one of the few well-documented studies of dispersal gradients in the field.

Cammack (1958a) in exemplary experiments giving a wealth of information, studied the pattern of infection by American corn rust *(Puccinia polysora)* on two successive crops of susceptible *Zea mais* at Ibadan, Nigeria. Each planting was about 1·6 hectares, isolated by 750 metres from other maize. Plants were spaced at ½ by 1 metre. A Hirst spore trap was operated at the centre with its orifice 3 metres above ground level, and near an anemometer and wind-direction vane. As a rust source, 16 mature plants infected with rust (covered in the glasshouse before being brought out) were transplanted to a central square and uncovered when rooted. Exposure started 45 days after planting the first receptor crop, and 56 days after the second. At each planting date the background infection was also assessed by two plots, each of 250 plants, 370 metres from the main experiment, and 250 metres up the prevailing wind of other maize plots.

Resulting infections were recorded in two ways: on concentric circles around the source, and on radii up and down the direction of the prevailing wind.

On concentric circles. All plants lying on the circumference of concentric circles at 5, 10, 20, and 40 metres radius from the infection centre, were tagged and examined for rust at 10, 20 and 30 days after uncovering the infectors. In Nigeria, rust uredosori regenerate in about 9 days, so the

10th day count must give the primary gradient. Numbers of infected plants were counted, and on each plant the total number of uredosori was counted on the 3rd, 5th, and 7th leaves above the second node (Figure 43).

On radii up and down wind. Mean uredosorus number on plants per 5-metre lengths of straight lines up wind and down wind were counted on the 10th and 30th days after uncovering the infectors (Figure 44).

The values for 'log mean pustule number' (log y), and for 'log distance from the source in metres' (log x) were read from Cammack's graphs and used to calculate the gradients in the two plantings at different dates (Table XXXII).

TABLE XXXII

DISPERSAL GRADIENTS OF *Puccinia polysora* ON MAIZE IN NIGERIA[a]

	1 d_p		2 d_{pw}		3 $d_{p.uw}$		4 D_p	
	b	S.E.	b	S.E.	b	S.E.	b	S.E.
1st planting								
Day 10	$-2 \cdot 618 \pm 0 \cdot 451$ (2)[b]		$-1 \cdot 861 \pm 0 \cdot 205$ (6)		$-2 \cdot 043 \pm 0 \cdot 342$ (2)		$-1 \cdot 583 \pm 0 \cdot 477$ (2)	
Day 20	$(-1 \cdot 060$	$0 \cdot 261)$ (2)	—		—		$(-0 \cdot 047$	$0 \cdot 258)$ (2)
Day 30	$(-0 \cdot 313$	$0 \cdot 009)$ (2)	$(-0 \cdot 077$	$0 \cdot 079)$ (6)	$(-0 \cdot 149$	$0 \cdot 066)$ (5)	$(+0 \cdot 742$	$0 \cdot 010)$ (2)
2nd planting								
Day 10	$-2 \cdot 019 \pm 0 \cdot 211$ (2)		$-1 \cdot 527 \pm 0 \cdot 198$ (6)		$-1 \cdot 932 \pm 0 \cdot 264$ (6)		$-0 \cdot 957 \pm 0 \cdot 236$ (2)	
Day 20	$(-0 \cdot 200$	$0 \cdot 056)$ (2)	—		—		$(+0 \cdot 861$	$0 \cdot 060)$ (2)
Day 30	$(-0 \cdot 043$	$0 \cdot 016)$ (2)	$(-0 \cdot 016$	$0 \cdot 017)$ (6)	$(-0 \cdot 157$	$0 \cdot 038)$ (6)	$(+1 \cdot 018$	$0 \cdot 035)$ (2)

[a] From Cammack (1958a)
[b] Residual degrees of freedom in parentheses after S.E.

GRADIENT DOWN WIND OF SOURCE, 10TH DAY, FIRST PLANTING (d_{pw})

By computing the linear regression for d_{pw}, we find $b = -1 \cdot 861 \pm 0 \cdot 205$. Alternatively a similar value can be obtained graphically: either plot log x and log y on linear graph paper (taking care that both axes have the same scale), or look up the antilogs and plot on log.log paper. Then draw the best fitting straight line through the points by eye and find the angle the line makes with the x-axis. I found the angle approximately 60°. From tables, tan 60° = 1·73 (this will be negative as there is an inverse relation between x and y). The linear regression coefficient is thus $b = -1 \cdot 73$, little different from the computed result.

At 10 days after first infection all lesions should be primary or derived from background sources. A measure of background (d_B) is given by the two distant control plots which had 3·2% plants naturally infected at 14 days after 'inoculation' date. Cammack noted that 'doublet analysis' (Plank,

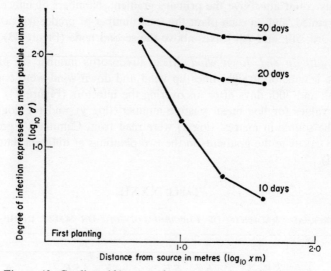

Figure 43. Gradient (d_p) assessed on concentric circles around point source of *Puccinia polysora* (first planting) at 10 days after exposure to inoculum, and showing secondary flattening after 20 and 30 days. (From Cammack (1958); reproduced by permission of the author.)

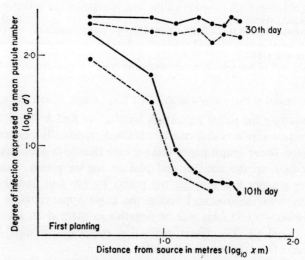

Figure 44. Gradient (d_p) assessed along straight lines up wind (broken lines), and down wind (continuous lines) of a point source of *Puccinia polysora*, and showing secondary flattening after 30 days. (From Cammack (1958a); reproduced by permission of the author.)

1946) gave strong evidence of secondary or neighbour spread in the control plots on the 20th day, but not as early as the 10th day. Spore trapping showed relatively few airborne uredospores over the plot at the date of exposure, when they averaged 20 per m^3 of air, but increased to nearly 2000 per m^3 30 days later. These facts are consistent with the idea that the gradient on the inoculated plot on the 10th day was primarily derived from the experimental source. The 3·2% of the plants infected by background tended to slightly mask the effect of the experimental source by flattening the gradient. Another clue to background infection is probably given by the flattening of the gradient at the greater distance in Figure 44 by the superimposition of a constant on the primary gradient. A small background contamination will not affect the gradient near the source, but will prevent the line from tailing off.

GRADIENT 'UP WIND' OF SOURCE, 10TH DAY, FIRST PLANTING ($d_{p.uw}$)

From Figure 44 we can also find that the gradient in the prevailing up-wind direction has a somewhat steeper slope ($b = -2·043$ by computer) than down wind (Table XXXII, column 3), and that at all distances the amount of disease is somewhat less than in the down-wind direction. Evidently the effect of the prevailing wind in all these experiments was small.

THIRTIETH-DAY GRADIENTS

At 30 days after infection the gradients both down and up wind are nearly flat, with b less than unity (Table XXXII, columns 2 and 3), but the amount of disease is greater at all points.

SECOND PLANTING EXPERIMENT

In accord with the local practice of taking two maize crops a year, a second planting was made during the peak period of airborne uredospores, when concentrations exceeded 300/m^3 from the time of inoculation. Both down wind ($b = -1·527$) and up wind ($b = -1·932$) gradients were flatter than on the first planting. The speculation that flattening was due to the large background contamination from the peak spore concentration is supported by the isolated control plots, which became 100% infected only 61 days after planting, compared with 79 days for the 1st planting. (For other gradients measured by Cammack in these experiments *see* Gregory (1968).

To summarize Cammack's results, the primary gradient in all directions around the point source for the 1st planting was: $b = -2·62$, a value considerably greater than that of -2, which would be required by an inverse square law. The 2nd planting had $b = -2·02$ on the 10th day. The background point can be estimated graphically at about 30 metres. This surprisingly short active range for the 16 heavily infected plants put out as a source is consistent with the steep gradient and the fairly heavy background contamination indicated by the control plots. As Cammack points out, later flattening was mainly by secondary infection on the first planting, but was dominated by background on the 2nd.

INTERPRETATIONS

Some general principles can be extracted from the fragmentary data on dispersal gradients on rusts, potato blight, *Rhizoctonia*, and potato leaf roll and virus Y, reviewed here and in Gregory (1968).

In general, values observed for b differed greatly, up to an occasional $b = -4$ for d_p, but values of $b = -1$, -2, or -3 are common. When coefficients near zero occur we may suspect secondary spread (which may be detectable statistically), background contamination, or proximity to a large area source.

Source geometry is confirmed as playing a large part in determining b, and for a line source the value of b for the corresponding point source is increased by $+1$. Line, strip and area sources give successively flatter curves (infection falling off less rapidly with increasing distance from the source) than point or near-point sources. Strip and area sources also produce more disease than point sources at all distances.

Effect of wind direction. Values for b down wind are nearly all smaller than for the mean of all wind directions, but the differences may be small, and will depend on the constancy of wind direction during emission. If wind direction fluctuates widely the value of b for d_{pw} will approach that for d_p.

Effect of terminal velocity on gradient. Stepanov's experiments have the merit that they give estimates of Q_0. In one test, Stepanov (1962) liberated simultaneously spores of three sizes (Table XXXIII) at 0·8 to 1·2 metres above ground level into a wind of 2·02 to 2·21 metres per second. Spores were trapped on the ground at 10, 20 and 40 metres down wind. As in Stepanov's 1935 experiments (*see* Chapter VI) gradients with larger and smaller spores were similar.

As expected from the sizes and terminal velocities of the different spores, the fraction of Q_0 deposited per unit area at any distance (10 metres is chosen as representative in Table XXXIII) was greatest with the largest spores *(Lycopodium)* and least with the smallest spores *(Bovista)*, and therefore the smaller spores will have the wider dispersal. But the gradients over the short distances studied are not in the expected order, and as refer-

TABLE XXXIII

DEPOSITION AND GRADIENTS WITH SPORES OF DIFFERENT SIZE[a]

Organism	Spore diameter (μ)	Terminal velocity (cm/s)	d_{pw}/Q_o at 10 m	b	S.E.	rdf
Lycopodium	32	1·76	$1\cdot20\times10^{-5}$	$-1\cdot700$	$\pm0\cdot273$	2
Tilletia	19	1·41	$0\cdot82\times10^{-5}$	$-2\cdot083$	$0\cdot256$	2
Bovista	4	0·24	$0\cdot63\times10^{-5}$	$-1\cdot853$	$0\cdot378$	2

[a] From Stepanov (1962)

ence to their standard errors shows, clearly did not differ from one another significantly.

A similar conclusion can be drawn from the experiments by Sreeramulu & Ramalingam (1961) who also estimated Q_o. The gradients from the heavier spores tend to be slightly steeper than those from the lighter, but here again the standard errors suggest that there is little real difference (Table XXXIV). But in contrast to the similarity of these gradients the fraction of total spores liberated that were deposited within the sampling area differed greatly for the two sizes: Podaxis had only about one-twentieth the recovery of Lycopodium.

TABLE XXXIV

GRADIENTS WITH Lycopodium AND Podaxis[a]

Experiment	Lycopodium clavatum			Podaxis pistillaris		
	b	S.E.	rdf	b	S.E.	rdf
7	$-1\cdot607$	$\pm0\cdot242$	5	$-1\cdot614$	$\pm0\cdot156$	5
8	$-1\cdot865$	$0\cdot141$	5	$-1\cdot696$	$0\cdot084$	5
9	$-1\cdot174$	$0\cdot342$	4	$-1\cdot145$	$0\cdot259$	4
10	$-1\cdot799$	$0\cdot198$	4	$-1\cdot522$	$0\cdot166$	4
11	$-1\cdot462$	$0\cdot175$	4	$-1\cdot437$	$0\cdot196$	4

[a] From Sreeramulu & Ramalingham (1961)

In spite of the similarity of gradients from spores of different sizes, we must not be misled into drawing conclusions about Q_x (the proportion of spores liberated remaining in suspension at distance x) from the extrapolation of b. Steep gradients are often interpreted to mean that the source disseminates only over a short distance. This inference may sometimes be justified, but the proportion of Lycopodium recovered within 30 metres in Sreeramulu and Ramalingam's tests suggests that it is possible for steep gradients to be associated with enhanced distant dispersal (compare Experiments 2 and 4 in Table XXXV).

TABLE XXXV

DISPERSAL GRADIENT AND RECOVERY OF *Lycopodium* WITHIN 30 METRES
OF SOURCE IN EXPERIMENTS BY SREERAMULU & RAMALINGAM (23)

Experiment	Recovery (%)	b	S.E.	rdf
2	11·4	−2·093	±0·027	4
3	12·1	−1·653	0·257	5
4	45·5	−1·609	0·109	4
5	10·4	−2·119	0·297	5
9	8·1	−1·961	0·374	3
12	9·5	−1·648	0·186	4
13	5·4	−1·369	0·177	4

For this reason I doubt whether a 'horizon' beyond which infection is negligible, can be deduced from gradients as suggested by Plank (1960). In general the proportion of the total effective emission remaining near the source is large, but this cannot be deduced from the gradient. Spore cloud and surface deposition are inconstantly related, like the height of a man and the length of his shadow. Only experiments where Q_0 (the total effective emission) is known can give grounds for drawing conclusions about spread beyond the immediate experimental area.

LOCATING A SOURCE

Tracing up a gradient can sometimes locate a source (Stakman & Hamilton, 1939; Snell, 1941; Bonde & Schultz, 1943; Gregory & Lacey, 1964). However, three types of anomalous gradient must be borne in mind. With an elevated source the gradient may lead the search in the wrong direction over the skip-distance (Figure 42). Topography or meteorological conditions may influence the diffusion of a particulate cloud (Zogg, 1949; van Arsdel, 1958). Field margins may show abnormal amounts of deposit, caused for example by filtration effects (Hirst & Stedman, 1971).

RELEVANCE TO SPRAYING TRIALS

'The study of disease gradients bears on the design of field experiments with epidemic plant diseases. Current designs for replicated and factorial trials have been developed largely for tests with varieties and fertilizers in which it is fair to assume that, provided narrow guard strips between plots are discarded, the treatments given to one plot does not affect its neighbours. With airborne organisms attacking the shoot system of plants this assumption is invalid. For instance, a small plot of potatoes sprayed with a copper fungicide for the control of blight might be exposed to many thousands of times the intensity of contamination that it would have experienced if surrounded only by treated plants' (Gregory, 1945). This can lead to what van der Plank (1963) refers to as 'the cryptic error in field experiments'. The error is tolerable in experiments designed to compare fungicides, but invalidates those designed to measure the effect of control in practice.

The sharp cut-off commonly observed at the boundary between sprayed and control plots ($b = -\infty$) has been interpreted to mean that the control plot is no danger to a neighbouring sprayed plot. The fallacy of this deduction has been elaborated by van der Plank (1963). The conclusion that should be drawn from this observation is that, for the time being, the spray has abolished one of the pre-requisites for an infection gradient – the presence of a population of susceptible host tissue (Principle 2). Later, if the disease eventually attacks sprayed plots and strong gradients are not observed, it may be again concluded that the controls are not quantitatively invalidating the experiment. This deduction may also be fallacious if: (1) new susceptible sites arise (by renewed growth, or erosion of protectant deposit) at a time when the combined source strength of the unsprayed plots is great enough to saturate randomly-arising sites at all distances across the sprayed plot; or (2) if the gradient near the control plot is flattened by the area source effect. When background is negligible, and unsprayed plots are narrow in comparison with sprayed plots, strong and typical gradients can arise on sprayed plots (Gregory, 1968, p. 204). An experiment can generate its own 'internal background'; only when contamination from distant sources outside the experimental area dominates the scene (i.e. background point reached in a few metres) can the effect of the control plots be neglected.

BACKGROUND AND ISOLATION

Using the log.log diagram, two parts to the gradient can often be distinguished: a steep initial decline (flattened at first if the source is a wide area), merging into a fluctuating horizontal line due to background infection. The gradient is the sum of the two lines. Background derives from the tails of the distributions of innumerable sources too far away to show a gradient on the field under consideration.

In isolation practice our problem is to find the distance (x_B) where the primary dispersal gradient cuts the background level (Figure 41), or if there is no background (perhaps an unusual case), to decide what is an acceptable level of infection from the source alone. The log.log diagram shows the background level more clearly than a diagram on a linear scale.

TOPOGRAPHICAL MODIFICATION OF GRADIENTS

Diffusion has been treated so far assuming nearly ideal conditions, but the literature contains information on the effects of topographical features which modify the gradient.

Pollen of wind-pollinated plants is distributed in typical gradients, as shown for example by Roemer (1932), Jensen & Bøgh (1942), Jones & Newell (1946), and Bateman (1947b). During strong winds in open fields, Jensen & Bøgh's catches of pollen of rye, ryegrass, cocksfoot, timothy, sugar-beet and mangold, on sticky microscope slides, showed a steady decrease with distance up to 1200 metres from the source field. But they found that hedgerows and plantations protected ryegrass, cocksfoot and mangolds, in proportion to the height of the obstacle. In tests near to the source a protection corresponding to an isolation distance of about 200 metres of open ground was obtained behind hedges, even so far down wind as 5 to 10 times their height.

Hirst & Stedman (1971) trapped sugar-beet pollen in a wheat field, and recorded a greater deposition in the 10–20 metres at the up-wind crop edge, where pollen entering the stand was rapidly filtered out on the plant surfaces, compared with parts of the field remote from the edge, where pollen concentration was restored by diffusion (boundary layer exchange?) from the wind above the top of the crop.

XII

Long-distance Dispersal

Dispersal of microbes over longer distances of a few kilometres and more, is an ever-present, world-wide phenomenon. Its experimental study is almost non-existent; but there is circumstantial and observational evidence of its magnitude, some of which was reviewed by J. J. Christensen (1942).

CONTROVERSY ON THE IMPORTANCE OF THE AIR SPORA

The discovery of air dispersal in any group of microbes has usually been followed by controversy between opposing specialists, some maximizing and some minimizing the significance of the phenomenon.

Views current on air hygiene affect the design and organization of hospitals, and traces of the controversy have appeared in Chapter XV.

In plant pathology the two schools of thought have flourished simultaneously. Butler (1917) held that spores of plant pathogens could be transported for short distances by air or rain-splash, but claimed that 'the distance to which spores may be carried in the air has often been exaggerated in the past, and is much less than might be expected'. The discontinuous spread of plant pathogens, he considered, was likely to occur on seeds, plants, and horticultural produce; but 'infection by spores carried through the air from remote centres is not a contingency which needs to be taken seriously into account'. Naumov (1934) held that human activity accounted for most long-distance transport of fungus pathogens and that, in the absence of host plants, fungi were dispersed with extreme slowness. *Endothia parasitica* in North America could not cross, in a period of 10 years, a 45–60 km wide tract that was free from chestnut trees; the spread of a disease of palm trees was estimated at only 4–5 km per annum.

The literature contains many instances of fungus pathogens failing to infect hosts at distances of a few metres (e.g. H. W. Long, 1914), or to colonize apparently favourable environments. Not all of these necessarily

263

represent the failure of transport: genetic differences in host population or the pre-establishment of competitors, may explain many puzzling failures.

By contrast, Stover (1962) considered wind transport feasible for asco-spores of banana leaf spot fungus, *Mycosphaerella musicola*, by high alti-tude winds from a large area source in Australia to Africa and the Caribbean direct. The possibility of Transocean spread is also raised tentatively by Hirst & Hurst (1967, p. 309) for tobacco blue mould, and by Bowden *et al.* (1971) for coffee rust.

Diverse views are also held about Bryophytes and Pteridophytes; how-ever, most authors accept the possibility of their dispersal over great distances, although the minimizing view has been held by some (*see* Pettersson, 1940, p. 22).

The maximizing view, if held too strongly, may lead to fatalism and to neglect of local hygiene and of reasonable precautions when moving plants from place to place. However, evidence presented in earlier chapters stresses the overriding importance of local sources. The minimizing view, if applied to certain diseases, may lead to over-reliance on local hygiene and neglect of protection by chemical and genetic measures. It may also increase the danger of introducing a vigorous organism by neglect of quarantine precautions.

Long-distance dispersal will be discussed under two main headings: (1) diffusion theories extrapolating from short-range experiments; and (2) observations on distant dispersal of inorganic and radioactive particles, or rust fungi, and of other microbes. The problem is complicated by the curvature of air-mass trajectories, by the possibility that spores may be re-concentrated within cumulus and other clouds, by unpredictable removal in precipitation, and by loss of viability.

THEORETICAL DISCUSSION

Presumably the tropopause limits the vertical expansion of the spore cloud, and an extensive temperature inversion at a lower level may have the same effect. After the spore cloud has travelled perhaps 20 km or less, diffusion will become two-dimensional, and concentration may be expected to decrease more slowly with increasing distance than it would when the spore cloud was nearer the source where its diffusion was three dimensional. In the limiting case of a land mass acting as a long strip or area source, the decrease of concentration at 20 km or more out to sea can be expected to depend only on depletion of Q_x (the fraction of the cloud remaining in suspension).

Sutton's theory appears satisfactorily to describe dispersal of microbes in air up to the limit of distance studied experimentally, but there is doubt about its use for distances greater than 1 km or heights above 30 metres. This is emphasized by Pasquill (1956, 1962b), who dispersed flourescent dusts and sampled, both on the ground and by aircraft, at distances of up to 64 km and at heights of up to 1220 metres. In these tests the height of the cloud was much less than the width: the width did not increase uniformly with distance and the angle subtended at the release point by the cloud at the greatest distance was only about half that subtended by the cloud at 1 or 2 km. At distances greater than those considered in Chapter XVI, the difference between turbulence in a vertical horizontal direction evidently becomes important (Chamberlain, 1966a). Further diffusion is limited by the tropopause, if not by the top of the turbulent layers of the atmosphere.

The calculation for Q_x referred to in Chapter XVI, can be used to define the 'escape fraction', the fraction of source output available for long-distance dispersal. This is illustrated in Figure 45, based on Q_x from Figure 33. The curve for Q_x can be regarded as consisting of three parts: (1) The initial skip distance over which $Q_x = Q_0$ (the length of this part depends mainly on height of liberation); (2) the main deposition region where there is a linear relation between $\log x$ and $\log Q_x$; and (3) a tail of reduced deposition where part of the cloud has sufficient altitude to decrease its chance of encounter with the ground surface; this represents the 'escape fraction' and its magnitude can be estimated by finding the value of Q_x at the distance where extrapolation of the linear part of the curve would have led to $Q_x = 0$.

The concept of 'escape fraction' illustrated here may resolve the dilemma posed by the apparent incompatability of heavy deposition near the source with transport to great distances. The magnitude of the escape fraction increases with smaller particles, greater liberation height, and increased turbulence.

OBSERVATIONS

THE WINDSCALE ACCIDENT

Samples taken after the accidental emission of about 20 000 curies of [131]iodine from a stack 122 metres high at Windscale atomic pile on 10 October 1957, provided detailed records of ground contamination by the main plume up to 290 km distance (Booker, 1958). Plotted on a log.log scale, the radioactivity deposition curves, whether measured on herbage or in milk, are relatively flat up to 15 km – a fact consonant with the height of emission. From about 28 km onwards the slope is similar to that of d_w for $m = 1.75$, but if the Sutton formulae are used over long distances it must be remembered that, as in the Windscale accident, the

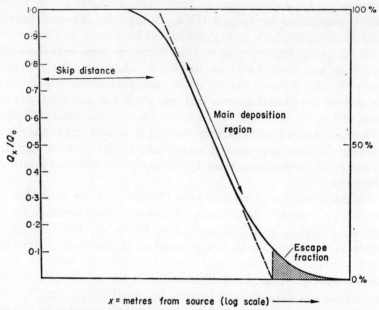

Figure 45. Diagram (based on Figure 33) illustrating skip distance, main deposition region, and escape fraction.

trajectory of the cloud is not likely to be in a straight line over the Earth's surface.

RE-COLONIZATION OF KRAKATOA

The volcanic island of Krakatoa lies between Java and Sumatra, and is almost encircled by land. The eruption of August 1883 blew away the mountainous two-thirds of the island, leaving a hole in the sea-bed 300 metres deep, and covered the remainder of the island with lava and ashes. A column of fine dust rose to a height estimated at 27 km, and was carried westwards by the prevailing wind. Eventually this dust circled the earth repeatedly, spreading over the whole tropical and temperate zones (Symons, 1888). Although this world circulation of the dust is relevant to long-distance dispersal of microbes, the story of re-colonization of the island after the destruction of living things, which is usually considered to have been complete, is of even greater significance.

Three years after the eruption, the only flowering plants found by Treub (1888) were two species of Compositae and two grasses. There were also eleven species of ferns, and the volcanic dust was colonized by a film of blue-green algae (Cyanophyceae). All these were thought to have been carried in by wind, the nearest land being 40 km away.

Records of subsequent visits are summarized by Ridley (1930) who concluded that of the 144 species of flowering plants then reported, 24% were wind distributed, 42% were seaborne on floating tree trunks, and most of the remainder had probably been carried by birds. In addition there were forty-eight species of Pteridophytes and nineteen of Bryophytes, all potentially windborne. Boedijn (1940, *and see* Leeuwen, 1936), who believed that most of the fungi present had been carried to Krakatoa by wind, was impressed by the paucity of lichens, of which he found only 13 species (0·1% of the world's list), compared with sixty-one species of Pteridophytes and 263 of flowering plants. Moreover, none of these lichens inhabited rocks: all were epiphytes which probably arrived on driftwood. Evidently lichens are poorly equipped for wind transport in comparison with Myxomycetes, which were represented by twenty-eight species (7% of the world's list). Only two rusts were recorded (0·02%), but these obligate parasites must needs wait for their flowering-plant hosts.

EXOGENOUS POLLEN AND SPORES

Wind-dispersed pollen, being very durable and well-studied morphologically, is of great interest as marking long-distance transport. Death in transit may nevertheless decrease its biological significance.

Bessey (1883) found quantities of pine pollen on prairie land 600 km or so from a probable source. In steppe country in Voronezh Oblast in 1949 thunder rain precipitated a massive deposit of *Pinus sylvestris* pollen which must have travelled at least 10 to 12 km (Meshkov, 1950). In the month of June, Dyakowska (1948) found pine and spruce pollen falling on the coasts of Greenland, 600 to 1000 km from the nearest pine or spruce trees. Still further north, in Spitsbergen, Polunin (1955a) noted a deposition of pine and spruce pollen which must have been equivalent to about 200 grains per square metre per day in July and August. Records of long distance transport by wind are compiled by Potter & Rowley (1960).

Mosses are excellent accumulators of contemporary pollen. In moss samples from Spitsbergen, Środoń (1960) found a substantial proportion of *Pinus sylvestris* pollen which must have come from Scandinavian forests 750 km to the south. Sporadic pollens of other genera came from 1500 km south. Środoń also cites pollen analysis of arctic surface samples by Kuprianova [1951] who found tree pollen from 2100 km distance. Ritchie & Lichti-Federovich (1967) studied exotic pollen in arctic Canada. At Resolute, 74° N, 60% of the pollen must have been blown at least 1500 km.

Hesselman (1919) traced the transport of tree pollens from Scandinavian forests across the Gulf of Bothnia. Pollen was trapped in open petri dishes on two lightships, *Västra Banken* and *Finngrundet*, situated respectively 30 and 55 km from land, from 16 May to 25 June 1918 (Table XXXVI).

TABLE XXXVI

POLLEN TRAPPED ON LIGHTSHIPS IN GULF OF BOTHNIA, 16 MAY TO 25 JUNE,
1918
(Hesselman, 1919)

*Number of pollen grains per cm^2 of
petri dish, for all 34 days when both
stations reporting*

	Västra Banken 30 km	Finngrundet 55 km
Spruce (*Picea*)	271	141
Pine (*Pinus*)	222	64
Birch (*Betula*)	68	36

TRISTAN DA CUNHA ISLANDS

The long-distance record is probably the discovery of 33 pollen grains of
Nothofagus and 25 of *Ephedra* in peat from the Tristan da Cunha islands,
4500 km from the nearest source in South America (Hafsten, 1960). Maher
(1964) confirmed that *Ephedra* pollen can be dispersed upwards of 1000 km.

Such pollens may have been dead on arrival. But it appears that some
cryptogamic spores survive the journey. Wace & Dickson (1965; *and see*
Wace, 1961) consider that the high proportion of species of ferns, mosses
and liverworts in the Tristan flora can only be accounted for by admitting
the efficiency of long-range dispersal of their spores, mainly from South
America and the sub-Antarctic.

BRYOPHYTE SPORE DISPERSAL: THE MOSS *Aloina*

An instance of long-distance transport of moss spores is reported by
Pettersson (1940), whose investigations of plant spores in rain water were
described in Chapter XIII. At Tvärminne, Finland, during 22–23 July,
1936 (with persistent rains and light, mainly easterly winds), 104 cm^3 of
rain water were collected during 15 hours. This sample proved particularly
rich in the spores of Bryophytes, yielding 300 plants of *Marchantia polymorpha* and *Metzgeria* sp. Most remarkable, however, was the occurrence
of spores which yielded 278 plants of *Aloina brevirostris* and 2 which were
identified as *A. rigida*. These are small, annual or biennial mosses of dry
calcareous soil, and belong to a genus hitherto unrecorded in Finland.
There are a few records of *A. brevirostris* in Eastern Europe, but their
presence in such quantities in rain over Finland suggested that the spores
must have come from a rich and extensive source area. This, Pettersson
suggested, must lie in Siberia, in the region of the River Yenisei – a distance
of at least 2000 km east of Tvärminne. It is true that this conclusion has
been questioned by Persson (1944), who thinks that they must have come

from a nearer source such as European Russia or possibly Sweden; and by Bergeron (1944), who examined air-mass trajectories for the day in question and reached a similar conclusion. Yet Pettersson found that *Aloina* spores were being deposited in rain to the number of 10 000 per square metre and, although their origin is unknown, they must clearly have travelled a very long distance.

MAIZE RUST *(Puccinia sorghi)* IN SWITZERLAND

While ample qualitative evidence exists of mass transport of microbes by wind over long distances, there are few quantitative studies. The work of Zogg (1949) on one of the maize rusts in the Upper Rhine Valley is of exceptional interest, both for the distance studied and for the complex topography of the area.

In the level area where the Rhine flows into Lake Constance, *Puccinia sorghi* produces its aecidial stage on the wild *Oxalis stricta* in spring. In early summer, nearby maize becomes infected and a uredospore focus is established from which the fungus is spread by wind up the narrow Rhine Valley. At various dates in the 1945 and 1947 seasons, Zogg measured the incidence of infection for 66 km (up as far as Chur). Three striking facts about the gradients observed are: (1) the general decrease in the number of uredosori per plant with increasing distance from the source; (2) the flattening of the gradient as a result of secondary spread later in the season; and (3) the great irregularity of the gradient, because the incidence of the rust decreased locally wherever the valley widens, and increased again where it narrows: a feature which Zogg attributes, no doubt rightly, to the nozzle effect of the valley increasing spore concentration, though other ecoclimatic factors may play a part. In the terminology adopted in this book, both area dose *(A.D.)* and efficiency of deposition *(E)* would be increased where the valley narrows.

The diffusion theory developed in Chapter XVI referred to dispersal over a level plain, and the modifications imposed on a spore cloud by dispersal up an alpine valley are particularly interesting. In comparison with the distance travelled, the focus of infection near Lake Constance would appear as a point source, though some flattening of the gradient near the source would be expected. As the winds blow characteristically either up or down the valley, it seems appropriate to consider d_w (deposition down wind of source). Plotting Zogg's data on a log.log scale we find the linear regression (calculated as $\log y = 5 \cdot 356 - 1 \cdot 818 \log x$) is compatible for the slope of d_w predicted (Chapter XVI). We take this to indicate that in the valley *diffusion* by atmospheric turbulence proceeds much as elsewhere, but that *deposition* is subject to pronounced fluctuations associated with valley width.

CEREAL RUSTS

In the early summer of 1938, Stakman & Hamilton (1939) studied the deposition of uredospores of stem rust of wheat *(Puccinia graminis tritici)* at various points to the north of ripening fields of winter wheat in the southern United States, which was then acting as a vast area source of uredospores. At this time of year it could be assumed that spores were not being produced locally on the spring-wheat area in the northern United States, but that infection would follow arrival of the spore cloud borne on southerly winds. Table XXXVII, compiled from Stakman & Hamilton's data, indicates the amount of deposition, first in the source area, and then at various points farther north.

TABLE XXXVII

STAKMAN & HAMILTON'S (1939) DATA FOR LONG-DISTANCE DISSEMINATION OF *Puccinia graminis* (DEPOSITION ON GROUND, 24–25 MAY, 1938)

Place	Approximate distance from source	Uredospores per 930 cm^2 (sq ft) per 48 hours
Dallas, Texas	(source area)	129 216
Oklahoma	300 km	6 288
Falls City, Nebraska	560 km	7 680
Beatrice, Nebraska	840 km	1 968
Madison, Wisconsin	970 km	192

Another series, taken during 13–14 June, gave the following numbers of uredospores deposited per 930 cm^2 in 48 hours: Kansas 336 000, Nebraska 54 336, Iowa 21 360, South Dakota 12 624, Minnesota 32 256, North Dakota 1344.

Such long-distance transport of cereal rusts is not merely an occasional risk. On the contrary, it is clear from work done over a vast area that an annual double transcontinental migration through the atmosphere is an essential condition for the development of the rust epidemics which regularly attack cereal crops in North America. Moreover, wind dispersal is relatively unselective, and what happens to rust fungi no doubt happens also to countless other organisms whose spores travel on a global scale.

It has been possible to demonstrate this phenomenon for cereal rusts because of the concerted study of a disease of a major food crop by a generation of scientists in plant pathology laboratories scattered over North America, and also because of the strange life-cycle of these rusts, which makes it possible to obtain clear evidence. The evidence derived from spore trapping, field records of outbreaks, geographical distribution of the physiological races, and meterological data is summarized by Craigie

(1945), and Stakman & Harrar (1957, pp. 221–32). To simplify a complex story, *Puccinia graminis* and *P. recondita (= P. triticina)*, for various reasons, do not survive the cold winters of the northern part of the continent or the hot and dry summers of the southern parts. Spring-sown wheat in the northern United States and Canada receives rust-spore showers annually from rusted autumn-sown wheat in Mexico and Texas. In some years it comes by a succession of short jumps, with intervening stops where inoculum multiplies locally, whereas in other years infection spreads suddenly from the south – for distances of a thousand or more kilometres when atmospheric pressure distribution produces suitable winds. Similarly, winter wheat in the south becomes infected during autumn by spore showers from the north. Large-scale movement east and west across the North American Continent is relatively infrequent. Yellow rust of wheat *(Puccinia glumarum = P. striiformis)* seems to be spread much more locally than the other rusts of wheat, perhaps because its uredospores are more easily killed by exposure.

A somewhat similar annual flow of airborne cereal rust spores has been demonstrated in other parts of the world. For example, India has an annual flow from the hills to the plains: Mehta (1940, 1952) found that rust failed to survive the long hot summers of the Indian plains, yet rusts on wheat and barley caused heavy annual loss there. The explanation lies in the over-summering of rusts on cereal crops and self-sown plants at 2000 metres or more in the hills, whence inoculum is carried by winds to start infection foci here and there on the plains in early winter. Upper-air currents and katabatic winds both play a part in the dissemination. The early dissemination of rust from inoculum coming from central Nepal and from the Nilgiris of Palni Hills to the Indo-Gangetic plain, is the cause of annual devastating outbreaks.

Russian work (summarized by Chester, 1946) suggested transport of rust uredospores over hundreds of kilometres from the west across the Sea of Azov, and also from Manchuria up the Amur Valley in Eastern Siberia. On the other hand, the Irkutsk wheat-growing area west of Lake Baikal seems for practical purposes to be isolated from the rest of the world by deserts, mountains and tundra. *Puccinia graminis* does not occur in the Irkutsk region, and *P. recondita* survives because it can form aecidia on *Isopyrum fumarioides*, a common weed of arable land which serves as an alternate host. The wheat-growing areas of Australia and the Argentine also seem to be autonomous to the extent that airborne spores from outside do not affect either the annual rust epidemic cycle or the population of rust races present.

In Europe similar studies have been fragmentary, but migration of *Puccinia graminis* from south to north was established by Ogilvie & Thorpe

(1961), and confirmed with a wealth of meteorological data by Hirst *et al.* (1967a), who measured concentrations of hundreds of uredospores per cubic metre at an altitude of 3000 metres during migration. Zadoks (1967) indicates two migration tracks: one in western Europe spreading from North Africa, the Iberian Peninsula and southern France; and another originating in the Balkans and extending to Russia and Scandinavia.

For the cereal rusts we thus have a picture of free interchange of uredospores over long distances, sometimes in the form of an annual immigration, or even a return trip each year; but isolation of several thousand kilometres limits this dispersal process. Whereas no other kind of microbe has been so fully traced in its atmospheric dispersal as have the cereal rusts, circumstantial evidence points to long distance dispersal of many crop pathogens. For example, Stephens & Woolman (1922) found 200 to 500 spores of wheat bunt *(Tilletia caries)* adhering to leaves of cottonwood 100 miles distant from the nearest possible source in Oregon. Rust uredospores are relatively large, and we would expect many smaller-spored organisms to be at least as well equipped for long-distance dispersal, e.g. the coloured spores of the agarics and Myxomycetes.

ANIMAL VIRUSES

Fowl pest. Evidence for wind dispersal of some viruses causing disease in animals is summarized in Chapter XV: that such wind transport can be effective over long distances is now becoming apparent. C. V. Smith (1964) concluded that windborne dispersal of infective material accounted for 85% of fowl pest outbreaks in eastern England in the springs of 1950 and 1952, with a range of over 50 km.

Foot-and-mouth disease virus, which is not endemic in Britain, was assumed to spread only between pigs, sheep or cattle in close contact, or by fomites. This assumption was shaken when the Committee of Inquiry on Foot-and-Mouth Disease 1968 Report (1969) mapped secondary outbreaks in a fan-shaped pattern down wind of a primary, and provided evidence of wind transport of the virus to a distance of nearly 50 km from the source.

Hurst (1968) restrospectively examined meteorological records during outbreaks in Britain over the past 30 years. Some *isolated* outbreaks could not be related to winds from Continental Europe. But many epidemics started as *multiple simultaneous* outbreaks in eastern areas, associated with winds from the Continent, pointing strongly to windborne inoculum. Moreover, as shown by L. P. Smith & Hugh-Jones (1969), spread of the disease by wind is associated particularly with night rain. Rain might act by precipitating windborne virus on fodder (*see* Chamberlain, 1970), but if infection is mainly by the respiratory route (*see* Norris & Harper, 1970), then it is possible that rain splash may act as a launching mechanism (Gregory, 1971).

SEA CROSSINGS

The McGill workers showed conclusively that the upper air contains an appreciable spore load in all the places which they examined (Chapter XII). Even in the high-arctic winter, the evidence proves that samples of a few cubic metres of air may or may not be sterile. *Cladosporium* appears to dominate the upper-air spora, often together with many yeasts and bacteria. Possibly the purest air is to be found near sea level in mid-ocean. The few observations suggest that samples taken on board ship are collected in a purified layer of the atmosphere, and that higher up over the ocean surface the spore concentration is greater than at sea level. The troposphere is always more or less contaminated with microbes. From Erdtman's (1937) results a ship in the North Atlantic in spring would be in a region of about one pollen grain per 100 cubic metres, whereas Pady & Kapica (1955), at 3000 metres over the same ocean, recorded up to 25 pollen grains (with moss spores) per cubic metre.

On a flight from New Zealand to Australia, at about 1000 metres above the Tasman Sea, Newman (1948) exposed sticky slides (behind a leading wire in the hope of improving trapping efficiency by breaking the stagnant layer). At a position 1100 km off Australia, he estimated pollen grains at 0·73, and fungus spores at 0·70 per cubic metre; at 340 km from Australia he found pollen to be 8·75 and fungus spores 16·8 per cubic metre, which is about 100 times as numerous as the concentration of pollen grains and fern spores at ship's mast level on the North Atlantic crossing recorded by Erdtman (cf. p. 165). Newman's values for fungus spores are somewhat similar to those of the McGill University workers for the upper air over the Atlantic. It seems likely that Erdtman's samples, taken on board ship, were from the zone of surface air which had been largely cleaned by rain-wash, sedimentation, and contact with the ocean, and only partially replenished from the stock in the air mass overhead, where aircraft samples were taken.

Two series of tests have traced the diffusion of microbial clouds at various distances from land over the ocean. With three aircraft flying at 152, 1066 and 1981 metres altitude respectively, Fulton (1966b) sampled microbial populations of air masses during flights seaward from Houston, Texas, for a distance of up to 640 km into the Gulf of Mexico. Gelatine foam filters were exposed in isokinetic samplers (p. 194): after exposure, the filters were dissolved in saline, and microbial numbers determined by cultivating aliquots of the suspension on various media.

At the lowest altitude, as expected, microbial concentrations decreased steadily the further from land. At higher altitudes concentrations varied irregularly, and failed to decrease steadily with increasing distance from land. No satisfactory explanation was provided for this irregular spatial distribution at the time. However, it might be explicable if the full records could be re-examined in light of a further series of tests by Hirst and his associates.

Hirst & Hurst (1967) also sampled from aircraft over the North Sea. By contrast with Fulton's technique, which reveals microbes cultivable on the media employed, Hirst used a microscopic method for counting fungi and pollens, living and dead alike, after impaction on isokinetic suction impactors. One flight took samples between Yorkshire and the Skagerrak, around mid-day with the wind between south and west in July 1964. The flight first encountered the expected daytime *Cladosporium* and pollen cloud coming off the land; this, again as expected, decreased in concentration soon after leaving the coast. Unexpectedly, concentrations increased again to a maximum at 400 to 500 km from the English coast. In between, at 100 to 200 km, and again further still from land at 500 to 600 km, they found maxima of the moist air spore types (Chapter X), such as *Sporobolomyces* and the hyaline ascospores, which had almost certainly been liberated into the air at night.

Evidently the aircraft, starting in that day's spore cloud, flew next into the previous night's cloud, then on into the previous day's cloud, and finally, near the Danish coast, into that of the preceding night (Figure 46).

If washed by rain, this supra-marine cloud would deposit an estimated 1000 spores and pollen grains per square centimetre of sea surface, a quantity of the same order as the summer concentration over land referred to in Chapter X.

EXTENSION OF GEOGRAPHICAL RANGE

Although there is no theoretical limit to the dispersal of windborne spores, the geographical distribution of some plant pathogens is narrower than their host crops (*see* Holliday, 1971). Often spread is restricted by physical barriers mentioned earlier, but when a virulent windborne pathogen is introduced into an area previously uninfested, the rapidity of its spread may be as spectacular as its social consequences are far reaching, but here we enter the realm of history. Both aspects were exemplified when *Phytophthora infestans* reached the European potato crop in the 1840's (*see* Bourke, 1964; Woodham-Smith, 1962). The process continues as some modern examples show.

Puccinia polysora is probably indigenous on maize in the New World, because American varieties sustain little damage, as though a balance had

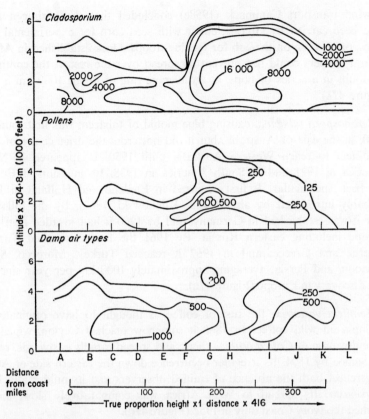

Figure 46. Isospore diagrams of a vertical section over the North Sea between Yorkshire and Skagerrak on 16 July 1964, showing concentrations of *Cladosporium*, pollen and 'damp air types' related to height and distance from the English coast. (Reproduced by permission of J. M. Hirst.)

been established by selection. In 1949 the fungus suddenly appeared in Sierra Leone, where African varieties of maize were highly susceptible and the attack was devastating. Once established, it spread rapidly, evidently by wind, reaching all other parts of West Africa by 1951, Congo and East Africa by 1952, Rhodesia and Madagascar by 1953, and the Islands of the Indian Ocean by 1955. While this was happening another focus was developing, starting in Malaya in 1948 and spreading to Siam and the Philippines, reaching Queensland in 1959 and Fiji in 1961 (van Eijnatten, 1965; Cammack, 1958b). Evidently until about 1949 the fungus in America remained isolated from vast areas of highly susceptible maize in Africa by the 5600 km of the Atlantic Ocean which formed an impassable barrier to wind transport Cammack (1958c) concluded that the pathogen must have been carried by aircraft, either with seed corn for experimental purposes, or on corn-on-the-cob for feeding troops. Once established in Africa, natural barriers could not prevent its spread over the rest of the continent, no doubt in a series of hops, at an average rate of about 1000 km a year (Figure 47A).

Peronospora tabacina, causing blue mould of tobacco, was first found in 1890, in the east of Australia, but it did not cross the drier centre of that Continent to reach Western Australia until 1950. It appeared in North America in 1921 and in South America in 1938. Its invasion of Europe has been spectacular. It first occurred in England and Holland in 1958, probably introduced by aircraft (*see* Klinkowski, 1962). By the following year it was established in Germany, and by 1960 it had invaded nearly all Europe including eastern Russia. By 1961 the fungus reached Tunisia, Algeria, and Greece, and in 1962 it reached Turkey, Morocco, Syria, Lebanon, and Persia, averaging approximately 1000 km per year since its first discovery in Europe (Figure 47B).

Hemileia vastatrix, the rust of coffee, is thought to have originated in Ethiopia on wild coffee. In 1868 it somehow reached Ceylon, wiped out the cultivation of *Coffea arabica* there, and spread rapidly eastwards, reaching Samoa by 1894. Its irregular occurrence down the eastern side of Africa may reflect both the absence of trained observers and its milder attack on *C. robusta*. Its spread to West Africa was comparatively slow, and it reached the Ivory Coast only in 1952 (Figure 47C).

The New World coffee-growing areas remained free until it was found in 1970 attacking a wide area in the State of Bahia, Brazil, and since then its range has been increasing at a pace typical of other windborne pathogens. The disease is generally believed to have been introduced into Brazil by human agency. Like maize rust, coffee rust seems to have been unable to cross the doldrums of the Atlantic near the equator in 20 years since it reached the West Coast of Africa. However it may be significant that the disease was recorded in Bahia only four years after its first record in Angola, from where, as pointed out by Bowden *et al.* (1971), it could easily be injected by convection into the southeast trade winds to make the crossing in 4 or 5 days. The extent of wind transfer of coffee rust spores is controversial, but it may prove to be a pathogen whose control has been impeded by adoption of the minimizing view of wind dispersal (Figure 47C).

Wind dispersal appears to spread a pathogen for hundreds of kilometres per annum, until some climatic or physical barrier is reached. Then

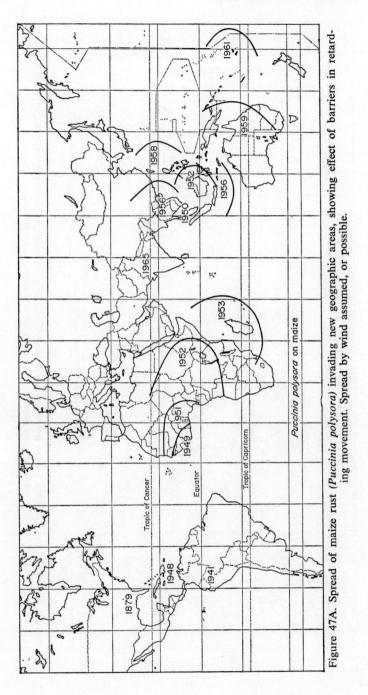

Figure 47A. Spread of maize rust (*Puccinia polysora*) invading new geographic areas, showing effect of barriers in retarding movement. Spread by wind assumed, or possible.

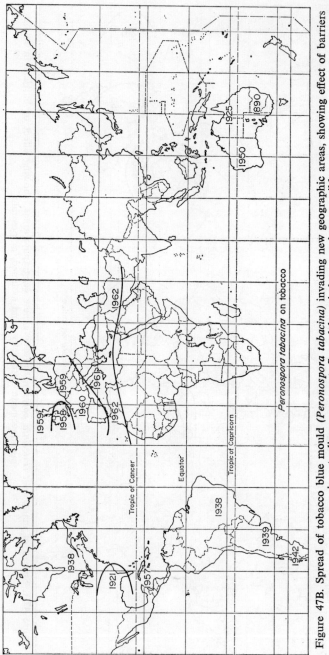

Figure 47B. Spread of tobacco blue mould (*Peronospora tabacina*) invading new geographic areas, showing effect of barriers in retarding movement. Spread by wind assumed, or possible.

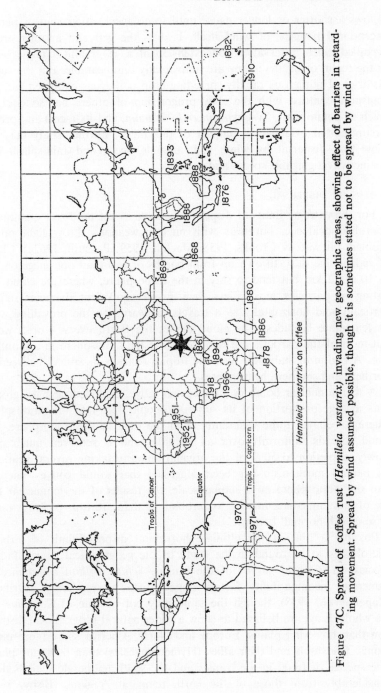

Figure 47C. Spread of coffee rust (*Hemileia vastatrix*) invading new geographic areas, showing effect of barriers in retarding movement. Spread by wind assumed possible, though it is sometimes stated not to be spread by wind.

follows a shorter or longer pause until some agent, often man or perhaps exceptional patterns of wind itself, bridges the gap and another area is occupied. Such observations on extension of a geographical range belong to the realm of history: they are individually unrepeatable. But the similar histories of numerous rapidly-spreading crop pathogens equipped for wind transport contrast with the slow progression of others not so equipped (such as potato wart, *Synchytrium endobioticum*, and its races) and provide circumstantial evidence in support of this thesis. Further study may well show that dispersal by wind, soil, animals, seeds and rain-splash have distinctly recognizable speeds and patterns.

LIMITS OF DISPERSAL

For the extreme limits of dispersal we have as yet no direct quantitative microbial evidence, but tests with nuclear weapons throw light on the problem (Stewart & Crooks, 1958; Machta, 1959; Pasquill, 1962a). Clouds of radioactive dust, thrown up from bombs of less than one megaton, rise to 10 to 12 km but tend to stay in the troposphere, where the cloud consisting of particles mostly less than 1 μm equivalent diameter diffuses vertically and horizontally as it travels eastwards on the prevailing winds of temperate latitudes. The cloud circles the Earth every 4 or 5 weeks, but most of the particles are removed from the atmosphere in a month or two. A region of stable air over the tropics restricts its spread between the northern and southern hemispheres.

A thermonuclear bomb in the megaton range, exploded near the ground, puts a large proportion of its sub-microscopic dust into the stratosphere, where it behaves quite differently from the dust in the troposphere. The cloud spreads uniformly over all latitudes and is removed much more slowly, depletion to 50% of the original quantity taking several years. As no natural mechanism has been suggested that would convey microbes into the stratosphere on a similar scale, the results of thermonuclear tests are of less immediate interest to aerobiology than are tests of lower power or so-called 'normal' bombs.

Possibly a hint at the ultimate horizontal dispersal-limit of airborne microbes is to be found in the study of the geographical distribution of species. In a world survey of the fungus, *Schizophyllum commune*, the numerous incompatibility factors appeared to be randomly distributed (Raper *et al.*, 1958), though the same may not be true of *Coprinus*. On the whole, fungi are believed to show a wider natural geographical distribution than flowering plants. Europe and North America share more mushrooms, toadstools and their allies (Hymenomycetes) than flowering plants. The species of fungi in the tropics and the south temperate regions differ considerably from those of the north temperate regions (Bisby, 1943).

Many saprophytic species of fungi, especially soil moulds, tend to be cosmopolitan. How far wind has operated in their transport, and how far other means, especially man, are responsible for the world distribution of these organisms, is unknown; but the limits of the rust fungi give us a useful yardstick, showing where dispersal is limited by survival.

XVIII

Aerobiology

Aerobiology seemed at first to fall naturally into two parts: extramural aerobiology which is mainly the concern of plant pathologists; and intramural aerobiology which concerned hygienists (Moulton, 1942). Today the dividing line could be drawn somewhat differently, to separate the aerobiology of the hardy organisms, well adapted naturally for wind transport, from that of tender organisms which become airborne only incidentally in nature, or deliberately in experiments.

The literature of aerobiology is scattered, but much of it can be traced through the bibliography on pp. 307 to 344 of this book. Much literature is also summarized in the various conference reports:

Aerobiology Symposium, American Association for the Advancement of Science, Chicago, 1942 (Edited: Moulton, 1942).

Conference on Airborne Infection, Miami Beach, Florida, 1960 (Edited: McDermott, 1961).

First International Conference on Aerobiology, Berkeley, California (Naval Biological Laboratory, 1963).

Atmospheric Biology Conference, Minneapolis, 1964 (Edited: Tsuchiya & Brown, 1965).

Second International Conference on Aerobiology (Airborne Infection), Chicago, 1966 (Edited: Lepper & Wolfe, 1966).

Third International Symposium on Aerobiology, University of Sussex, England, 1969 (Edited: Silver, 1970).

Extensive reference lists are given in the following works: Cunningham (1873), Heald (1913), Gardner (1918), Sernander (1927), Stepanov (1935), Rempe (1937), Pettersson (1940), Craigie (1940), duBuy *et al.* (1945), Gregory (1945), Stakman & Christensen (1946), Wolfenbarger (1946, 1959), Jacobs (1951), Maunsell (1954a), Wells (1955), Werff (1958), Hirst (1959), Dimmick & Akers (1969).

It now remains to review the conclusions of each of the preceding chapters in the light of the whole, to consider their implications, and to attempt a synthesis, based mainly with work on the 'hardy' microbes because they are easier to study.

THE PHENOMENA

TAKE-OFF

Airborne microbes, whether carried singly, in groups, or on 'rafts', are heavier than air. In still air they fall by gravity with constant terminal velocities ranging from 0·01 to 1500 mm/s according to their size and density. This falling would lead to their sedimentation out of the air if other forces did not oppose gravity. Two atmospheric processes tend to prevent sedimentation through the quiet boundary layer: turbulent diffusion in the wind carries the spore cloud horizontally, at the same time diffusing it both horizontally and vertically; and thermal convection can carry a spore cloud to great heights in the atmosphere. Modes by which spores cross the boundary layer of air near the Earth's surface and reach the turbulent wind layer, are therefore of prime importance in the dispersal of microbes. Energy for take-off may be supplied by the organism itself or may come from external sources supported by a wide variety of mechanisms, such as wind or rain-splash. Factors controlling the take-off mechanism also control the occurrence of spores in the air. The more an organism is specialized towards one dispersal route, the more unfit it becomes for dispersal by another route. For many purposes, knowledge of the take-off mechanism is important. The regular occurrence of a microbe in the air spora implies an efficient launching mechanism. Understanding this mechanism may be important in control.

Rain-splash dispersal results in a local scatter, because the larger splash droplets, which are less easily carried by the wind, pick up more spores than smaller droplets.

DIFFUSION

Frictional turbulence will diffuse the spore cloud to the top of the dust horizon, convection will take it higher. The concentration of spore clouds approximates to the expected logarithmic decrease with height, but equili-

L

brium is never reached in the atmosphere. Often there is a zone of increased concentration at an altitude of two or three thousand metres.

Dilution of the spore cloud as it travels horizontally results from eddy diffusion. Of the various meteorologists and physicists who have formulated eddy diffusion, the most useful come from the work of O. G. Sutton. Experimental values obtained in microbiological work for the parameters of Sutton's diffusion equations are useful in predicting spore concentrations at various distances from the source.

Concentration of the spore cloud decreases not only by diffusion but also because particles are lost by various deposition processes – of which impaction, turbulent deposition, and sedimentation by gravity, are the most important in dry air. The diffusion process may be virtually terminated when the spore cloud is washed out by rain.

DEPOSITION

Deposition from the atmosphere under natural conditions outdoors has been studied experimentally. Under uniform conditions, particle size has an important effect on the amount deposited from a spore cloud of given concentration at ground level. When liberation is near ground level, loss by deposition is great. We cannot yet choose between two theories of deposition to the ground, one of which depends on the wind speed and the other on the distance traversed by the cloud. Near the source, unexpected high deposition values have been observed. Elevating the point of liberation reduces the amount of deposition near the source. Topographical features which affect wind speed also affect the number of spores deposited, and may play a part in the development of a plant disease epidemic – complicated by the fact that they may simultaneously modify the ecoclimate in a direction favouring or inhibiting infection.

Sooner or later, if they have not been already deposited by other means, airborne microbes are removed from the atmosphere by collection in rain, snow or hail. If the carrying droplet later evaporates below the cloud base, there could possibly be an actual local increase in concentration high above ground. Rain, hail and snow bring down a large microbial deposit and are factors which make for non-uniform deposition, in contrast with the more uniform diffusion which results from turbulence.

THE AIR SPORA

The microbial content of the air is difficult to enumerate because, although microscopic, the particles are often large enough to demand attention to the aerodynamic design of the sampling equipment. Choice of the correct apparatus must be determined by the range of organisms to be sampled. Throughout this book emphasis has been placed on methods, because methods commonly determine results. Single bacterial cells in aerosols are small enough to be handled in the manner of a gas, without regard to their inertia; but larger organisms (and bacteria on 'rafts') impact on surfaces, stick on corners, slip out of streamlines, and settle under gravity. These aerodynamic effects must be allowed for in the design of apparatus for reasonably accurate sampling. All these effects aid retention of inhaled particles in the biological air sampler – the vertebrate respiratory tract.

The basic study of the air spora must be by visual methods under the light microscope or electron microscope. Crude as visual identification is, it is based on the only common property shared by all microbes, visibility, and no other method can reveal the whole range of the air spora and disclose what numbers and kinds are in the air awaiting identification. More precise methods can then be applied to the study of limited groups.

As yet the air spora is imperfectly explored, but sampling has already shown that bacteria, algae, yeasts, spores of fungi, mosses and ferns, pollens and protozoa, commonly occur in the air. The air spora near the ground is extremely variable, both from time to time and from place to place. It changes with season, and with weather; the concentration of constituent types often changing several thousand-fold in the course of an hour or two. It changes regularly in composition and concentration throughout the day and night, partly reflecting launching mechanisms. Damp night air has its characteristic spora, but at dawn the night spora disappears suddenly; where it goes we do not know. Rain removes the dry-air spora and substitutes a different one.

Visual methods indicate that fungus spores are normally in the majority over other components of the air spora – probably outnumbering bacteria. However, we lack methods for the complete enumeration of the bacteria in outdoor air, and improved techniques may reveal the presence of far more bacteria than we can recognize at present. Survival of the windborne microbes is not measured by the visual methods, which are applicable especially to the 'hardy' group of microbes. Loss of viability must act like a further dilution factor on the 'tender' organisms which are the main interest of microbiologists.

Vegetation, living or dead, is the main source of the air spora, but most bacteria come from blown soil or splashed water. The source of the yeasts, which are sometimes recorded in large numbers, is still obscure – unless they prove to be Sporobolomycetes. Air near the ground at times contains from tens of thousands to hundreds of thousands of microbes per cubic metre. The air spora near the ground is commonly dominated by contributions from local and intermediate sources; but local effects are smoothed out at high altitudes, over the oceans and in polar regions. A map of microbial wind dispersal would show a mosaic of intermingling local centres of distribution, greyed by a background derived from the 'escape fraction'. On a large scale, concentrations would be patchy, with layers of enhanced concentration below clouds, or empty volumes newly cleansed by rain.

Spore concentration over land usually decreases with altitude, though not always regularly. Often a large proportion of the air spora must occur at altitudes higher than 10 metres. Far out to sea, a few metres above sea level, the microbial content of the air is usually small; but, at an altitude of a few thousand metres over the ocean, the air often contains a few bacteria and from tens to hundreds of spores per cubic metre. Current information suggests that, over the oceans, concentrations are greater in the upper air than near sea level. Clearing the lowest zone of the atmosphere of microbes is most evident over the ocean. For above land the spore cloud near the ground is not only fed from above by diffusion, and to a minor extent by gravity, but it may also be fed from below, from new sources of soil and vegetation. Even over the Arctic in winter, the air is only relatively sterile; samples of a few cubic metres may or may not contain viable microbes. Wind transport of microbes is a process that is evidently going on continuously on a world-wide scale.

AN AEROPLANKTON?

Many unsolved problems still remain. Is there a biotic zone at the height of a few thousand metres? Or is the lower air often cleared by rain, with microbes reconcentrating transiently at the base of a cloud? Is there a true aerial plankton in the sense of a population permanently living and reproducing at great heights, as suggested by R. C. McLean (1935, 1943)? This would seem improbable unless a microbial population is permanently balanced over tropical regions by rising air currents and descending cloud droplets. Probably a negative answer must be given to the hypothesis of an aerial plankton.

Do air masses retain for a long period a characteristic polar or tropical air spora, or do they rapidly receive and give up the spora of the land over which they pass?

GRADIENTS

Spores deposited on the ground, or on vegetation near a source, show a pronounced gradient (following the concentration gradient of the wind-blown spore cloud). This can be estimated by making assumptions based on observed data, and can be used in predicting the danger of contamination by foreign pollens, plant pathogens, and so on. In practice the ideal gradient is often modified by topography: sometimes a decreased wind velocity decreases spore deposition.

In the past it may have been too easily assumed that the dispersion of an organism around its origin resembles a normal frequency distribution. This may be true when dispersion is due to motile surface-animals, actively flying insects, and possibly even rain splash. But wind dispersal is not 'normal' (in a statistical sense) around the point of origin; by contrast, it is a hollow curve. This throws light on a paradoxical situation: when wind-dispersed microbes are liberated near ground level, the gradient near the source is very steep, and relatively short isolation distances give good protection. Pollens, and plant pathogens with large dry spores liberated at ground level may have 90% of their spores deposited within about 100 metres. Yet spores can travel immense distances, and spores found over the oceans clearly represent the tails of the distributions of the sources on the up-wind continent. The equation for Q_x has the curious property that the farther a spore has travelled, and the longer it has survived deposition, the farther it is likely to travel.

Although most spores are deposited near their source, some are readily transported to great distances. Transport over long distance plays a regular role in some crop disease epidemics, and presumably in the movement of many other organisms. Mountain ranges, oceans, and deserts, may all be effective barriers to dispersal. Although conditions in the upper air, especially in cloud, may not be unfavourable to survival, loss of viability rather than failure of the transport mechanism limits the range of many organisms. Enhanced viability may be expected among microbes that survive exposure to ultraviolet radiation. Because of the overwhelming importance of unsuspected nearby sources, it is often difficult to be sure that an organism observed has come from a great distance.

Numerically, fungal spores commonly dominate the air spora. This is because the fungi (not themselves microbes) excel other groups in the variety of spore-launching devices, culminating in the superbly efficient, but still obscure, basidiospore drop-excretion mechanism. We now know that basidiospores form an important part of the air spora, often carrying electrical charges, and probably outnumbering even *Cladosporium* (which is the dominant mould in most parts of the world). The tardy recognition of the abundance of basidiospores is partly explained by their inefficient collection by standard sampling methods, and partly by the unfamiliarity of most microbiologists with the spores of the higher fungi (cf. Plate 7).

BIOLOGICAL WARFARE

It seems that microbial weapons have not been used on any large scale by man against man, but the example of myxomatosis in rabbits should convince sceptics of what might happen when all conditions are favourable for epidemic spread. The topic is shrouded in official secrecy, but information already released suggests that if deliberate dissemination of pathogens or toxins were ever attempted, contamination of air might be one of the dangers to be anticipated. Rosebury *et al.* (1947), from their comprehensive analysis of the principles of bacterial warfare, consider that the air-borne group of pathogens contains the most important infective agents for war use (*see also* Rosebury, 1947, 1949).

A report on *Chemical and Bacteriological (Biological) Weapons and the Effects of Their Possible Use* (United Nations, 1969) points out that the delivery of chemical and biological agents via the atmosphere is the means most likely to be chosen. 'Infections through the respiratory tract by means of aerosols is by far the most likely route to be used in warfare', using a means of dissemination which produces most particles under 5 μm diameter resulting in better suspension, inhalation and rapid drying down to droplet nuclei. By reason of its power of multiplication a living organism would be a more efficient weapon than an equal weight of chemical. But like poison gases, microbes discharged into the atmosphere are at the mercy of meteorological factors; inversion conditions at night being especially favourable for producing dense concentrations.

The report cites a field trial in which 200 kg of a harmless tracer consisting of zinc cadmium sulphate particles 2 μm in diameter were liberated from a ship at sea while travelling for 260 km on a course parallel with the coastline at 16 km off shore. It was estimated that a microbial pathogen, if following the same pattern of distribution as the harmless tracer, would have effectively contaminated from 5000 to 20 000 km², depending on viability in air. In another experiment with 600 litres of *Bacillus globigii* (another harmless tracer with particularly resistant spores) liberated at sea, bacteria were recovered at monitoring stations over an area of 250 km² and reaching to 30 km inland. Fortunately most human and animal pathogens are much less hardy in aerial suspension than *B. globigii* (*see* Chapter XI).

Pathogens considered for use in causing death or incapacity in man or animals by inhalation include: influenza, dengue, Venezuelan equine encephalitis; foot-and-mouth disease viruses; Q-fever *Rickettsia*; anthrax, plague, tularemia, and brucellosis bacteria; aspergillosis and coccidioidomycosis fungi (United Nations Report, 1969). Even yellow fever has been transmitted experimentally by aerosol inhalation, dispensing with the mosquito vector.

Detection and warning of microbial attack would be more difficult than with chemical agents. The use of non-specific particle counters and protein detectors seems unrealistic in view of the abundance, complexity and great fluctuation of the air spora discussed in Chapter X. Identification of specific microbes by normal methods would take several days, but time could be saved by special techniques discussed by Westwood (1970).

The medical, social and psychological implications of biological agents as weapons for mass destruction of man, animals and plants, are discussed in a report by the World Health Organization (1970).

ISOLATION, QUARANTINE AND GEOGRAPHICAL DISTRIBUTION

The prodigious reproductive powers of microbes always excites comment and is particularly remarkable in the sporulation of fungi. The wastage of spores must be enormous. The world is more or less fully populated with fungi and, within reasonable geographic and climatic bounds, a fresh substratum will rapidly select its normal colonizers from the ample supply of suitable spores brought to it. In crop pathology we study a system that is temporarily unbalanced by human activities – large, reasonably pure stands of susceptible 'artificial' crop plants which are renewed at regular intervals, are especially subject to attack by parasites. Dispersal gradients become obvious under these conditions, and our isolation and quarantine methods can play a major role in limiting the development of a plant pathogen.

Willis (1940) claims that 'nothing in the distribution of plants would lead anyone to suppose that the "mechanisms for dispersal" have produced for the plants that possess them any wider dispersal than usual'. In the unbalanced condition necessary for agriculture, this claim obviously fails; it is comparatively easy to limit the dispersal of *Sclerotium cepivorum* and *Synchytrium endobioticum* which have poor dispersal mechanisms, difficult with *Phytophthora infestans*, and impossible with *Puccinia graminis* whose dispersal mechanism is excellent. Willis's claim refers to floras which man has not greatly disturbed, and, it seems, mainly to higher plants.

The rules of M. W. Beijerinck and L. G. M. Baas-Becking, as given by Overeem (1937), state that 'as far as microbes are concerned; *"Everything is everywhere"* and from this *"everything"*, *"the milieu selects"* '. This is an exaggeration, a microbiological half-truth that is useful as a corrective to narrow parochialism; but if it were universally true, aerobiology would not be interesting. For example, the rubber tree is attacked by two serious pathogens, *Oidium heveae* and *Dothidella ulei*, which fortunately are at present limited in their distribution, the former to Asia and the latter to the Americas. Holliday (1971) lists plant pathogens whose distribution is narrower than that of their host crops.

Many saprophytic microbes have a world-wide distribution, and the soil bacteria and moulds tend to be similar in all continents. Floras of the different areas are remarkably uniform in their coprophilous fungi, and in their green and blue-green algae – a testimony to the effectiveness of various dispersal processes where similar environments, existing over long periods, have given time for equilibrium to be established. With local disturbances from time to time, sites become available in the midst of a population of organisms in apparent equilibrium. Man is the most active disturber of this equilibrium, and human activity therefore frequently operates in the borderline region of a microbial concentration gradient. In this region we can attempt to interfere with the dispersal process.

Normally in a field population, multiplication and elimination proceed simultaneously. Altering the balance between these two processes sometimes leads to the disappearance of an organism or disease. Isolation of allergic patients from allergens outdoors may need great distances, because the threshold may be so low that one pollen grain is enough to provoke an attack. Isolation and quarantine in medicine scarcely falls within the province of this book; intramural aerobiology is mainly the concern of hygiene and public-health workers. But the use of distance to control cross pollination and the spread of crop pathogens outdoors is obviously relevant.

Where the geographical distribution of a pathogen is more restricted than that of its host crop, the existence of natural barriers, for practical purposes uncrossable by airborne plant pathogens, make control of some crop diseases by official action a feasible procedure. For example, Zadoks (1967) regards Europe as a single area in this context, bounded by the Atlantic, the Sahara, the Russo-Siberian steppes and the Arctic. New pathogens enter this area by natural migration from the east or southeast or by human introduction. Within Europe wind dispersal of pathogenic fungi occurs in all directions.

Where natural wind transport fails, other methods may dominate dispersal. With freshwater and soil microbes, transport by birds may be important. But of all animals, man is the most dangerous, because his opportunities for carrying exotic microbes are greatest. The major portion of the load carried across international boundaries by the world's highly organized transport system consists of unprocessed plant material. There are two dangers to be guarded against in this connection. An important crop disease may be excluded from a country where the pathogen does not occur by: (1) prohibiting imports of possible host plants; (2) inspection in the country of origin or on arrival; and (3) disinfection (supplemented by local eradication measures should the pathogen gain a temporary hold). The more insidious danger, however, is in an organism which has settled down in its original home as an insignificant parasite on a host with which it is in balanced relationship, suddenly being transported to a new country and finding a highly susceptible host crop where it can cause devastating losses, e.g. *Puccinia polysora* on African maize (cf. p. 274), and *Endothia parasitica*, a trivial parasite on oriental chestnuts, but devastating when introduced into the United States of America from Asia (McCubbin, 1944b, 1954).

MEDICAL MYCOLOGY AND ALLERGY

The winds, which had been suspected from antiquity of bearing epidemic diseases, have been gradually exonerated so far as most human and animal epidemics are concerned. Malaria and yellow fever come through the air, but only by the activity of their insect vectors. The old scourges of cities are now known to be spread by vectors, water, milk or contact, and are no longer attributed to the winds. Air is not entirely blameless, however. Droplets expelled from the mouth and nose spread disease in confined places; wind carries pollen and other allergens. Some of the less well-known fungus diseases of man, such as coccidioidomycosis and histoplasmosis are clearly airborne (Rooks, 1954; Hoggan *et al.*, 1956; Furcolow & Horr, 1956; Fiese, 1958; Levine, 1969), but we lack detailed information on their launching mechanisms. Foot-and-mouth disease virus is now shown to be windborne. Air emerges as one of the major routes for transport of microbes, pollens, and many crop pathogens.

Outdoor airborne allergens appear to originate from vegetation above ground, but not significantly from the soil (though the allergenic potentialities of various soils themselves would be worth testing). To a large extent, therefore their origin is an agricultural problem, aggravated by our needs to cultivate plants in pure stands. Fortunately, the interests of farmers and allergic patients are to some extent identical. The farmer, unless he is a seed grower, does not require his grass to scatter protein-rich pollen uselessly, and the development of non-flowering strains of grasses (from seed raised in other countries where conditions such as length of day permit flowering) is beginning to attract the attention of farmers and seed-merchants (Peterson *et al*, 1958); nor does the farmer want his wheat straw weakened by *Alternaria* or *Cladosporium*.

Improved methods for measuring spore concentrations of the atmosphere bring forward many new organisms for test as potential allergens, and may throw light on some of the seasonal asthmas – especially those of later summer whose etiology is unknown (Maunsell, 1958; Ganderton, 1968). Air of alpine valleys such as Davos, traditionally beneficial for asthmatics may escape some spore pollution as a result of shielding (R. R. Davies, 1969b).

Not all effects of the air spora are on the debit side of the health account.

Calculations show that microbial deposition could add a few kilogrammes per hectare of combined nitrogen to the land surface: a quantity negligible in terms of agriculture, but useful to pioneer organisms colonizing new sites. Possibly the air spora contributes vitamins such as B12 (Parker, 1968).

The purest air appears to be just above the ocean surface, though life aboard ship has its allergic hazards. Microbial pollution of the atmosphere is comparatively short lived because its size range makes for rapid removal.

PALYNOLOGY

Pollen statistics, pollen analysis, and more recently palynology (a term introduced by Hyde & Williams, 1944), are names given to a group of studies, including investigation of the ecology, vegetation and pre-history of the Quaternary Period by examining pollens preserved in peat and other deposits. This is possible by virtue of the highly resistant sporopollenin of the pollen exine (see Brooks et al., 1971). Aeropalynology is reviewed by Hyde (1969), and palynologists have contributed much to the development of aerobiology, being well aware of the complications introduced by the occurrence of wind-blown pollen from distant sources (e.g. Buell, 1947; Potter & Rowley, 1960). The problem is one of sampling in a given locality, so as to eliminate uneven distribution from the dominating influence of one nearby source. The influence of meteorological factors is discussed by Schmidt (1967).

We still need to know how the total deposit at one point is made up of a few local distributions, plus the tails of the distributions of many distant sources (i.e. the problem of measuring 'background' as discussed in Chapter XVI). Furthermore, active concentration within a cumulus cloud may sometimes reverse the diffusion process, leading to the kind of incident which Pettersson recorded with a heavy shower of *Aloina* spores. How far reconcentration in the air needs to be taken into account, and how far surface obstacles may enhance local deposition, are matters for further study.

The proportion of particles becoming resuspended in air, after having once been deposited, is normally negligible, but the possibility interests palynologists, and Tauber (1967) showed it may sometimes be locally significant.

Necessarily, palynologists have usually preferred deposition traps: for example the recessed sampler of Tauber (1965), and the opportunistic use of moss polsters as samplers by Carrol (1943), Benninghoff (1960), and King & Kapp (1963), among others.

Spores and pollen preserved in glacial ice in mountainous and polar regions offer scope for investigation that is still almost unexplored (cf. Vareschi, 1942).

EVOLUTION

It seems that in each locality or habitat, mutation and recombination are apt to act, through selection, to evolve special local divergent populations. But, simultaneously, dispersal mechanisms tend to counteract this process by encouraging outbreeding and so increasing uniformity. How the two processes balance is a genetical problem depending on external factors and on the breeding systems involved. Aerobiology contributes quantitative information on the size of the breeding group, which must be determined partly by the characteristics of the dispersal gradient. Wind-borne genes are not distributed 'normally' (in a statistical sense) around their point of origin but in a characteristic hollow curve, which has the interesting property of involving greater frequencies near the source, and again at great distance (at the expense of intermediate distances) than if the genes followed the normal frequency distribution.

Over the greater part of the Earth's surface the ecologically dominant flowering plants and conifers belong to wind-pollinated (anemophilous) species. Temperate and tropical grasslands, coniferous forests and deciduous forests, and some semi-desert lands, are dominated by species which reach up to shed pollen into the turbulent boundary layer. Tropical rain forest, on the other hand, is dominated by insect-pollinated plants. Probably the number of anemophilous *individuals* in many times greater than the number of entomophilous *individuals* in the world, yet only about one-tenth of the *species* of flowering plants are anemophilous. By contrast, tropical rain forest is noted for its extraordinary number and diversity of species of flowering plants. Perhaps it is because of meteorological factors such as rain or the difficulty of access for turbulent wind that these plants are mostly insect pollinated, and clearly the two phenomena are inextricably linked. The wider possibilities of gene dissemination in wind-pollinated plants has tended towards relative uniformity over wide areas, while the statistically more 'normal' and localized character of insect pollination has favoured specialization and speciation (cf. Raper *et al.*, 1958).

GENE DISPERSAL

The role of wind dispersal is evidently two-fold. Its role in leading to colonization is clear enough with seed plants and with many microbes, and the establishment of infection is an example of this process in pathogens.

A second important role, obvious enough in wind-pollinated flowering plants but curiously overlooked in microbes, is that of gene transmission,

Theories of gene dispersion in populations at first assumed that there is a random scatter around the source. In such a distribution the gradient in any direction would have the form of one-half of the normal frequency curve. S. Wright (1943, 1946) studied genetic effects of isolation distance, and his methods were applied by J. W. Wright (1952) to compare dispersion distances of pollens of various forest trees with a view to delimiting a 'neighbourhood' for race formation. Simple sticky-slide traps were exposed at various distances around isolated trees, and pollen counts were used to find the standard deviation of the scatter. Observed values for the standard deviation were as follows: ash 17–46 metres, Douglas fir 18 metres, poplar and elm 300 metres or more, spruce 40 metres, Atlas cedar 73 metres, Lebanon cedar 43 metres, and pinyon pine 17 metres. Dispersal data were well fitted by the Gregory formulae, but not by theories which assume that the trajectory of each grain can be calculated from the rate and distance of fall, and the wind velocity (*and see* Chamberlain, 1967a).

Bateman (1950) questioned whether gene dispersal is 'statistically' normal, and showed by his regression method that many observed distributions, including those of fungus spores, passively borne insects, pollen and wind-dispersed seeds, were highly leptokurtic, i.e. the peak and tails of the distribution are exaggerated at the expense of the shoulders. Compared with a normal frequency distribution having the same standard deviation ('same over-all degree of inbreeding'), the leptokurtosis characteristic of passive airborne dispersal produces more breeding between close relatives and simultaneously more breeding between distant relatives (*see also* Parker-Rhodes, 1951).

In the fungi the process of gene dispersal is probably a major function of spore dispersal (Gregory, 1952). So far as well-established fungi are concerned, the immense output of spores probably does little to promote the extension of the range of a species. When some accidental or cyclic effect offers a suitable environment within the range, the area involved rapidly becomes colonized and the average number of mycelia of the species is at least maintained. From this we need not necessarily deduce, as has been usual hitherto, that all the rest of the spores are functionless. The higher fungi possess one characteristic that is unmatched in other organisms – their ability to form vegetative hyphal anastomoses which lead to a mixing of cytoplasm and nuclei from spores of different origin – and it is tempting to speculate, as Transeau (1949) did with *Coprinus variegatus*, that far more spores germinate and fuse with an already-established mycelium than ever themselves succeed in establishing a new mycelium (cf. Savile, 1964; Burnett & Partington, 1957).

The work of H. M. Hansen & Smith (1932) on heterocaryosis and subsequent investigators, show that wild mycelia may contain genetically different nuclei. Some of these may be due to mutations in mycelia derived from a single nucleus, but experimental evidence shows that heterocaryons can also be produced by artificial mixing of suitable, genetically different mycelia, and it seems highly probable that this mixing also occurs in nature. The value to a species of storing an adequate supply of mutant genes to be drawn upon in future, and so giving plasticity under varying conditions, has been stressed by workers on fungal genetics, including Craigie (1942) and Whitehouse (1949).

Our speculation would suggest that besides increasing the range and colonizing new substrata within the range, spore dispersal of the higher fungi has another function: dispersing genes and transmitting novelties, arising from mutation, re-combination and para-sexual recombination, between one established mycelium and another. The shift of sexual reproduction from the sedentary spores of the Phycomycetes, which lack hyphal fusions, to the dispersal-spore form in the Ascomycetes and Basidiomycetes, may be a phase in the evolution of this habit. On this hypothesis, a perennial mycelium would be the locus of activity and multiplication of individual nuclei – some perhaps descended from the original spore-colonizer, others descended from outside immigrant sources. The mycelium would remain, whereas nuclei would come and go. The established mycelium would resemble a city rather than an individual. Thus, besides their function as colonizers, spores may perhaps act as a sort of unreliable air-mail service, transmitting genes between established mycelia.

Dispersal spores of some fungi, such as the Gasteromycetes are difficult to germinate. If we reject, as first choice, the hypothesis that they are functionless, we must suppose that there exist special conditions under which germination occurs naturally. The experiments of Ferguson (1902; see Lösel, 1964), who showed that mushroom spores would germinate readily when in contact with living hyphae of the same species, are suggestive in this connection. Other examples are known of spores that are stimulated to germinate by hyphae of the same species, an instance being the spermatia (pycniospores) of *Puccinia helianthi* (Craigie, 1933). Experiment may show that this phenomenon of gene interchange plays a bigger part than has hitherto been considered possible. If this speculation is justified, the production and dissemination of novelty must be a major activity of the fungi.

Ascospores and basidiospores are the spore forms most likely to contain genetic novelty, and they are most commonly dispersed by wind to potentially new environments. The conidia of fungi, which are ordinarily dispersed by wind, splash or insects, were well named 'repeating spores' in the older literature, functioning as they do for rapid exploitation of the same environment as the parent. However, G. Pontecorvo's discovery of para-sexual recombination shows that genetic novelty can also arise from conidial forms in several ways.

BEYOND THE ATMOSPHERE

Aerobiological technique has much to contribute to research beyond the Earth's atmosphere.

To counter the idea that living organisms evolved from non-living matter on this planet, Arrhenius (1908) put forward the hypothesis that space is permeated by spores. On this hypothesis, spores might be carried to great heights, for example in the Earth's stratosphere by volcanic activity, and then driven off into space by electrical repulsion. If this were so, a planet might leave a trail of dust in space, and large particles might move in the sun's gravitational field; but the smallest bacterial spores (0.2 μm) would move away from the sun with high velocities under radiation pressure, even crossing interstellar space and entering planetary atmospheres. Arrhenius's hypothesis now appears improbable. Conditions in space seem inimical to the survival of life (Oparin, 1957), though Halvorson & Srinivasan (1964) point out that bacterial endospores are extremely resistant to adverse environments and might survive in space if protected from ionizing radiation by inclusion within a dust particle.

The techniques of aerobiology would probably not solve the novel problems that would arise in sampling in Space, but present knowledge will need thorough exploitation when we begin to probe the *atmospheres* of other planets (*see* discussions in Quimby, 1964; Bruch, 1967). Many problems will arise, such as: (1) the necessity of studying any atmospheric spore flora of another planet; (2) the moral obligation to avoid contaminating the atmosphere of another planet; and (3) the practical necessity to avoid contaminating our own atmosphere with completely unknown organisms which might be carried on a returning space-vehicle (Alexander, 1969). We must explore any atmospheric spora of other planets less inefficiently than we have done our own, and for this much developmental work remains to be done.

FUTURE STUDY OF OUR ATMOSPHERE

Our knowledge of the terrestrial air spora is fragmentary in the extreme. The air has never been systematically explored simultaneously in different parts of the world by comparable methods. There is a heap of accumulated data, from which Chapters X to XVII attempt to sort out a few principles. Here and there are intriguing suggestions of phenomena; but many of the data are uninterpretable, and we need a fresh study of aerobiology as part of a vast terrestrial process.

Before starting, aims must be clearly defined, and the needs of three separate analyses formulated: concentration in air, concentration in precipitation, and surface deposition. Methods must then be worked out, based on visual examination of the whole air spora, and supplemented by cultural and other measures of viability, for the taxonomically diverse components. Equipment must be tested and calibrated in wind-tunnels and in the open air, including sampling during flight at high altitudes.

In Britain outdoor air has been monitored almost continuously by the Allergy Department of the Wright-Fleming Institute for Microbiology, London, for over 15 years, and for nearly as long at St David's Hospital, Cardiff. Many shorter sets of continuous records with the Hirst trap for purposes of allergy or plant pathology exist in other parts of the world.

At present the use made of the information is largely retrospective; at best we have yesterday's spore and pollen count, but commonly the delays are longer. For the future we must look for a television microscope built into the sampler to give a picture of the air spora as the sample is taken.

When methods have been developed, international co-operation will be necessary to organize a chain of routine sampling stations and to co-ordinate programmes. Routine sampling stations should cover a wide variety of climates, of topographic, altitudinal and ecological environments, and also the special interests of cities, agriculture, and medicine. An international aerobiological research institute is needed, with laboratories in temperate and arctic regions and, above all, in the tropics and on remote oceanic islands, about which we know least.

After exploring the stratosphere, we shall be better able to tackle outer space. Time is short; already the Moon is at our doorstep, and before we know we shall have Mars and Venus on our hands.

Appendix 1

VISUAL IDENTIFICATION

The morphological characters of many pollen grains are known in considerable detail, and identification can often be carried to species level. The beginner who is faced with the problem of visual identification of the catch is helped by published illustrations, including the two volumes by Erdtman (1952, 1957), and one by Hyde & Adams (1958); also by the journal *Pollen et spores* published in Paris. The shells of thecate amoebae, which occasionally occur in suction trap slides, are illustrated by Bonnet & Thomas (1960).

For many other groups of organisms little help can be had from books, and any laboratory doing visual scanning of atmospheric spore deposits must assemble a collection of reference slides of specimens in the standard mountant adopted, and prepared from reliably identified specimens collected in the field. (Fungus spores from cultures are often abnormal and unlike the forms which occur on spore traps.) For the basic collection the spores should be mounted unstained, because the natural colour greatly aids categorization or identification.

The reference collection will also be useful in identifying contaminants. For example, among strange objects turning up in histological preparations from medical and veterinary pathology laboratories we have recognized: grass pollen, pine pollen, and spores of *Cladosporium* and Basidiomycetes – all doubtless arriving through the laboratory window. The reference collection is also relevant to the work of forensic laboratories. An atlas of the air spora is a project which needs encouragement. The need is exemplified by the recent discussion on 'brochosomes' (Wiffen & Heard, 1969; Neville, 1970; and D. S. Smith, 1970).

Plates 5 and 6 illustrate in colour some typical airborne spores, etc., at the uniform magnification of 1000 diameters to help the beginner who may have been confused by many published illustrations which portray plant spores at widely different magnifications. Plate 8 illustrates some typical anemophilous pollen in monochrome on the same scale. Fresh pollen grains mostly appear a pale yellow in microscopic preparations.

The paintings, by Mrs Maureen E. Lacey (née Bunce) attempt to depict the object as it appears under the microscope, with a minimum of interpretation. As far as possible they are from specimens collected in the field (the actual source of each is indicated by a key symbol). All were mounted unstained in glycerine jelly.

The colour blocks used in the first edition of this book having been destroyed, Mrs Lacey has painted Plates 6 and 7 afresh, and a few additions and deletions have been made.

Key to Source

c from culture   h from hay
s from field specimen   t from trap slide

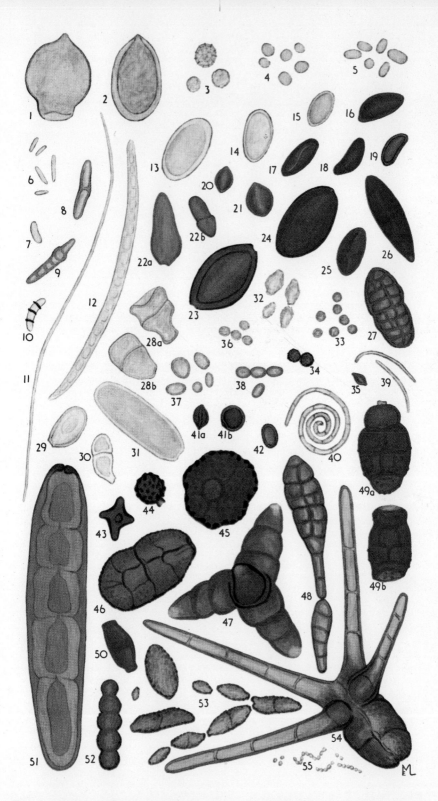

Plate 6. Spores of Phycomycetes, Ascomycetes, Fungi Imperfecti, and Actinomycetes. All magnifications, ×1000 diameters.

PLATE 6

Spores of Phycomycetes, Ascomycetes, and Fungi Imperfecti.

1   *Entomphthora muscae*, conidium, s
2   *Peronospora parasitica*, sporangium, s
3   *Mucor spinosus*, sporangiospores, c
4   *Absidia corymbifera*, sporangiospores, c
5   *Absidia ramosa*, sporangiospores, c
6   'Hyaline rods', t
7   *Tubercularia vulgaris* (= conidium of *Nectria cinnabarina*), s
8   *Nectria cinnabarina*, ascospore, s
9   Unknown ascospore, t
10 Unknown ascospore, t
11 *Claviceps purpurea*, ascospore, s
12 *Ophiobolus graminis*, ascospore, s
13 *Helvella crispa*, ascospore, s
14 *Humaria granulata*, ascospore, s
15 *Pyronema confluens*, ascospore, s
16 *Bulgaria inquinans*, ascospore, s
17 *Xylaria polymorpha*, ascospore, s
18 *Hypoxylon coccineum*, ascospore, s
19 *Hypoxylon multiforme*, ascospore, s
20 *Chaetomium indicum*, ascospore, c
21 *Chaetomium globosum*, ascospore, c
22 *Venturia inaequalis*: (a) conidium; (b) ascospore, s
23 *Melanospora zamiae*, ascospore, s
24 *Sordaria fimicola*, ascospore, c
25 *Daldinia concentrica*, ascospore, s
26 *Rosellinia aquila*, ascospore, s
27 *Pleospora herbarum*, ascospore, s
28 *Polythrincium trifolii*, conidia: (a) t, (b) s
29 *Botrytis* sp., conidium, c
30 *Trichothecium roseum*, conidium, c
31 *Erysiphe (graminis?)*, conidium, t
32 *Aspergillus glaucus* (series), conidia, c
33 *Aspergillus fumigatus*, conidia, c
34 *Aspergillus niger*, conidia, c
35 *Aspergillus nidulans*, ascospore, c
36 *Penicillium cyclopium*, conidia, c
37 *Penicillium digitatum*, conidia, c
38 *Penicillium chrysogenum*, conidia, c
39 *Diatrype stigma*, conidia, s
40 *Helicomyces* sp., conidium, t
41 *Papularia arundinis*, conidia: (a) face view, (b) edge view, c
42 *Cryptostroma corticale*, conidium, s
43 *Humicola stellata*, conidium, c
44 *Humicola lanuginosa*, conidium, h
45 *Epicoccum* sp., conidium, t

46 *Stemphylium* sp., conidium, t
47 *Asterosporium* sp., conidium, t
48 *Alternaria* sp., conidia, t
49 *Pithomyces chartarum*, conidia: (a) rounded, (b) collapsed, t
50 *Bispora monilioides*, conidium, s
51 *Helminthosporium* sp., conidium, t
52 *Torula herbarum*, conidium, t
53 *Cladosporium* sp., conidia, t
54 *Tetraploa aristata*, conidium, s
55 *Streptomyces* sp., spores, h

Magnification: all × 1000 diameters (1 mm = 1 μm)

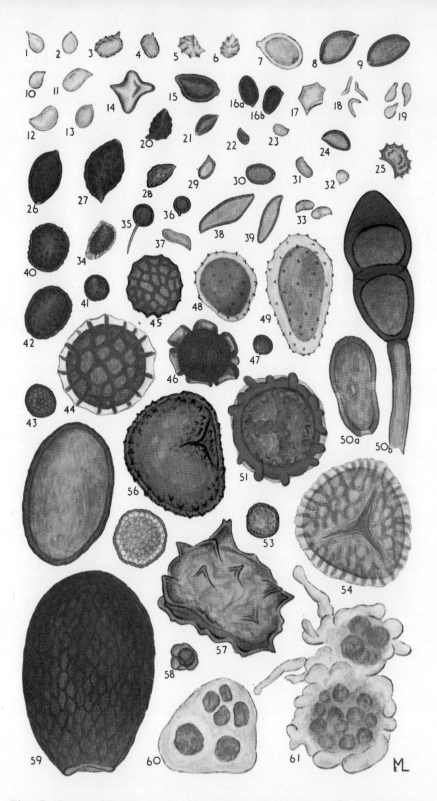

Plate 7. Spores of Basidiomycetes, Myxomycetes, Pteridophytes, Bryophytes, Protozoa, etc. All magnifications, ×1000 diameters.

PLATE 7

Spores of Basidiomycetes, Myxomycetes, Pteridophytes, Bryophytes, Protista [Protozoa], etc.

1   *Collybia maculata*, basidiospore, s
2   *Tricholomopsis (Tricholoma) rutilans*, basidiospore, s
3   *Russula nigricans*, basidiospore, s
4   *Russula vesca*, basidiospore, s
5   *Lactarius blennius*, basidiospore, s
6   *Lactarius rufus*, basidiospore, s
7   *Amanita (Amanitopsis) fulva*, basidiospore, s
8   *Pholiota (Naucoria) myosotis*, basidiospore, s
9   *Bolbitius vitellinus*, basidiospore, s
10 *Hygrophorus niveus*, basidiospore, s
11 *Armillaria mellea*, basidiospore, s
12 *Amanita rubescens*, basidiospore, s
13 *Inocybe geophylla*, basidiospore, s
14 *Nolanea staurospora*, basidiospore, s
15 *Cortinarius elatior*, basidiospore, s
16 *Coprinus atramentarius*, basidiospore: (a) profile, (b) face view, s
17 *Entoloma rhodopolium*, basidiospore, s
18 *Tilletiopsis* sp., basidiospores, t
19 *Sporobolomyces* sp., basidiospores, c
20 *Lacrymaria (Hypholoma) velutina*, basidiospore, s
21 *Hypholoma fasciculare*, basidiospore, s
22 *Psathyrella (Hypholoma) hydrophila*, basidiospore, s
23 *Fistulina hepatica*, basidiospore, s
24 *Stropharia aeruginosa*, basidiospore, s
25 *Thelephora terrestris*, basidiospore, s
26 *Panaeolus sphinctrinus*, basidiospore, s
27 *Panaeolina (Psilocybe) foenisecii*, basidiospore, s
28 *Phaeolepiota (Pholiota) spectabilis*, basidiospore, s
29 *Pholiota squarrosa*, basidiospore, s
30 *Merulius (Serpula) lacrymans*, basidiospore, s
31 *Crepidotus mollis*, basidiospore, s
32 *Fomes annosus*, basidiospore, s
33 *Stereum purpureum*, basidiospore, s
34 *Ganoderma applanatum*, basidiospore, s
35 *Bovista plumbea*, basidiospore, s
36 *Calvatia gigantea*, basidiospore, s
37 *Boletus elegans*, basidiospore, s
38 *Boletus scaber*, basidiospore, s
39 *Boletus chrysenteron*, basidiospore, s
40 *Badhamia utricularis*, myxomycete spore, s
41 *Fuligo septica*, myxomycete spore, s
42 *Leocarpus fragilis*, myxomycete spore, s
43 *Reticularia lycoperdon*, myxomycete spore, s
44 *Tilletia holci*, smut spore, s

45   *Tilletia caries*, smut spore, s
46   *Urocystis agropyri*, smut spore, s
47   *Ustilago avenae*, smut spore, s
48   *Triphragmium ulmariae*, uredospore, s
49   *Melampsoridium betulinum*, uredospore, s
50   *Puccinia graminis:* (a) teleutospore; (b) uredospore, s
51   *Selaginella pulcherrima*, spore, s
52   *Funaria hygrometrica*, moss spore, s
53   *Barbula fallax*, moss spore, s
54   *Lycopodium*, club-moss spore, s
55   *Dryopteris filix-mas*, fern spore, s
56   *Pteridium aquilinum*, fern spore, s
57   *Phyllitis scolopendrium*, fern spore, s
58   *Tetramitus* sp., amoeboid cyst, t
59   *Thecamoeba* sp., t
60   *Gloeocapsa* sp., algal group, t
61   *Cladonia* sp., lichen soredium, s

Magnification: all × 1000 diameters (1 mm = 1 μm)

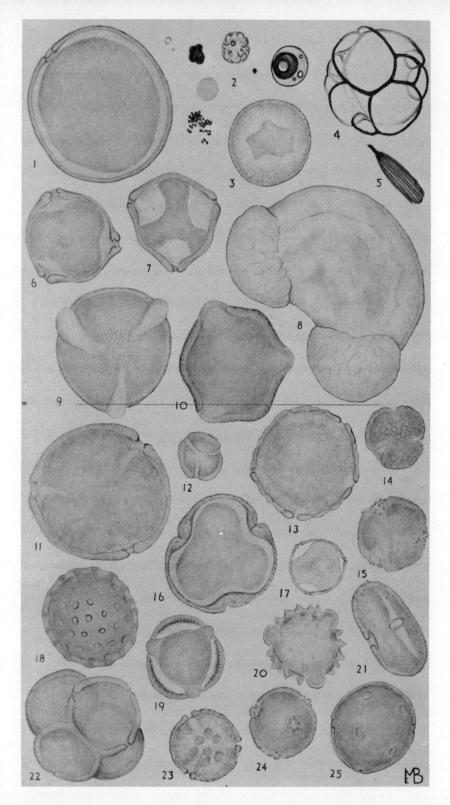

Plate 8. Pollen grains and miscellaneous particles. All magnifications, ×1000 diameters.

PLATE 8

Pollen grains and miscellaneous particles.

1   *Phleum pratense*, grass pollen, s
2   'Fly-ash' spheres, t
3   *Taxus baccata*, yew pollen, s
4   'Cenosphere', t
5   Insect scale, t
6   *Betula verrucosa*, birch pollen, s
7   *Corylus avellana*, hazel pollen, s
8   *Pinus sylvestris*, pine pollen, s
9   *Acer pseudoplatanus*, sycamore pollen, s
10   *Quercus robur*, oak pollen, s
11   *Fagus sylvatica*, beech pollen, s
12   *Castanea sativa*, sweet chestnut pollen, s
13   *Ulmus* sp., elm pollen, s
14   *Salix caprea*, willow pollen, s
15   *Platanus* sp., plane pollen, s
16   *Tilia* sp., lime pollen, s
17   *Urtica dioica*, nettle pollen, s
18   *Chenopodium album*, fat-hen pollen, s
19   *Artemisia vulgaris*, mugwort pollen, s
20   *Solidago* sp., golden-rod pollen, s
21   *Anthriscus sylvestris*, cow-parsley pollen, s
22   *Calluna vulgaris*, ling (heather) pollen, s
23   *Rumex acetosa*, sorrel pollen, s
24   *Thalictrum* sp., rue pollen, s
25   *Plantago* sp., plantain pollen, s

Magnification: all × 1000 diameters (1 mm = 1 μm)

Appendix 2   Conversion Factors

Measures of Weight

Avoirdupois to Metric

1 dram (dr)	27·344 grains	1·772 grams
1 ounce (oz)	16 drams	28·3 grams
1 pound (lb)	16 ounces	0·454 kilogram
1 stone (st)	14 pounds	6·350 kilograms
1 quarter (qr)	2 stones	12·701 kilograms
1 hundredweight (cwt)	4 quarters	50·802 kilograms
1 (long) ton	20 hundredweight	1·016 tonnes

Metric to Avoirdupois

1 milligram (mg)		0·015 grain
1 gram (gm)	1000 milligrams	0·564 dram
1 kilogram (kg)	1000 grams	2·205 pounds
1 quintal (q)	100 kilograms	220·5 pounds
1 tonne	1000 kilograms	0·984 ton

U.S. Weights to Metric

1 pound	16 ounces	453·592 grams
1 cental	100 pounds	45·359 kilograms
1 (short) ton	20 centals	0·907 tonne

Metric to U.S. Weights

1 quintal (q)	100 kilograms	2·205 centals
1 tonne	1000 kilograms	1·102 (short) tons

Measures of Length

British to Metric

1 inch (in)		25·400 millimetres
1 foot (ft)	12 inches	30·480 centimetres
1 yard (yd)	3 feet	0·914 metre
1 mile	1760 yards	1·609 kilometres

Metric to British

1 micron (n)	1/1000 mm (1/1 000 000 m)	1/25 400 inch
1 millimetre (mm)		0·039 inch
1 centimetre (cm)	10 mm	0·394 inch
1 decimetre (dm)	10 cm	3·937 inches
1 metre (m)	10 dm	1·094 yards
		3·281 feet
		39·370 inches
1 kilometre (km)	1000 m	0·621 mile

Measures of Area (Based on 1 *metre* = 39·370 *inches)*

British to Metric

1 square inch (sq in)		6·452 sq centimetres
1 square foot (sq ft)	144 sq in	0·093 sq metre
1 square yard (sq yd)	9 sq ft	0·836 sq metre
1 acre	4840 sq yd	0·405 hectare
1 square mile	640 acres	2·590 sq kilometres
		258·998 hectares

Metric to British

1 square millimetre (mm²)		0·00155 sq inch
1 square centimetre (cm²)	100 mm²	0·155 sq inch
1 square decimetre (dm²)	100 cm²	0·108 sq foot
1 square metre (m²)	100 dm²	1·196 sq yards
1 hectare (ha)	10 000 m²	2·471 acres
1 square kilometre (km²)	100 ha	0·386 sq mile

Measures of Volume

British to Metric

1 cubic inch (cu in)		16·387 cu centimetres
1 cubic foot (cu ft)	1728 cu in	28·317 cu decimetres
1 cubic yard (cu yd)	27 cu ft	0·765 cu metre
1 bushel (bu)	2219·3 cu in	0·364 cu metre

Metric to British

1 cubic centimetre (cm³ = ml)		0·061 cu inch / 0·035 cu foot
1 cubic decimetre (cu dm)	1000 cm³	1·308 cu yards
1 cubic metre (cu m)	1000 dm³	2·750 bushels

Measures of Capacity **I.** *Based on* 1 *Imperial gallon (British)* = 4·546 *litres*
(used for both liquid and dry measure)

British to Metric

1 pint (pt)		0·568 litre
1 quart (qt)	2 pints	1·136 litres
1 gallon (gal)	4 quarts	4·546 litres
1 peck (pk)	2 gallons	9·092 litres
1 bushel (bu)	4 pecks	36·368 litres

Metric to British

1 millilitre (ml = cm³)		0·0610 cu inch
1 centilitre (cl)	10 cm³	0·0176 pint
1 decilitre (dl)	10 cl	0·176 pint
1 litre (l)	10 dl	1·760 pints

II. *Based on* 1 *U.S. gallon (liquid measure)* = 3·785 *litres*

U.S. to Metric

1 pint (pt)		0·473 litre
1 quart (qt)	2 pints	0·946 litre
1 gallon (gal)	4 quarts	3·785 litres

Metric to U.S.

1 millilitre (ml = cm³)		0·0610 cu inch
1 centilitre (cl)	10 cm³	0·021 pint

| 1 decilitre (dl) | 10 cl | 0·211 pint |
| 1 litre (l) | 10 dl | 1·057 quart |

Note: 1 British pint, quart, or gallon = 1·201 U.S. (liquid) pints, quarts, or gallons respectively.

1 U.S. (liquid) pint, quart, or gallon = 0·833 British pint, quart, or gallon, respectively.

III. *Based on* 1 *U.S. quart (dry measure)* = 1·1012 *litres*

U.S. (dry measure) to Metric

1 pint (pt)	33·600 cu in	0·5506 litre
1 quart (qt)	2 pints	1·101 litres
1 peck (pk)	8 quarts	8·810 litres
1 bushel (bu)	4 pecks	35·238 litres

Metric to U.S. (dry measure)

1 litre (l)		0·908 quart
1 dekalitre (dkl)	10 l	0·284 bushel
1 hectolitre (hl)	10 dkl	2·838 bushels

Temperature

0° Centigrade (Celsius) = 32° Fahrenheit

The following formulae connect the two major thermometric scales:

Fahrenheit to Centigrade: $°C = 5/9 (°F = 32)$

Centigrade (Celsius) to Fahrenheit: $°F = (9/5 °C) + 32$

Bibliography

ABBOTT, A. C. (1889). Bacteriological study of hail. *Johns Hopkins Hosp. Bull.*, 1, 56.

ADAMS, K. F. (1964). Year to year variation in the fungus spore content of the atmosphere. *Acta allerg.*, 19, 11–50.

— & HYDE, H. A. (1965). Pollen grains and fungus spores indoors and out at Cardiff. *J. Palynology*, 1, 67–69.

— — & WILLIAMS, D. A. (1968). Woodlands as a source of allergens: with special reference to basidiospores. *Acta allerg.*, 23, 265–81.

AINSWORTH, G. C. & AUSTWICK, P. K. C. (1959). *Fungal diseases of animals.* Commonwealth Agricultural Bureaux: Farnham Royal, England, 148 pp.

AIRY, H. (1874). Microscopic examination of air. *Nature, Lond.*, 9, 439–40.

AJELLO, L., MADDY, K., CRESELIUS, G., HUGENHOLTZ, P. G. & HALL, L. B. (1965). Recovery of *Coccidioides immitis* from the air. *Sabouraudia*, 4, 92–95.

AKERS, A. B. & WON, W. D. (1969). Assay of living airborne micro-organisms. *See* Dimmick, R. L. & Akers, A. B. (Eds.), 1969.

ALEXANDER, M. (1969). Possible contamination of earth by lunar or martian life. *Nature, Lond.*, 222, 432–33.

ALVAREZ, J. C. & CASTRO, J. F. (1952). Quantitative studies of airborne fungi of Havana in each of the twenty-four hours of the day. *J. Allergy*, 23, 259–64.

ANDERSEN, A. A. (1958). New sampler for the collection, sizing, and enumeration of viable airborne particles. *J. Bact.*, 76, 471–84.

— & ANDERSEN, M. R. (1962). A monitor for airborne bacteria. *Appl. Microbiol.*, 10, 181–84.

ANDERSEN, J. D. & COX. C. S. (1967). Microbial survival. *See* Gregory, P. H. & Monteith, J. L. (Eds.), 1967.

— DARK, F. A. & PETO, S. (1968). The effect of aerosolization upon survival and potassium retention by various bacteria. *J. gen. Microbiol.*, 52, 99–105.

ANDERSEN, S. T. (1967). Tree-pollen rain in a mixed deciduous forest in South Jutland (Denmark). *Rev. Palaeobot. Palynol.*, 3, 267–75.

ANDREWES, F. W. (1902). Examination of the atmosphere of the Central London Railway, in *London County Council, Report to the Parliamentary Committee*, No. 615, 21 pp.

ANTWEILER, H. (1958). Über die Funktion des Flimmerepithels der Luftwege, insbesondere unter Staubbelastung. *Beitr. Silikose-Forsch. Sonderband*, **3**, 509–13.

APPERT, N. (1810). *L'art de conserver pendant plusieurs années toutes les substances animales et végétales*. Paris, 116 pp.

ARRHENIUS, S. (1908). *Worlds in the Making: the Evolution of the Universe* (Transl. H. Borns). Harper, London, 230 pp.

ARSDEL, E. P. VAN (1958). Smoke movement clarifies spread of blister rust to distant white pines. *Bull. Am. meteorol. Soc.*, **39**, 442–43.

ASAI, G. N. (1960). Intra- and inter-regional movement of black stem rust in the upper Mississippi Valley. *Phytopathology*, **50**, 535–41.

ASHWOOD-SMITH, M. J., COPELAND, J. & WILCOCKSON, J. (1967). Sunlight and frozen bacteria. *Nature, Lond.*, **214**, 33–35.

AUGUSTINE, R. & HAYWARD, B. J. (1962). Grass pollen antigens: IV The isolation of some of the principal allergens of *Phleum pratense* and *Dactylis glomerata* and their sensitivity spectra in patients. *Immunology*, **5**, 424–60.

AUSTWICK, P. K. C. (1969). Effects of inhalation of spores on animals. Filtration Soc. Symposium on *Filtration in Medical and Health Engineering*, April 14–16, 1969, pp. 65–69. Filtration Society, London.

BAILEY, R. H. (1966). Studies on the dispersal of lichen soredia. *J. Linn. Soc. (Bot.)*, **59**, 479–90.

BANASZAK, F. F., THEIDE, W. H. & FINK, J. N. (1970). Hypersensitivity pneumonitis due to contamination of an air conditioner. *New England J. Med.*, **283**, 271–76.

BARGHOORN, E. S. (1960). Palynological studies of organic sediments and of coated slides. *Scientific studies on Fletcher's Ice Island, T–3, 1952–1955*, Vol. III (Geophysics Research Papers, No. 63, Geophysics Research Directorate, Air Force Cambridge Research Center, Air Research and Development Command, United States Air Force, Bedford, Massachusetts), pp. 86–91.

BARUAH, H. K. (1961). The air spora of a cowshed. *J. gen. Microbiol.*, **25**, 483–91.

BASSI, A. (1835/1958). *Del mal del segno*. Transl. P. J. Yarrow. *Phytopath. Class.*, No. 10, American Phytopathological Society, 49 pp.

BATEMAN, A. J. (1947a). Contamination of seed crops. I Insect Pollination. *J. Genet.*, **48**, 257–75.

— (1947b). Contamination of seed crops. II Wind Pollination. *Heredity*, **1**, 235–46.

— (1947c). Contamination of seed crops. III Relation with isolation distance. *Heredity*, **1**, 303–36.

— (1950). Is gene dispersal normal? *Heredity*, **4**, 353–63.

BELASCO, J. E. (1952). Characteristics of air masses over the British Isles. *Geophys. Mem., Lond.*, **11**, No. 87, 34 pp.

BELIN, L., FALSEN, E., HOBORN, J. & ANDRÉ, J. (1970). Enzyme sensitization in consumers of enzyme-containing washing powder. *Lancet*, 1970, ii, 1153–1157.

BELLI, C. M. (1901). Chemische, mikroskopische und bakteriologische Untersuchungen über den Hagel. *Hyg. Rdsch.*, **11**, 1181–7.

BENNINGHOFF, W. S. (1966). Pollen spectra from bryophyte polsters, Inverness Mud Lake Bog, Cheboygan, Michigan. *Pap. Mich. Acad. Sci.*, **46**, 41–60.

BERGERON, T. (1944). On some meteorological conditions for the dissemination of spores, pollen, etc., and a supposed wind transport of *Aloina* spores from the region of Lower Yenisey to southwestern Finland in July 1936. *Svensk. bot. Tidskr.*, **38**, 269–92.

BERRY, C. M. (1941). An electrostatic method for collecting bacteria from air. *Publ. Hlth. Rep. Wash.*, **56**, (Part 2), 2044–51.

BESSEY, C. E. (1883). Remarkable fall of pine pollen. *Am. Nat.*, **17**, 658.

BEST, A. C. (1950). The size distribution of raindrops. *Q. Jl R. met. Soc.*, **76**, 16–36.

— (1952). The evaporation of raindrops. *Q. Jl R. met. Soc.*, **78**, 200–25.

BILGRAMI, K. S. (1963). Studies on conidial dispersal of some pathogenic species of *Phyllosticta*. *Naturwissenschaften*, **50**, 360.

BISBY, G. R. (1935). Are living spores to be found over the ocean? *Mycologia*, **27**, 84–5.

— (1943). Geographical distribution of fungi. *Bot. Rev.*, **9**, 466–88.

BLACK, S. H. & GERHARDT, P. (1962). Permeability of bacterial spores. IV Water content, uptake and distribution. *J. Bact.*, **83**, 960–7.

BLACKLEY, C. H. (1873). *Experimental Researches on the Causes and Nature of Catarrhus Aestivus (Hay fever or hay asthma)*. Ballière, Tindall & Cox, London, 202 pp. (Reprinted: Dawson, London, 1959).

BLACKMAN, G. E. (1942). Statistical and ecological studies in the distribution of species in plant communities. I Dispersion as a factor in the study of changes of plant populations. *Ann. Bot. Lond.*, **6**, 351–70.

BLANCHARD, D. C. (1950). The behaviour of water drops at terminal velocity in air. *Trans. Am. geophys. Un.*, **31**, 836–42.

— & WOODCOCK, A. H. (1957). Bubble formation in the sea and its meteorological significance. *Tellus*, **9**, 145–158.

BODMER, H. (1922). Über den Windpollen. *Natur u. Tech. Zürich*, **3**, 66.

BOEDIJN, K. B. (1940). The mycetozoa, fungi and lichens of the Krakatau group. *Bull. Jard. Bot. Buitenz, III*, **16**, 358–429.

BONDE, R. & SCHULTZ, E. S. (1943). Potato refuse piles as a factor in the dissemination of late blight. *Bull. Maine agric. Exp. Sta.* No. 416. (*Am. Potato J.*, **20**, 112–18.)

BONNER, G., MATRUCHOT, L. & COMBES, R. (1911). Recherches sur la dissémination des germes microscopiques dans l'atmosphère. *C.R. Acad. Sci., Paris*, **152**, 652–9.

BONNIER, L. & THOMAS, R. (1960). *Faune terreste et d'eau douce des Pyrénées-Orientales*. Fasc. 5. *Thecamoebiens du sol*. Hermann, Paris, 182 figures.

BOOKER, D. V. (1958). *Physical measurements of activity in samples from Windscale*. Atomic Energy Research Establishment Rept. HP/R 2607, H.M.S.O., London. 16 pp.

BOSANQUET, C. H. & PEARSON, J. L. (1936). The spread of smoke and gases from chimneys. *Trans. Faraday Soc.*, **32**, 1249–63.

BOURDILLON, R. B. & COLEBROOK, L. (1946). Air hygiene in dressing-rooms for burns or major wounds. *Lancet*, 1946, i, 561–5, 601–5.

— LIDWELL, O. M. & LOVELOCK, J. E. (1948). *Studies in air hygiene*. Med. Res. Council Special Rept. Series, No. 262, H.M.S.O., London, 356 pp.

— LIDWELL, O. M., LOVELOCK, J. E. & RAYMOND, W. F. (1948). Airborne bacteria found in factories and other places: suggested limits of bacterial contamination. *See* Bourdillon, Lidwell & Lovelock (1948), 257–63.

— LIDWELL, O. M. & THOMAS, J. C. (1941). A slit sampler for collecting and counting airborne bacteria. *J. Hyg., Camb.,* **41**, 197–224.

BOURKE, P. M. A. (1964). Emergence of potato blight. *Nature, Lond.,* **203**, 805–8.

BOWDEN, J., GREGORY, P. H. & JOHNSON, C. G. (1971). Possible wind transport of coffee leaf rust across the Atlantic Ocean. *Nature, Lond.,* **229**, 500–1.

BRACHMAN, P. S., KAUFMANN, A. F. & DALLDORF, F. G. (1966). Industrial inhalation anthrax. *Bact. Rev.,* **30**, 646–57.

BRANDES, E. W. (1919). Banana wilt. *Phytopathology,* **9**, 330–89.

BRODIE, H. J. (1951). The splash-cup dispersal mechanism in plants. *Can. J. Bot.,* **29**, 224–34.

— (1957). Raindrops as plant dispersal agents. *Indiana Acad. Sci.,* **66**, 65–73.

— & GREGORY, P. H. (1953). The action of wind in the dispersal of spores from cup-shaped plant structures. *Can. J. Bot.,* **31**, 402–10.

BRONSWIJK, J. E. M. H. VAN & SINHA, R. N. (1971). Tyroglyphid mites (Acari) and house dust allergy. *J. Allergy,* **41**, 31–52.

BROOK, P. J. (1959). A volumetric spore trap for sampling pastures. *N.Z. Jl agric. Res.,* **2**, 690–93.

BROOKS, J., GRANT, P. R., MUIR, M. D., GIJZEL, P. VAN & SHAW, G. (1971). *Sporopollenin.* Academic Press, London & New York, 718 pp.

BROWN, J. G. (1942). Wind dissemination of angular leaf spot of cotton. *Phytopathology,* **32**, 81–90.

BROWN, M., WEINTROUB, D. & SIMPSON, M. W. (1947). Timber as a source of sporotrichosis infection. In *Sporotrichosis infection on mines of the Witwatersrand,* Transvaal Chamber of Mines, Johannesburg, pp. 5–33.

BROWN, R. M., LARSON, D. A. & BOLD, H. C. (1964). Airborne algae: their abundance and heterogeneity. *Science, N.Y.,* **143**, 583–585.

BROWN, W. D. (1951). *Parachutes.* Pitman, London, 322 pp.

BROWNE, J. G. (1930). Living micro-organisms in the air of the arid south west. *Science, N.Y.,* **72**, 322-3.

BRUCH, C. W. (1967). Microbes in the upper atmosphere and beyond. *See* Gregory & Monteith (1967), pp. 345–374.

BRUN, R. J., LEWIS, W., PERKINS, P. J. & SERAFINI, J. S. (1955). *Impingement of cloud droplets on a cylinder and procedure for measuring liquid-water content and droplet sizes in supercooled clouds by multirotating cylinder method.* U.S. Nat. Adv. Comm. Aeronautics, Report 1215(1), pp. 1–43.

BRUNT, D. (1934). *Physical and dynamical meteorology.* Cambridge University Press, England, 411 pp.

BUCHBINDER, L., SOLOWEY, M. & SOLOTOROVSKY, M. (1945). Comparative quantitative studies of bacteria in air of enclosed spaces. Air pollution survey of New York City. Part I. Report of New York City Air Pollution Survey. *Heat. Pip. Air. Condit.* (ASHVE Journal Section), 1945, 389–97.

BUELL, M. F. (1947). Mass dissemination of pine pollen. *J. Elisha Mitchell scient. Soc.,* **63**, 163–7.

BUJWID, O. (1888). Die Bakterien in Hagelkörnern. *Zentbl. Bakt.,* **3**, 1–2.

BULLER, A. H. R. (1909–1950). *Researches on Fungi.* Longmans, England. Vol. I, 1909; Vol. II, 1922; Vol. III, 1924; Vol. IV, 1931; Vol. V, 1933; Vol. VI, 1934; University of Toronto Press, Toronto, Vol. VII, 1950.

— (1915). Micheli and the discovery of reproduction in fungi. *Trans. R. Soc. Can.* (Ser. 3), **9**, Sect. IV, 1–25.

— & LOWE, C. W. (1911). Upon the number of micro-organisms in the air of Winnipeg. *Trans. R. Soc. Can.* (Ser. 3), **4**, Sect. IV, 41–58.

BULLOCH, W. (1938). *The History of Bacteriology*. Clarendon Press, Oxford, 422 pp.

BURGES, A. (1950). The downward movement of fungal spores in sandy soil. *Trans. Br. mycol. Soc.*, **33**, 142–147.

BURNETT, J. H. & PARTINGTON, M. (1957). Spatial distribution of fungal mating-type factors. *Proc. R. phys. Soc. Edinb.*, **26**, 61–68.

BURRILL, T. J. (1907). Bitter rot of apples: botanical investigations. *Illinois agr. Exp. Sta. Bull.*, **118**, 555–608.

BURRILL, T. J. & BARRETT, J. T. (1909). Ear rots of corn. *Illinois agric. Exp. Sta. Bull.*, **133**, 65–109.

BUSSE, J. (1926). Kiefernpollenflug und forstliche Saatgutanerkennung. *Tharandt. Forstl. Jb.*, **77**, 225–31.

BUTLER, E. J. (1917). The dissemination of parasitic fungi and international legislation. *Mem. Dept. Agric. India, Bot.*, **9**, 1–73.

BYWATER, J. (1959). Infection of peas by *Fusarium solani* var. *martii* forma 2 and the spread of the pathogen. *Trans. Br. mycol. Soc.*, **42**, 201–212.

CADHAM, F. T. (1924). Asthma due to grain rusts. *J. Am. med. Ass.*, **83**, 27.

CALDER, K. L. (1952). Some recent British work on the problem of diffusion in the lower atmosphere. Chapter 91 in *Air Pollution* (Edited by L. C. McCabe), Proc. U.S. Tech. Conf. on Air Pollution, pp. 787–93, New York.

CALPOUZOS, L. (1955). Studies on the Sigatoka disease of bananas and its fungus pathogen. *Atkins Garden and Research Laboratory, Cuba*, 70 pp.

CAMMACK, R. H. (1955). Seasonal changes in three common constituents of the air spora of Southern Nigeria. *Nature, London*, **176**, 1270–2.

— (1958a). Factors affecting infection gradients from a point source of *Puccinia polysora* in a plot of *Zea mays*. *Ann. appl. Biol.*, **46**, 186–97.

— (1958b). Studies on *Puccinia polysora* Underw. I The world distribution of forms of *P. polysora*. *Trans. Br. mycol. Soc.*, **41**, 89–94.

— (1958c). Studies on *Puccinia polysora* Underw. II A consideration of the method of introduction of *P. polysora* into Africa. *Trans. Br. mycol. Soc.*, **42**, 27–32.

CARLILE, M. J. (1970). The photoresponses of Fungi. In P. Halldal (Ed.) *The Photobiology of Micro-organisms*, Wiley, pp. 309–344.

CARNELLEY, T. & HALDANE, J. S. (1887). The air of sewers. *Proc. R. Soc.*, **42**, 501–522.

— HALDANE, J. S. & ANDERSON, A. M. (1887). The carbonic acid, organic matter and micro-organisms in air, more especially in dwellings and schools. *Phil. Trans. R. Soc. Ser. B*, **178**, 61–111.

CARROL, G. (1943). The use of bryophyte polsters and mats in the study of recent pollen deposition. *Am. J. Bot.*, **30**, 361–366.

CARTER, M. V. (1961). Calibration of whirling-arm air sampler. *Rep. Rothamst. exp. Sta. for 1960*, p. 125.

— (1965). Ascospore deposition in *Eutypa armeniacae*. *Aust. J. agric. Res.*, **16**, 825–36.

— MOLLER, W. J. & PADY, S. M. (1970a). Factors affecting uredospore production and dispersal in *Tranzschelia discolor*. *Aust. J. agric. Res.*, **21**, 905–14.

— YAP, A. S. & PADY, S. M. (1970b). Factors affecting uredospore liberation in *Puccinia antirrhini*. *Aust. J. agric. Res.*, **21**, 921–25.

CATON, P. G. F. (1966). A study of raindrop-size distribution in the free atmosphere. *Q. Jl R. met. Soc.*, **92**, 15–30.

CAWOOD, W. (1936). The movement of dust or smoke particles in a temperature gradient. *Trans. Faraday Soc.*, **32**, 1068–73.

CHAMBERLAIN, A. C. (1953). Experiments on the deposition of Iodine-131 vapour onto surfaces from an airstream. *Phil. Mag.* Ser. 7, **44**, 1145–53.

— (1956). *Aspects of travel and deposition of aerosol and vapour clouds.* Atomic Energy Research Establishment Report, HP/R 1261, H.M.S.O., London, 35 pp.

— (1966a). Transport of *Lycopodium* spores and other small particles to rough surfaces. *Proc. R. Soc. A.*, **296**, 45–70.

— (1966b). Transport of gases to and from grass and grass-like surfaces. *Proc. R. Soc. A.*, **190**, 236–65.

— (1967a). Deposition of particles to natural surfaces. *See* Gregory & Monteith (1967), pp. 138–64.

— (1967b). Cross-pollination between fields of sugar beet. *Q. Jl R. met. Soc.*, **93**, 509–15.

— (1970). Deposition and uptake by cattle of airborne particles. *Nature, London*, **225**, 99–100.

CHANNELL, S., BLYTH, W., LLOYD, M., WEIR, D. M., AMOS, W. M. G., LITTLE-WOOD, A. P., RIDDLE, H. F. V. & GRANT, I. W. B. (1969). Allergic alveolitis in maltworkers. *Q. Jl Med.*, **38**, 351–76.

CHATIGNY, M. A. & CLINGER, D. I. (1969). Contamination control in aerobiology. *See* Dimmick & Akers (1969), pp. 194–263.

CHATTERJEE, G. (1931). A note on an apparatus for catching spores from the upper air. *Indian J. agric. Sci.*, **1**, 306–8.

CHAUVIN, R. & LAVIE, P. (1956). Recherches sur la substance antibiotique du pollen. *Annls. Inst. Pasteur, Paris*, **90**, 523–7.

CHERRY, E. & PEET, C. E. (1966). An efficient device for the rapid collection of fungal spores from infected plants. *Phytopathology*, **56**, 1102–3.

CHESTER, K. S. (1946). *The nature and prevention of the cereal rusts as exemplified in the leaf rust of wheat.* Chronica Botanica, Waltham, Massachussetts, 269 pp.

CHRISTENSEN, C. M. (1950). Intramural dissemination of spores of *Hormodendrum resinae. J. Allergy*, **21**, 409–13.

CHRISTENSEN, J. J. (1942). Long distance dissemination of plant pathogens. *See* Moulton (1942), pp. 78–87.

CHRISTOFF, A. (1934). Sposobŭ za khvaschane na raznasyanitê chrezŭ vêtŭra spori. *Rev. Inst. Rech. agron. Bulg.*, **6**, 41–48.

CLARK, H. E. (1951). An atmospheric pollen survey of four centres in the North Island, New Zealand, 1949–50. *N.Z. Jl Sci. Technol. B*, **33**, 73–91.

COCKE, E. C. (1937). Calculating pollen concentration of the air. *J. Allergy*, **8**, 601–6.

— (1938). A method for determining pollen concentration of the air. *J. Allergy*, **9**, 458–63.

COLEBROOK, L. & CAWSTON, W. C. (1948). Microbic content of the air on the roof of a city hospital, at street level, and in the wards. *See* Bourdillon, Lidwell & Lovelock (1948), pp. 233–41.

COLWELL, R. N. (1951). The use of radioactive isotopes in determining spore distribution patterns. *Am. J. Bot.*, **38**, 511–23.

COMMITTEE ON APPARATUS IN AEROBIOLOGY (1941). Techniques for appraising airborne populations of micro-organisms, pollen and insects. *Phyto-pathology*, **31**, 201–25.

COMMITTEE OF INQUIRY ON FOOT-AND-MOUTH DISEASE, 1968 (1969). *Report, Part 2*, Cmnd. 4225, H.M.S.O., London, 135 pp.

CONE, C. D. (1962). The soaring flight of birds. *Scient. Am.*, **206**, 130–40.

CONVERSE, J. L. & REED, R. E. (1966). Experimental epidemiology of coccidioi-domycosis. *Bact. Rev.*, **30**, 678–94.

CORNET, G. (1889). Die Verbreitung der Tuberkelbacillen ausserhalb des Körpers. *Z. Hyg. InfecktKrankh.*, **5**, 191–331.

COTTER, R. U. (1931). Black stem rust combed from the air by fliers. *Yearb. Agric. U.S. Dep. Agric. 1931*, 116–18.

COX, C. S. (1968). The aerosol survival and cause of death of *Escherichia coli*. *J. gen. Microbiol.*, **54**, 169–175.

— & BALDWIN, F. (1967). The toxic effect of oxygen upon the aerosol survival of *Escherichia coli* B., *J. gen. Microbiol.*, **49**, 115–17.

COX, R. A. & PENKETT, S. A. (1971). Oxidation of atmospheric SO_2 by products of the ozone–olefin reaction. *Nature, Lond.*, **230**, 321–2.

CRAIGIE, J. H. (1933). Union of pycniospores and haploid hyphae in *Puccinia helianthi* Schw., *Nature, Lond.*, **131**, 25.

— (1940). Aerial dissemination of plant pathogens. *Proc. VI Pacific Sci. Congr., 1939*, **4**, 753–767.

— (1942). Heterothallism in the rust fungi and its significance. *Trans. R. Soc. Can.* (Ser. 3), Sect. V, **36**, 19–40.

— (1945). Epidemiology of stem rust in Western Canada. *Scient. Agric.*, **25**, 285–401.

CRAWLEY, W. E., CAMPBELL, A. G. & DREW SMITH, J. (1962). Movement of spores of *Pithomyces chartarum* on leaves of rye-grass. *Nature, Lond.*, **193**, 295.

CRISTIANI, H. (1893). Analyse bactériologique de l'air des hauteurs, puisé pendant un voyage en ballon. *Annls. Inst. Pasteur, Paris*, **7**, 665–71.

CUNNINGHAM, D. D. (1873). *Microscopic Examinations of Air*. Government Printer, Calcutta, 58 pp.

DADE, H. A. (1927). Economic significance of cacao pod diseases and factors determining their incidence and control. *Dep. Agric. Gold Coast, Bull.*, No. 6, 59 pp.

DANIELSEN, E. F. (1964). Extrusion of stratopheric air into troposphere. *See* Tsuchiya & Brown (1964).

DARLING, C. A. & SIPLE, P. A. (1941). Bacteria of Antarctica. *J. Bact.*, **42**, 83–98.

DARLOW, H. M. & BALE, W. R. (1959). Infective hazards of water-closets. *Lancet*, i, 1196–1200.

DARWIN, C. (1846). An account of the fine dust which often falls on vessels in the Atlantic Ocean. *Q. Jl geol. Soc. Lond.*, **2**, 26–30. (Reprinted in various editions of Findlay's *Memoir Descriptive of the North Atlantic Ocean*; and *Directory for the North Atlantic Ocean*.)

DAVIES, C. N. (1947). The sedimentation of small suspended particles. *Trans. Instn. Chem. Engrs. Lond.*, **25** (Suppl. Symposium on Particle Size Analysis, p. 25).

— (1952). The separation of airborne dust and particles. *Proc. Inst. Mech. Engrs. Lond. B*, **1**, 185–99.

— (1960). Deposition of *Lycopodium* spores upon glass slides exposed in a wind tunnel. *Br. J. appl. Phys.*, **11**, 535–8.

— (1966). Deposition of aerosols from turbulent flow through pipes. *Proc. R. Soc. A*, **289**, 235–46.

— (1968). The entry of aerosols into sampling tubes and heads. *Br. J. appl. Phys. Ser. 2*, **1**, 921–32.

— AYLWARD, M. & LEACEY, D. (1951). Impingement of dust from air jets. *A.M.A. Arch. Industr. Hyg.*, **4**, 354–97.

— & PEETZ, C. V. (1956). Impingement of particles on a transverse cylinder. *Proc. R. Soc. A*, **234**, 269–5.

DAVIES, R. R. (1957). A study of airborne *Cladosporium*. *Trans. Br. mycol. Soc.*, **40**, 409–14.

— (1959). Detachment of conidia by cloud droplets. *Nature, Lond.*, **183**, 1695.

— (1961). Wettability and the capture, carriage and deposition of particles by raindrops. *Nature, Lond.*, **191**, 616–7.

— (1969a). Aerobiology and the relief of asthma in an alpine valley. *Acta allerg.*, **24**, 377–95.

— (1969b). Climate and topography in relation to aero-allergens at Davos and London. *Acta allerg.*, **24**, 396–409.

— (1971). Air sampling for fungi, pollens and bacteria. In C. Booth (Ed.), *Methods in Microbiology*, Vol. 4, pp. 367–404, Academic Press, London & New York.

— DENNY, M. J. & NEWTON, L. M. (1963). A comparison between the summer and autumn air sporas at London and Liverpool. *Acta allerg.*, **18**, 131–47.

— & NOBLE, W. C. (1962). Dispersal of bacteria on desquamated skin. *Lancet*, **11**, 1295–7.

DAVIS, D. R. & SECHLER, D. (1966). A simple, inexpensive, continuous sampling spore trap. *Pl. Dis. Reptr.*, **50**, 906–7.

DAWS, L. F. (1967). Movement of air streams indoors. *See* Gregory & Monteith (1967).

DEUTSCHEN WETTERDIENST (1963). *Aspirations-Psychrometer-Tafeln*. 4th Edition, F. Vierweg u. Sohn, Braunschweig. 194 pp.

DILLON WESTON, W. A. R. *See* Weston, W. A. R. Dillon.

DI MENNA, M. E. *See* Menna, M. E. Di.

DIMMICK, R. L. (1965). (On *Coccidioides immitis* spores.) *See* Tsuchiya & Brown (1965), 212.

— (1969). Stirred-settling aerosols and stirred-settling aerosol chambers. *See* Dimmick & Akers (1969), pp. 127–63.

— & AKERS, A. B. (Eds) (1969). *An Introduction to Experimental Aerobiology*. Wiley-Interscience, New York & London, 494 pp.

— & HATCH, M. T. (1969). Dynamic aerosols and dynamic aerosol chambers. *See* Dimmick & Akers (1969), pp. 177–93.

— HATCH, M. T. & NG, J. (1958). A particle-sizing method for aerosols and fine powders. *A.M.A. Arch. Industr. Health*, **18**, 23–29.

DIMOND, A. E. & HORSFALL, J. G. (1965). The theory of inoculum. In K. F. Baker and W. C. Snyder (Eds.), *Ecology of Soil-Borne Plant Pathogens*, John Murray, London, pp. 404–415.

DINGLE, A. (1957). Meteorological considerations in ragweed hay fever research. *Fed. Proc. Fedn. Am. Socs. exp. Biol.*, **16**, 615–27.

— GILL, C. C., WAGNER, W. H. & HEWSON, E. W. (1959). The emission, dispersion and deposition of ragweed pollen. *Adv. Geophys.*, 6, 367–387.

DIXON, P. A. (1961). Spore dispersal in *Chaetomium globosum* (Kunze). *Nature, Lond.*, 191, 1418–9.

— (1963). Spore liberation by water drops in some myxomycetes. *Trans. Br. mycol. Soc.*, 46, 615–9.

DOBBS, C. G. (1942a). Spore dispersal in the Mucorales. *Nature, Lond.*, 149, 583.

— (1942b). On the primary dispersal and isolation of fungal spores. *New Phytol.*, 41, 63–9.

DOBELL, C. (1932). *Antony van Leeuwenhoek and his 'Little Animals'*. Bale & Danielsson, London, 435 pp.

DOCTERS VAN LEEUWEN, W. M. *See* Leeuwen, Docters van.

DOUGLAS, H. W., COLLINS, A. E. & PARKINSON, D. (1959). Electric charge and other surface properties of some fungal spores. *Biochem. biophys. Acta*, 33, 535–8.

DOWDING, P. (1969). The dispersal and survival of spores of fungi causing bluestain in pine. *Trans. Br. mycol. Soc.*, 52, 125–37.

DRANSFIELD, M. (1966). The fungal air spora at Samaru, Northern Nigeria. *Trans. Br. mycol. Soc.*, 49, 121–32.

DRUETT, H. A. (1942). The microimpactor: an apparatus for sampling solid and liquid particulate clouds. *Porton Report, No. 2458, Serial No. 32*, 12 pp.

— (1964). Equilibrium temperature of a small sphere suspended in air and exposed to solar radiation. *Nature, Lond.*, 201, 611.

— (1967). The inhalation and retention of particles in the human respiratory system. *See* Gregory & Monteith (1969).

— & MAY, K. R. (1952). A wind tunnel for the study of airborne infections. *J. Hyg., Camb.*, 50, 69–81.

— & MAY, K. R. (1968). Unstable germicidal pollutant in rural air. *Nature, Lond.*, 220, 395–6.

— & PACKMAN, L. P. (1968). Sensitive microbiological detection of air pollution. *Nature, Lond.*, 218, 699.

D.S.I.R. (1931). *The investigation of atmospheric pollution*. 16th Report, 1930, p. 11.

DUBOIS, R. (1918). Sur la présence d'organismes vivants dans les grêlons, *Annls. Soc. Linn. Lyon* (n.s.), 64, 45–51.

DUBUY, H. G. & CRISP, L. R. (1944). A sieve device for sampling airborne micro-organisms. *Publ. Hlth. Rep., Wash.*, 59, 829–32.

— & HOLLAENDER, A. & LACKEY, M. D. (1945). A comparative study of sampling devices for airborne micro-organisms. *Publ. Hlth. Rep., Wash. Suppl.* 184, 40 pp.

DUGUID, J. P. (1946). The size and duration of air-carriage of respiratory droplets and droplet-nuclei. *J. Hyg. Camb.*, 44, 471–9.

DURHAM, O. C. (1942). Airborne fungus spores as allergens. *See* Moulton (1942).

— (1943). The volumetric incidence of atmospheric allergens. I Specific gravity of pollen grains. *J. Allergy*, 14, 455–61.

— (1944). The volumetric incidence of atmospheric allergens. II Simultaneous measurements by volumetric and gravity slide methods. Results with ragweed pollen and *Alternaria* spores. *J. Allergy*, 15, 226–35.

318 BIBLIOGRAPHY

— (1946). The volumetric incidence of atmospheric allergens. IV A proposed standard method of gravity sampling, counting and volumetric interpolation of results. *J. Allergy*, **17**, 79–86.

— (1947). The volumetric incidence of atmospheric allergens. V Spot testing in the evaluation of species. *J. Allergy*, **18**, 231–8.

DURRELL, L. W. & PARKER, J. H. (1920). The comparative resistance of varieties of oats to crown and stem rusts. *Res. Bull. Ia. agric. Exp. Sta.*, **62**, 1–56.

DWORIN, M. (1966). A study of atmospheric mould spores in Tucson, Arizona. *Ann. Allergy*, **24**, 31–6.

DYAKOWSKA, J. (1936). Researches on the rapidity of the falling down of pollen of some trees. *Bull. Int. Acad. Cracovie (Acad. Po. Sci.), Ser. B, Sci. Nat.*, **1**, 155–168.

— (1948). The pollen rain on the sea and on the coast of Greenland. *Bull. Int. Acad. Cracovie (Acad. Pol. Sci.), Ser. B, Sci. Nat.*, **1**, 25–33.

DYE, M. H. & VERNON, T. R. (1952). Airborne mould spores. *N.Z. Jl Sci. Technol.*, **34**, 118–27.

EDGERTON, H. E. & KILLIAN, J. R. (1939). *Flash! Seeing the Unseen by Ultra-High-Speed Photography*. Hale, Cushman & Flint, Boston, 203 pp.

EHRENBERG, C. G. (1849). *Passatstaub und Blutregen*. Kg. Akad. Wiss. Berlin, 192 pp.

— (1872a). Übersicht der Seit 1847 fortgesetzten Untersuchungen über das von der Atmosphäre unsichtbar getragene reich organische Leben. *Abhandl. Kg. Akad. Wiss. Berlin, Phys. Kl. 1871*, pp. 1–150.

— (1872b). Nachtrag zur Übersicht der organischen Atmosphärilien. *Abhandl. Kg. Akad. Wiss. Berlin, Phys. Kl. 1871*, pp. 233–75.

EIJNATTEN, C. L. M. VAN (1965). Towards the improvement of maize in Nigeria. *Meded. LandbHoogesch. Wageningen.*, **65**, (3), 67–85.

EKELÖF, E. (1907). Studien über den Bakteriengehalt der Luft und des Erdbodens der antarktischen Gegende, ausgeführt während der Swedischen Südpolar-Expedition, 1901–1904. *Z. Hyg. InfecktKrankh.*, **56**, 344–70.

ELLIS, F. P. & RAYMOND, W. F. (1948). Air hygiene in H.M. Ships under wartime conditions. *See* Bourdillon, Lidwell & Lovelock, 1948.

ELLISON, W. D. (1944). Studies in raindrop erosion. *Agric. Engng. St Joseph, Mich.*, **25**, 131–6; 181–2.

EMMONS, C. W. (1943). Coccidioidomycosis in wild rodents. A method of determining the extent of endemic areas. *Publ. Hlth. Rep. Wash.*, **58**, 1–5.

— (1955). Saprophytic sources of *Cryptococcus neoformans* associated with the pigeon *(Columba livia)*. *Am. J. Hyg.*, **62**, 227–32.

— BINFORD, C. H. & UTZ, J. P. (1970). *Medical Mycology*, 2nd edition. Lea & Febiger, Philadelphia, 508 pp.

ENGLISH, H. & GERHARDT, F. (1946). The effect of ultra-violet radiation on the viability of fungus spores and on the development of decay in sweet cherries. *Phytopathology*, **36**, 100–11.

ERDTMAN, G. (1937). Pollen grains recovered from the atmosphere over the Atlantic. *Acta Hort. Gothoburg.*, **12**, 185–96.

— (1943). *An Introduction to Pollen Analysis*. Chronica Botanica, Waltham, Massachussetts, 239 pp.

— (1952). *Pollen Morphology and Plant Taxonomy. Angiosperms. (An Introduction to Palynology, I)*. Almquist, Stockholm, 539 pp.

— (1957). *Pollen and Spore Morphology/Plant Taxonomy. Gymnosperms, Pteridophytes, Bryophytes. (An introduction to Palynology, II)*. (Illustrations.) Almquist, Stockholm, 151 pp.

ERRINGTON, F. P. & POWELL, E. O. (1969). A cyclone separator for aerosol sampling in the field. *J. Hyg. Camb.*, **67**, 387–99.

EVANS, E. (1961). Relationship between the physical characteristics of a banana leaf and the distribution of leafspot lesions (*Mycosphaerella musicola* Leach). *Trans. Br. mycol. Soc.*, **44**, 299.

EVANS, R. GWYN (1965). Sporobolomyces as a cause of respiratory allergy. *Acta allerg.*, **20**, 197–205.

EWART, G. (1965). On the mucus flow-rate of the human nose. *Acta otolaryngol. Suppl.*, **200**, 1–61.

FALCK, R. (1904). Die Sporenverbreitung bei den Basidiomyceten. *Beitr. Biol. Pfl.*, **9**, 1–82.

— (1927). Über die Grössen, Fallgeschwindigkeit und Schwebewarte der Pilzsporen und ihre Gruppierung mit Bezug auf die zu ihrer Verbreitung nötigen Temperaturströmungs-Geschwindigkeit. *Ber. Dtsch. Bot. Ges.*, **45**, 262–81.

FAULWETTER, R. F. (1917a). Dissemination of the angular leafspot of cotton. *J. agric. Res.*, **8**, 457–75.

— (1917b). Wind-blown rain, a factor in disease dissemination. *J. agric. Res.*, **10**, 639–48.

FEINBERG, S. M. (1935). Mold allergy: its importance in asthma and hay fever. *Wisconsin med. J.*, **34**, 254.

— DURHAM, O. C. & DRAGSTEDT, C. A. (1946). *Allergy in Practice*. 2nd edition, Yearbook Publishers, Chicago, pp. 216–84.

FERGUSON, M. C. (1902). A preliminary study of the germination of the spores of *Agaricus campestris* and other Basidiomycetous fungi. *Bull. U.S. Bur. Pl. Ind.*, No. 16, 40 pp.

FIESE, M. J. (1958). *Coccidioidomycosis*. Thomas Springfield, Illinois. 253 pp.

FINCHER, E. L. (1969). Aerobiology and hospital sepsis. *See* Dimmick & Akers (1969), pp. 407–436.

FINDEISEN, W. (1939). Das Verdampfen der Wolken- und Regentropfen. *Met. Zeit.*, **56**, 453.

FINNEY, D. J. (1947). Errors of estimation in inverse sampling. *Nature, Lond.*, **160**, 195–6.

FIRST, M. W. & SILVERMAN, L. (1953). Air sampling with membrane filters. *A.M.A. Arch. Industr. Hyg.*, **7**, 1–11.

FISCHER, B. (1886). Bacteriologische Untersuchungen auf einer Reise nach Westindien. *Z. Hyg. InfecktKrankh.*, **1**, 421–64.

FISCHER, E. & GÄUMANN, E. (1929). *Biologie der Pflanzenbewohnenden parasitischen Pilze*. Fischer, Jena, 428 pp.

FISHER, D. J. & RICHMOND, D. V. (1969). The electrostatic properties of some fungal spores. *J. gen. Microbiol.*, **57**, 51–60.

FISHER, R. A., CORBET, A. S. & WILLIAMS, C. B. (1943). The relationship between the number of species and the number of individuals in a random sample of an animal population. *J. Anim. Ecol.*, **12**, 42–58.

FLEMMING (1908). Über die Arten und die Verbreitung der lebensfähigen Mikroorganismen in der Atmosphäre. *Z. Hyg. InfecktKr.*, **58**, 345–85.

FLINDT, M. L. H. (1969). Pulmonary disease due to inhalation of derivatives of *Bacillus subtilis* containing proteolytic enzyme. *Lancet*, 1969, ii, 1177–81.

FLUGGE, C. (1897). Ueber Luftinfektion. *Z. Hyg. InfektKr.*, **25**, 179–224.

FONTANA, F. (1767/1932). Observations on the rust of grain. Translation by P. P. Pirone, *Phytopath Class.*, No. 2, Ithaca, New York, 1932, 40 pp.

FORBES, J. GRAHAM (1924). The atmosphere of the underground electric railways of London. *J. Hyg. Camb.*, **22**, 123–55.

FOUTIN, W. M. (1889). Die Bakteriologische Untersuchungen von Hagel. (Cited from *Zbl. Bakt.*, **7**, 372–4, 1890.)

FRACKER, S. B. & BRISCHLE, H. A. (1944). Measuring the local distribution of *Ribes. Ecology*, **25**, 283–303.

FRANKLAND, A. W. & HAY, M. J. (1951). Dry rot as a cause of allergic complaints. *Acta allerg.*, **4**, 186–200.

FRANKLAND, P. F. (1886). The distribution of micro-organisms in air. *Proc. R. Soc.*, **40**, 509–26.

— (1887). A new method for the quantitative estimation of the micro-organisms present in the atmosphere. *Phil. Trans. R. Soc. Ser. B.*, **178**, 113–152.

— & HART, T. G. (1887). Further experiments on the distribution of micro-organisms in air (by Hesse's method). *Proc. R. Soc.*, **42**, 267–82.

FREY, C. N. & KEITT, G. W. (1925). Studies of spore dissemination of *Venturia inaequalis* (Cke.) Wint. in relation to seasonal development of apple scab. *J. agric. Res.*, **30**, 529–40.

FUCHS, N. A. (1964). *The Mechanics of Aerosols.* Translation C. N. Davies (Ed.), Pergamon Press, Oxford.

FULTON, H. R. & COBLENTZ, W. H. (1928). The fungicidal action of ultra-violet radiation. *J. agric. Res.*, **38**, 159–68.

FULTON, J. D. (1966a). Micro-organisms of the upper atmosphere. III Relationship between altitude and micropopulation. *Appl. Microbiol.*, **14**, 237–40.

— (1966b).Micro-organisms of the upper atmosphere. IV Micro-organisms of a land air mass as it traverses an ocean. *Appl. Microbiol.*, **14**, 241–4.

— (1966c). Micro-organisms of the upper atmosphere. V Relationship between frontal activity and the micropopulation at altitude. *Appl. Microbiol.*, **14**, 245–50.

— & MITCHELL, R. B. (1966). Micro-organisms of the upper atmosphere. II Micro-organisms in two types of air masses at 690 metres over a city. *Appl. Microbiol.*, **14**, 232–236.

FURCOLOW, M. L. & HORR, W. H. (1956). Air and water in the natural history of *Histoplasma capsulatum*. In *Proceedings on Histoplasmosis.* Publ. Hlth. Monograph, 39. *Publ. Hlth. Serv. Publ. Wash.*, No. 465, 282–8.

GANDERTON, M. A. (1968). Phoma in the treatment of seasonal allergy due to *Leptosphaeria. Acta allerg.*, **23**, 173.

GARDNER, M. W. (1918). The mode of dissemination of fungus and bacterial diseases of plants. *Rep. Michigan Acad. Sci.*, **20**, 357–423.

GARRETT, S. D. (1960). Inoculum potential. *See* Horsfall & Dimond (1960), Vol. III, pp. 23–56.

GÄUMANN, E. (1950). *Principles of Plant Infection.* Crosby Lockwood, London.

GAZERT, H. (1912). Untersuchungen über Meeresbakterien und ihren Einfluss auf den Stoffwechsel in Meere. *Deutsche Sudpolar-Expedition, 1901–1903*, **7**, (3), 235–96.

GEIGER, R. (1965). *The Climate Near the Ground.* Harvard University Press, Cambridge, Massachussetts, 611 pp.

GERONE, P. J., COUCH, R. B., KEEFER, G. V., DOUGLAS, R. G., DERRENBACHER, E. B. & KNIGHT, V. (1966). Assessment of experimental and natural viral aerosols. *Bact. Rev.*, **30**, 576–84.

GILBERT, G. E. (1950). Volumetric and gravity slide tests for airborne ragweed and oak pollen at Columbus, Ohio. *Ohio. J. Sci.*, **50**, 60–70.

GISLÉN, T. (1948). Aerial plankton and its conditions of life. *Biol. Rev.*, **23**, 109–26.

GLYNNE, M. D. (1953). Production of spores by *Cercosporella herpotrichoides*. *Trans. Br. mycol. Soc.*, **36**, 46–51.

GOETZ, A. (1953). Application of molecular filter membranes to the analysis of aerosols. *Am. J. publ. Hlth.*, **43**, 150–9.

— (1965). Parameters for biocolloidal matter in the atmosphere. *See* Tsuchiya & Brown (1965), pp. 79–97.

GOLDBERG, L. J., WATKINS, H. M. S., BOERKE, E. E. & CHATIGNY, M. A. (1958). The use of a rotating drum for the study of aerosols over extended periods of time. *Am. J. Hyg.*, **68**, 85–93.

GOLDFARB, A. R. (1968). Separation of new skin reactive antigens from dwarf ragweed pollen. *J. Immun.*, **100**, 902–11.

GOLDSMITH, P. & BROWN, F. (1961). World-wide circulation of air within the stratosphere. *Nature, Lond.*, **191**, 1033–7.

GOODMAN, D. H., NORTHEY, W. T., LEATHERS, C. R. & SAVAGE, T. H. (1966). A study of airborne fungi in the Phoenix Arizona metropolitan area. *J. Allergy*, **38**, 56–62.

GORDON, F. B. (1970). The microflora of man in confined, controlled environments: review and analysis. In Silver (1970), pp. 167–190.

GORDON, M. A. & CUPP, H. B. (1953). Detection of *Histoplasma capsulatum* and other fungus spores in the environment by means of the membrane filter. *Mycologia*, **45**, 241–52.

GRAHAM FORBES, J. *See* Forbes, J. Graham.

GRAY, R. WHYTLAW-, *See* Whytlaw-Gray, R.

GOTTLIEB, D. (1950). The physiology of spore germination in fungi. *Bot. Rev.*, **16**, 228–57.

GREEN, H. L. & LANE, W. R. (1957). *Particulate Clouds, Dusts, Smokes and Mists.* 2nd edition, Spon, London.

GREENBURG, L., FIELD, F., REED, J. I. & ERHARDT, C. L. (1967). Asthma and temperature change. *Biometeorology*, **2**, (1), 3–6.

GREENE, V. W., PEDERSON, P. D., LUNDGREN, D. A. & HAGBERG, C. A. (1964). Microbiological exploration of stratosphere: results of six experimental flights. *See* Tsuchiya & Brown (1964), pp. 199–211.

GREGORY, P. H. (1945). The dispersion of airborne spores. *Trans. Br. mycol. Soc.*, **28**, 26–72.

— (1948). The multiple-infection transformation. *Ann appl. Biol.*, **35**, 412–7.

— (1949). The operation of the puff-ball mechanism of *Lycoperdon perlatum* by raindrops shown by ultra-high-speed Schlieren cinematography. *Trans. Br. mycol. Soc.*, **32**, 11–15.

— (1951). Deposition of airborne *Lycopodium* spores on cylinders. *Ann. appl. Biol.*, **38**, 357–76.

— (1952). Fungus spores. *Trans. Br. mycol. Soc.*, **35**, 1–18.

— (1954). The construction and use of a portable volumetric spore trap. *Trans. Br. mycol. Soc.*, **37**, 390–404.

— (1957). Electrostatic charges on spores of fungi in air. *Nature, Lond.*, **180**, 330.

— (1958). A correction. *Trans. Br. mycol. Soc.*, **41**, 202.

— (1966). The fungus spore: what it is and what it does. In M. F. Madelin (Ed.), *The Fungus Spore*, Butterworths, London, pp. 1–13.

— (1968). Interpreting plant disease dispersal gradients. *Ann. Rev. Phytopath.*, **6**, 189–212.

— (1971). The Leeuwenhoek Lecture, 1970. Airborne Microbes: their significance and distribution. *Proc. R. Soc. B.*, **177**, 469-83.

— GUTHRIE, E. J. & BUNCE, M. E. (1959). Experiments on splash dispersal of fungus spores. *J. gen. Microbiol.*, **20**, 328–54.

— HAMILTON, E. D. & SREERAMULU, T. (1955). Occurrence of the alga *Gloeocapsa* in the air. *Nature, Lond.*, **176**, 1270.

— & HENDEN, D. R. (1967). Seasonal occurrence of airborne *Ophiobolus graminis* ascospores. *Rep. Rothamsted exp. Sta. for 1966*, pp. 134–5.

— & HIRST, J. M. (1952). Possible role of basidiospores as airborne allergens. *Nature, Lond.*, **170**, 414.

— & HIRST, J. M. (1957). The summer air spora at Rothamsted in 1952. *J. gen. Microbiol.*, **17**, 135–52.

— HIRST, J. M. & LAST, F. T. (1953). Concentrations of basidiospores of the dry rot fungus *(Merulius lacrymans)* in the air of buildings. *Acta allerg.*, **6**, 168–74.

— & LACEY, M. E. (1963). Liberation of spores from mouldy hay. *Trans. Br. mycol. Soc.*, **46**, 73–80.

— & LACEY, M. E. (1964). The discovery of *Pithomyces chartarum* in Britain. *Trans. Br. mycol. Soc.*, **47**, 25–30.

— LONGHURST, T. J. & SREERAMULU, T. (1961). Dispersion and deposition of airborne *Lycopodium* and *Ganoderma* spores. *Ann appl. Biol.*, **49**, 645–58.

— & MONTEITH, J. L. (Eds.) (1967). *Airborne microbes. Symp. Soc. gen. Microbiol.*, **17**, Cambridge University Press, 385 pp.

— — & READ, D. R. (1949). The spatial distribution of insect-borne plant virus diseases. *Ann. appl. Biol.*, **36**, 475–82.

— & SREERAMULU, T. (1958). Air spora of an estuary. *Trans. Br. mycol. Soc.*, **41**, 145–56.

— & STEDMAN, O. J. (1953). Deposition of airborne *Lycopodium* spores on plane surfaces. *Ann. appl. Biol.*, **40**, 651–74.

— & STEDMAN, O. J. (1958). Spore dispersal in *Ophiobolus graminis* and other fungi of cereal foot rots. *Trans. Br. mycol. Soc.*, **41**, 449–56.

GREIG-SMITH, P. (1964). *Quantitative plant ecology*. Butterworths, London.

GUNN, R. & KINZER, G. D. (1949). The terminal velocity of fall for water droplets in stagnant air. *J. Met.*, **9**, 243–48.

HAAS, G. J. (1956). Use of the membrane filter in the brewing laboratory. *Wallerstein Labs. Commun.*, **19**, 7–20.

HAFSTEN, U. (1960). Pleistocene development of vegetation and climate in Tristan da Cunha and Gough Island. *Arbok Univ. Bergen, Mat. Natur. Ser. 1960*, No. 20, 48 pp.

HAHN, M. (1909). Die Bestimmung und meteorologische Verwertung der Keimzahl in den höheren Luftschichten. Nach vom Luftballon aus angestellten Beobachtungen. *Zbl. Bakt.* Abt. I, **51**, 97–114.

HALVORSON, H. O. & SRINIVASAN, V. R. (1964). Can spores survive space travel? *See* Tsuchiya & Brown (1964).

HAMILTON, E. D. (1957). A comparison of the pollen and fungus spore content of the air in two localities as a contribution to the study of respiratory allergy. *Ph.D. Thesis,* University of London.

— (1959). Studies in the air spora. *Acta allerg.*, **13**, 143–175.

HAMILTON, E. M. & JARVIS, W. D. (1962). The identification of atmospheric dust by use of the microscope. RD/P/R21. Power Generation Board, Central Electricity Generating Board, December 1962.

HAMMETT, K. R. W. & MANNERS, J. G. (1971). Conidium liberation in *Erysiphe graminis*. I Visual and statistical analysis of spore trap records. *Trans. Br. mycol. Soc.*, **56**, 387–401.

HANSEN, H. M. & SMITH, R. E. (1932). The mechanism of variation in imperfect fungi: *Botrytis cinerea. Phytopathology*, **22**, 953–64.

HARRINGTON, J. B., GILL, G. C. & WARR, B. R. (1959). High efficiency pollen samplers for use in clinical allergy. *J. Allergy*, **30**, 357–75.

HARRISON, F. C. (1898). Bacterial content of hailstones. *Bot. Gaz.*, **26**, 211–4.

HARVEY, R. (1970). Air spora studies at Cardiff. *Trans. Br. mycol. Soc.*, **54**, 251–4.

HARZ, C. O. (1904). Bakteriologische Untersuchungen der freien Atmosphäre mittels Luftballons, nebst Bemerkungen über den atmosphärischen Staub. *Jb. Dtsch. LuftschVerb*. 1904, 147–70.

HASKELL, R. J. & BARSS, H. P. (1939). Fred Campbell Meier, 1893–1938. *Phytopathology*, **29**, 293–302.

HATCH, T. F. & GROSS, P. (1964). *Pulmonary Deposition and Retention of Inhaled Aerosols*. Academic Press, New York & London.

HAUGEN, D. A., BARAD, M. L. & ANTANAITIS, P. (1961). Values of parameters appearing in Sutton's diffusion models. *J. Meteorol.*, **18**, 368–72.

HAYES. Comparison of the rotoslide and Durham samplers in a survey of airborne pollen. *Ann. Allergy*, **27**, 575–584.

HEALD, F. D. (1913). The dissemination of fungi causing disease. *Trans. Amer. Micr. Soc.*, **32**, 1–29.

— & GEORGE. D. C. (1918). The wind dissemination of the spores of bunt or stinking smut of wheat. *Bull. Wash. St. agric. Exp. Sta.*, No. 151, 23 pp.

HEISE, H. A. & HEISE, E. R. (1948). The distribution of ragweed pollen and *Alternaria* spores in the upper atmosphere. *J. Allergy*, **19**, 403–7.

— & HEISE, E. R. (1949). The influence of temperature variations and winds aloft on the distribution of pollens and molds in the upper atmosphere. *J. Allergy*, **20**, 378–82.

— & HEISE, E. R. (1950). Meteorological factors in the distribution of pollens and molds: a review and geographic influence. *Ann. Allergy*, **8**, 641–4.

— & HEISE, E. R. (1957). Effect of a city on the fall-out of pollens and molds. *J. Am. med. Ass.*, **163**, 803–4.

HENDERSON, D. W. (1952). An apparatus for the study of airborne infection. *J. Hyg. Camb.*, **50**, 53–68.

HERMANSEN, J. E., JOHANSEN, B., HANSEN, H. W. & CARSTENSEN, P. (1965). Notes on the trapping of powdery mildew conidia and uredospores by aircraft in Denmark in 1964. *Kgl. Vetr. & Landbøhojsk. Årsskr.*, 1965, 121–9.

HERXHEIMER, H., HYDE, H. A. & WILLIAMS, D. A. (1969). Allergic asthma caused by basidiospores. *Lancet*, 1969, ii, 131–2.

HESSE, E. (1914). Bakteriologische Untersuchungen auf einer Fahrt nach Island, Spitsbergen und Norwegen im Juli 1913. *Zbl. Bakt.*, Abt. I, **72**, 454–77.

HESSE, W. (1884). Ueber quantitative Bestimmung der in der Luft enthaltenen Mikroorganismen. *Mitth. Kaiserl. Gesundheitsamte*, **2**, 182–207.

— (1888). Bemerkungen zur quantitative Bestimmung der Mikroorganismen in der Luft. *Z. Hyg. InfektKrankh.*, **4**, 19–21.

HESSELMAN, H. (1919). Über die Verbreitungsfähigkeit des Waldbaumpollens. [In Swedish, German Summary] *Medd. Skogsförsöksanst. Stockh.*, **16**, 27–60.

HESSELTINE, C. W. (1960). Relationships of the Actinomycetales. *Mycologia*, **52**, 460–74.

HIRST, J. M. (1952). An automatic volumetric spore trap. *Ann. appl. Biol.*, **39**, 257–65.

— (1953). Changes in atmospheric spore content: diurnal periodicity and the effects of weather. *Trans. Br. mycol. Soc.*, **36**, 375–93.

— (1959). Spore liberation and dispersal. In C. S. Holton (Ed.), *Plant Pathology – Problems and Progress, 1908–1958*, University of Wisconsin Press, Madison, pp. 529–38.

— & HURST, G. W. (1967). Long-distance spore transport. *See* Gregory & Monteith (1967), pp. 307–44.

— & STEDMAN, O. J. (1961). The epidemiology of apple scab (*Venturia inaequalis* (Cke.) Wint.). I Frequency of airborne spores in orchards. *Ann. appl. Biol.*, **49**, 190–305.

— & STEDMAN, O. J. (1963). Dry liberation of fungus spores by raindrops. *J. gen. Microbiol.*, **33**, 335–44.

— & STEDMAN, O. J. (1971). Patterns of spore dispersal in crops. In T. F. Preece and C. H. Dickinson (Eds.) *Ecology of Leaf Surface Micro-organisms*, Academic Press, London & New York, pp. 229–37.

— STEDMAN, O. J. & HOGG, W. H. (1967). Long-distance spore transport: methods of measurement, vertical spore profiles and the detection of immigrant spores. *J. gen. Microbiol.*, **48**, 329–55.

— STEDMAN, O. J. & HURST, G. W. (1967). Long-distance spore transport: vertical sections of spore clouds over the sea. *J. gen. Microbiol.*, **48**, 357–77.

— STOREY, I. F., WARD, W. C. & WILCOX, H. J. (1955). The origin of apple scab epidemics in the Wisbech area in 1953 and 1954. *Plant Path.*, **4**, 91–6.

HOGGAN, M. D., RANSOM, J. P., PAPPAGIANIS, D., DANALD, G. E. & BELL, A. D. (1956). Isolation of *Coccidioides immitis* from the air. *Stanf. med. Bull.*, **14**, 190.

HOLLAENDER, A. (1942). Abiotic and sublethal effects of ultra-violet radiation on micro-organisms. *See* Moulton (1942), pp. 156–165.

— & DALLA VALLE, J. M. (1939). A simple device for sampling airborne bacteria. *Publ. Hlth. Rep. Wash.*, **54**, 574–77.

HOLLIDAY, P. (1971). Some tropical plant pathogenic fungi of limited distribution. *Rev. Plant Path.*, **50**, 337–48.

HOPKINS, J. C. (1959). A spore trap of the 'Vaseline' slide type. *Can. J. Bot.*, **37**, 1277–8.

HORNE, A. S. (1935). On the numerical distribution of micro-organisms in the air. *Proc. R. Soc. B.*, **117**, 154–74.

HORNICK, R. B. & EIGELSBACH, H. F. (1966). Aerogenic immunization of man with tularemia virus. *Bact. Rev.*, **30**, 532–8.

HORSFALL, J. G. & DIMOND, A. E. (Eds.) (1959–1960). *Plant Pathology: an Advanced Treatise.* Vol. I, 1959; Vol. II, 1960; Vol. III, 1960. Academic Press, New York and London.

HOTTLE, G. A. (1969). Aerosol immunization. *See* Dimmick & Akers (1969), pp. 375–89.

HUBERT, K. (1932). Beobachtungen über die Verbreitung des Gelbrostes bein künstlichen Feldinfektionen. *Fortschr. Landw.*, **7**, 195–8.

HUDSON, N. W. (1963). Raindrop size distribution in high intensity storms. *Rhodesia J. agric. Res.*, **1**, 6–11.

HUGH-JONES, M. E. & AUSTWICK, P. K. C. (1967). Epidemiological studies in bovine mycotic abortion. *Vet. Rec.*, **81**, 273–6.

HURST, G. W. (1968). Foot-and-mouth disease: the possibility of continental sources of the virus in England in epidemics of October 1967 and several other years. *Vet. Rec.*, **82**, 610–4.

HUSAIN, S. M. (1963). An automatic suction-impaction type spore trap and its use with onion blotch *Alternaria*. *Phytopathology*, **53**, 382–4.

HYDE, H. A. (1952). Grass pollen in Great Britain, *Acta allerg.*, **5**, 98–112.

— (1956). Tree pollen in Great Britain. *Acta allerg.*, **10**, 224–45.

— (1959a). Volumetric counts of pollen grains at Cardiff, 1954–57. *J. Allergy*, **30**, 219–34.

— (1959b). Weed pollen in Great Britain. *Acta allerg.*, **13**, 186–209.

— (1969). Aeropalynology in Britain – an outline. *New Phytol.*, **68**, 579–90.

— & ADAMS, K. F. (1958). *An Atlas of Airborne Pollen Grains.* Macmillan, London, 112 pp.

— & ADAMS, K. F. (1960). Airborne allergens at Cardiff, 1942–59. *Acta allerg. Suppl.* No. VII, pp. 159–69.

— & WILLIAMS, D. A. (1944). The right word. *Pollen Science Circ.*, No. 8, 28 October 1944.

— & WILLIAMS, D. A. (1945). Studies in atmospheric pollen. II Diurnal variation in the incidence of grass pollen. *New Phytol.*, **44**, 83–94.

— & WILLIAMS, D. A. (1950). Studies in atmospheric pollen. IV Pollen deposition in Great Britain, 1943. Part 1 The influence of situation and weather. *New Phytol.*, **49**, 398–406.

HYRE, R. A. (1949). Trapping sporangia of *Phytophthora infestans* as an aid in forecasting the development of late blight. *Phytopathology*, **39**, 10–11.

— (1950). Spore traps as an aid in forecasting several downy mildew type of diseases. *Pl. Dis. Reptr. Suppl.* No. 190, pp. 14–18.

IBACH, M. J., LARSH, H. W. & FURCOLOW, M. L. (1954). Isolation of *Histoplasma capsulatum* from the air. *Science, N.Y.*, **119**, 71.

INGOLD, C. T. (1939). *Spore Discharge in Land Plants.* Clarendon Press, Oxford, 178 pp.

— (1953). *Dispersal in Fungi.* Clarendon Press, Oxford, 179 pp.

— (1956). A gas phase in viable fungal spores. *Nature, Lond.*, **177**, 1242–3.

— (1957). Spore liberation in higher fungi. *Endeavour*, **16**, 78–83.

— (1960). Dispersal by air and water – the take-off. *See* Horsfall & Dimond (1960), Vol. III, pp. 137–68.

— (1961). The stalked spore-drop. *New Phytol.,* **60,** 181–3.

— (1965). *Spore Liberation.* Clarendon Press, Oxford, 210 pp.

— (1971). *Fungal Spores: Their Liberation and Dispersal.* Clarendon Press, Oxford, 302 pp.

JACOBS, W. C. (1951). Aerobiology. In *Compendium of Meteorology.* American Meteorological Society, Boston, pp. 1103–11.

JAGGER, J. (1958). Photoreactivation. *Bact. Rev.,* **22,** 99–142.

JANOWSKI, T. (1888). Ueber den Bakteriengehalt des Schnees. *Zbl. Bakt.,* **4,** 547–52.

JARVIS, W. R. (1962). The dispersal of spores of *Botrytis cinerea* Fr. in a raspberry plantation. *Trans. Br. mycol. Soc.,* **45,** 549–59.

JELLISON, W. L. (1969). *Adiaspiromycosis (= Haplomycosis).* Mountain Press, Missoula, Montana, 99 pp.

JENKINS, P. A. & PEPYS, J. (1965). Fog-fever. Precipitin (FLH) reactions to mouldy hay. *Vet. Rec.,* **77,** 464–66.

JENNISON, M. W. (1942). Atomizing of mouth and nose secretions into the air as revealed by high-speed photography. *See* Moulton (1942), pp. 106–128.

JENSEN, I. & BØGH, H. (1942). Om Forhold der har Indflydelse paa Krydsnings-faren hos vindestovende Kulturplanter. (With English summary.) *Tidsskr. Planteavl.,* **46,** 238–66.

JHOOTY, J. S. & McKEEN, W. E. (1965). The influence of host leaves on germina-tion of the asexual spores of *Sphaerotheca maculans* (Wallr. ex Fr.), *Can. J. Microbiol.,* **11,** 539–45.

JOHNSON, C. G. (1957). The distribution of insects in the air and the empirical relation of density to height. *J. Anim. Ecol.,* **26,** 479–94.

— & PENMAN, H. L. (1951). Relationship of aphid density to altitude. *Nature, Lond.,* **168,** 337.

JOHNSTONE, H. F., WINSCHE, W. E. & SMITH, L. W. (1949). The dispersion and deposition of aerosols. *Chem. Rev.,* **44,** 353–71.

JONES, M. D. & NEWELL, L. C. (1946). Pollination cycles and pollen dispersal in relation to grass improvement. *Res. Bull. Nebraska agric. Exp. Sta.,* No. 148, 1–43.

JUNGE, C. E. (1964). Large-scale distribution of micro-organisms in atmosphere. *See* Tsuchiya & Brown (1964), pp. 117–22.

KATAJIMA, H. (1951). Studies on the dissemination of peach anthracnose II. *Ann. phytopath. Soc. Japan,* **15,** 67–71.

KEITT, G. W. & JONES, L. K. (1926). Studies of the epidemiology and control of apple scab. *Wis. agric. Exp. Sta. Res. Bull.,* **73,** 1–104.

KELLY, C. D. & PADY, S. M. (1953). Microbiological studies of air over some non-arctic regions of Canada. *Can. J. Bot.,* **31,** 90–106.

— & PADY, S. M. (1954). Microbiological studies of air masses over Montreal during 1950 and 1951. *Can. J. Bot.,* **32,** 591–600.

— PADY, S. M. & POLUNIN, N. (1951). Aerobiological sampling methods from aircraft. *Can. J. Bot.,* **29,** 206–14.

KERLING, L. C. P. (1949). Attack of peas by *Mycosphaerella pinodes* (Berk. et Blox.) Stone. *Tijdschr. PlZiekt.,* **55,** 41–68.

KERNER VON MARILAUN, A. *See* Marilaun, Kerner von.

KIENTZLER, C. F., ARONS, A. B., BLANCHARD, D. C. & WOODCOCK, A. H. (1954).

Photographic investigation of the projection of droplets by bubbles bursting at a water surface. *Tellus*, 6, 1–7.

KING, J. E. & KAPP, R. O. (1963). Modern pollen rain studies in Eastern Ontario. *Can. J. Bot.*, 41, 243–52.

KITTREDGE, J. (1948). *Forest Influences*. McGraw Hill, New York, pp. 394.

KLINKOWSKI, M. (1962). Die europäische Pandemie von *Peronospora tabacina* Adam, dem Erreger des Blauschimmels des Tabaks. *Biol. Zbl.*, 81, 75–89.

KLYUZKO, S. O., KISHKO, Y. G. & VERSCHCHANSKIY, Y. I. (1960). (Bacterial aeroplankton of the upper layers of the atmosphere during the winter). Translated from Kiev, Vrachebnoye Delo, 1960, 1, 75–76. JPRS: 3732, Joint Publication Research Service, Washington, D.C.

KNIGHT, T. A. (1799). *Experiments on the Fecundation of Vegetables*.

KNOLL, F. (1932). Über die Fernverbreitung des Blütenstaubes durch den Wind. *Forsch. Fortschr. dtsch. Wiss.*, 8, 301–2.

KNUTH, P. (1906). *Handbook of Flower Pollination*. Translated by J. R. Ainsworth Davis, Clarendon Press, Oxford, 3 vols.

KNUTSON, K. W. & EIDE, C. J. (1961). Parasitic aggressiveness in *Phytophthora infestans*. *Phytopathology*, 51, 286–90.

KOBAYASHI, M., STAHMANN, M. A., RANKIN, J. & DICKIE, H. A. (1963). Antigens in mouldy hay as the cause of farmer's lung. *Proc. exp. Biol. Med.*, 113, 472–6.

KOELLREUTER, J. (1951). Morphologie und Biologie von *Rhabdospora ramealis*. *Phytopath. Z.*, 17, 129–60.

KOENIG, L. R. & SPYERS-DURAN, P. A. (1961). Simple methods of determining water drop sizes by means of photographic emulsions. *Rev. scient. Instrum.*, 32, 909–13.

KRAMER, C. L. & PADY, S. M. (1966). A new 24-hour spore sampler. *Phytopathology*, 56, 517–20.

— & PADY, S. M. (1968). Viability of airborne spores. *Mycologia*, 60, 448–9.

— PADY, S. M. & ROGERSON, C. T. (1959). Kansas aeromycology, III *Cladosporium*. *Trans. Kans. Acad. Sci.*, 62, 200–7.

— PADY, S. M., ROGERSON, C. T. & OUYE, L. G. (1959). Kansas aeromycology, II Material, methods and general results. *Trans. Kans. Acad. Sci.*, 62, 184–99.

KRUEGER, A. P., KOTAKA, S. & ANDRIESE, P. C. (1969). Atmospheric ions and aerosols. *See* Dimmick & Akers (1969), pp. 100–12.

KUNKEL, W. B. (1948). Magnitude and characters of errors produced by shape factors in Stokes' Law estimates of particle radius. *J. appl. Physics*, 19, 1056–8.

KURSANOV, L. I. (1933). *Mikologiya*, Sel'khozgiz, Moscow. (*Cited* by Stepanov, 1935.)

— (1971). *Thermoactinomyces sacchari* sp. nov., a thermophilic Actinomycete causing bagassosis. *J. gen. Microbiol.*, 66, 327–38.

LACEY, J. (1972). The microbiology of moist barley storage in unsealed silos. *Ann. appl. Biol.*, 69, 187–212.

— & LACEY, M. E. (1964). Spore concentrations in the air of farm buildings. *Trans. Br. mycol. Soc.*, 47, 547–52.

LACEY, M. E. (1962). The summer air spora of two contrasting adjacent rural sites. *J. gen. Microbiol.*, 29, 485–501.

LAMANNA, C. (1952). Biological role of spores. *Bact. Rev.*, **16**, 90–3.

LANDAHL, H. D. & HERRMANN, R. G. (1949). Sampling of liquid aerosols by wires, cylinders and slides, and the efficiency of impaction of the droplets. *J. Colloid Sci.*, **4**, 103–36.

LANGMUIR, A. D. (1961). Epidemiology of airborne infections. *Bact. Rev.*, **25**, 173–181.

LANGMUIR, I. (1948). The production of rain by a chain reaction in cumulus clouds at temperatures above freezing. *J. Meteorol.*, **5**, 175–92.

— & BLODGETT, K. B. (1949). Mathematical investigation of water droplet trajectories. *Rep. General Electric Res. Lab.*, No. R.L. 225 (Dec. 1944–July 1945), Schenectady, pp. 1–47.

LANIER, L. & ZELLER, C. (1968). Sur l'épidémiologie des spores filiformes: vitesse de chute des ascospores du *Lophodermium pinastri* (Schrad.) Chev., *C. R. Acad. Sci.*, **267** (D), 1574–7.

— ZELLER, C. & ZELLER, F. (1969). Contribution a l'étude du rouge crypto-gamique du pin sylvestre dû au *Lophodermium pinastri* (Schrad.) Chev.: déssémination des spores. *Ann. Sci. Forest.*, **26**, 321–43.

LARGE, E. C. (1940). *The Advance of the Fungi.* Jonathan Cape, London, 488 pp.

LAST, F. T. (1955). Seasonal incidence of Sporobolomyces on cereal leaves. *Trans. Br. mycol. Soc.*, **38**, 221–39.

LAWS, J. OTIS (1939). The camera 'stops' a moving object. *Soil Conserv.*, **5**, 101.

— (1940). Recent studies in raindrops and erosion. *Agric. Engng.*, St Joseph, Michigan, **21**, 431–3.

— (1941). Measurements of the fall velocities of waterdrops and raindrops. *Am. Geophys. Un. Trans.*, **22**, 709–21.

LEEUWEN, W. M. DOCTERS VAN (1936). Krakatau, 1883 to 1933. *Ann. Jard. Bot. Buitenz.*, **46–7**, 1–506.

LEPPER, M. H. & WOLFE, E. K. (Editors) (1966). Second International Conference on Aerobiology (Airborne Infection). *Bact. Rev.*, **30**, (3), 485–698.

LEVIN (1899). Les microbes dans les régions arctiques. *Ann. Inst. Pasteur, Paris*, **13**, 558–67.

LEVIN, Z. & HOBBS, P. V. (1971). Splashing of water drops on solid and wetted surfaces: hydrodynamics and charge separation. *Phil. Trans. R. Soc. Ser. A.*, **269**, 555–85.

LEVINE, H. B. (1969). Biological properties of fungal aerosols. *See* Dimmick & Akers (1969), pp. 340–6.

LEWIS, H. E., FOSTER, A. R., MULLEN, B. J., COX, R. N. & CLARK, R. P. (1969. Aerodynamics of the human microenvironment. *Lancet*, 1969, i, 1273–7.

LIBBY, W. F. (1956). Current research findings on radioactive fallout. *Proc. Nat. Acad. Sci., Wash.*, **42**, 945–62.

LIDDELL, H. F. & WOOTTEN, N. W. (1957). The detection and measurement of water droplets. *Q. Jl R. met. Soc.*, **83**, 263–6.

LIDWELL, O. M. (1948). Bacterial content of air in a dwelling house. *See* Bourdillon, Lidwell & Lovelock (1948), pp. 253–7.

— (1959). Impaction sampler for size grading airborne bacteria-carrying particles. *J. sci. Instrum.*, **36**, 3–7.

— (1967). Take-off of bacteria and viruses. *See* Gregory & Monteith (1967), pp. 116–37.

— (1970). Mikroorganismer: levende stof i luften. In Niels Jonassen (Ed.),

Termisk og Atmosfaerisk Indeklima, Polyteknisk Forlag, Lyngby, Denmark.

— & Towers, A. G. (1970). Uni-directional flow ventilation in patient isolation. *See* Silver (1970), pp. 109–30.

Lindner, G. (1899). Die Protozöenkeime im Regenwasser. *Biol. Zbl.*, **19**, 421–32.

Lingappa, B. T. & Lockwood, J. L. (1964). Activation of soil microflora by fungus spores in relation to soil fungistasis. *J. gen. Microbiol.*, **35**, 215–227.

Lipmann, M. (1961). A compact cascade impactor for field survey sampling. *Am. Industr. Hyg. Ass.*, **22**, 348–53.

Lister, J. (1868). An address on antiseptic system treatment in surgery. *Br. med. J.*, ii, 53–56.

Lloyd, A. B. (1969). Dispersal of Streptomycetes in air. *J. gen. Microbiol.*, **57**, 35–40.

Long, H. W. (1914). Influence of the host on the morphological characters of *Puccinia ellisiana* and *Puccinia andropogonis*. *J. agric. Res.*, **2**, 303–19.

Long, W. H. & Ahmad, S. (1947). The genus *Tylostoma* in India. *Farlowia*, **3**, 225–67.

Lortie, M. & Kuntz, J. E. (1963). Ascospore discharge and conidium release by *Nectria galligena* Bres. under field and laboratory conditions. *Can. J. Bot.*, **41**, 1203–10.

Lösel, D. M. (1964). The stimulation of spore-germination in *Agaricus bisporus* by living mycelium. *Ann. Bot.*, **28**, 541–54.

Luckiesh, M., Taylor, A. H. & Knowles, T. (1949). Sampling devices for determining the bacterial content of air. *Rev. sci. Instrum.*, **20**, 73–7.

Ludi, W. & Vareschi, V. (1936). Die Verbreitung, das Blühen und der Pollenniederschlag der Heufieberpflanzen im Hochtale von Davos. *Ber. Geobot. Forsch-Inst. Rübel*, 1935, 47–112.

Ludlam, F. H. (1967). The circulation of air, water and particles in the atmosphere. *See* Gregory & Monteith (1967), pp. 1–17.

— & Scorer, R. S. (1953). Convection in the atmosphere. *Q. Jl R. met. Soc.*, **79**, 317–41.

Lurie, H. I. & Way, M. (1957). The isolation of dermatophytes from the atmosphere of caves. *Mycologia*, **49**, 178–80.

Machta, L. (1959). Transport in the stratosphere and through the tropopause. *Adv. Geophys.*, **6**, 273–86.

MacLachlan, J. D. (1935). The dispersal of viable basidiospores of the Gymnosporangium rusts. *J. Arnold Arbor.*, **16**, 411–22.

MacQuiddy, E. L. (1935). Air studies at higher altitudes. *J. Allergy*, **6**, 123–7.

Maddox, R. L. (1870). On an apparatus for collecting atmospheric particles. *Monthly Microsc. J.*, **3**, 286–90.

— (1871). Observations on the use of the aeroconioscope, or air-dust collecting apparatus. *Monthly Microsc. J.*, **5**, 45–9.

Magarvey, R. H. & Hoskins, J. (1968). Entrainment of small particles by a large sphere. *Nature, Lond.*, **218**, 460.

Maguire, B. (1963). The passive dispersal of small aquatic organisms and their colonization of isolated bodies of water. *Ecol. Monogr.*, **33**, 161–85.

Maher, L. J. (1964). *Ephedra* pollen in sediments of the Great Lakes region. *Ecology*, **45**, 391–5.

MALIK, M. M. S. & BATTS, C. C. V. (1960). The determination of the reaction of barley varieties to loose smut. *Ann. appl. Biol.*, **48**, 39–50.

MANDELL, M. (1967). Proof of genus specific mold allergens. *Ann. Allergy*, **25**, 48–50.

MARILAUN, A. KERNER VON (1895). *The Natural History of Plants.* Transl. F. W. Oliver, 2 vols., Blackie, London.

MARSHALL WARD, H. *See* Ward, H. Marshall.

MARTIN, W. J. (1943). A simple technique for isolating spores of various fungi from exposed slides in aerobiological work. *Phytopathology*, **33**, 75–6.

MASON, B. J. (1957). *The Physics of Clouds.* Clarendon Press, Oxford. 481 pp.

MASON-WILLIAMS, A. & BENSON-EVANS, K. (1958). A preliminary investigation into the bacterial and botanical flora of caves in South Wales. *Cave Res. Gp. of Gt. Britain*, No. 8, pp. 1–70.

MATHUR, R. S. (1951). Function of spore matrix in *Colletotrichum lindemuthianum. Indian Phytopath.*, **4**, 152–5.

MAUNSELL, K. (1954a). Respiratory allergy to fungus spores. *Progr. Allergy*, **4**, 457–520.

— (1954b). Concentration of airborne spores in dwellings under normal conditions and under repair. *Int. Archs Allergy, Basel*, **5**, 373–6.

— (1958). The seasonal variations in allergic bronchial asthma in relation to the concentration of pollen and fungal spores in the air in 1954, 1955 and 1956. *Acta allerg.*, **12**, 257–76.

— WRAITH, D. G. & CUNNINGTON, A. M. (1968). Mites and house-dust allergy in bronchial asthma. *Lancet*, **1**, 1267–70.

MAY, F. G. (1958). The washout of *Lycopodium* spores by rain. *Q. Jl R. Met. Soc.*, **84**, 451–8.

MAY, K. R. (1945). The cascade impactor: an instrument for sampling coarse aerosols. *J. sci. Instrum.*, **22**, 187–95.

— (1956). A cascade impactor with moving slides. *Arch. Industr. Hlth.*, **13**, 481–8.

— (1961).Fog-droplet sampling using a modified impactor technique. *Q. Jl R. met. Soc.*, **87**, 535–48.

— (1964). Calibration of a modified Andersen bacterial aerosol sampler. *Appl. Microbiol.*, **12**, 37–43.

— (1966). A multi-stage liquid impinger. *Bact. Rev.*, **30**, 559–70.

— (1967). Physical aspects of sampling airborne microbes. *See* Gregory & Monteith (1967).

— & CLIFFORD, R. (1967). The impaction of aerosol particles on cylinders, spheres, ribbons and discs. *Ann. occup. Hyg.*, **10**, 83–95.

— & DRUETT, H. A. (1953). The pre-impinger: a selective aerosol sampler. *Brit. J. industr. Med.*, **10**, 142–51.

— & DRUETT, H. A. (1968). A microthread technique for studying the viability of microbes in a simulated airborne state. *J. gen. Microbiol.*, **51**, 353–66.

— DRUETT, H. A. & PACKMAN, L. P. (1969). Toxicity of open air to a variety of micro-organisms. *Nature, Lond.*, **221**, 1146–7.

— & HARPER, G. J. (1957). The efficiency of various liquid impinger samplers in bacterial aerosols. *Brit. J. industr. Med.*, **14**, 287–97.

MAYNE, W. WILSON (1932). Annual report of the Coffee Scientific Officer, 1931–32. *Bull. Mysore Coffee exp. Stat.*, **7**, 1–32.

McCALLAN, S. E. A. (1944). Evaluating fungicides by means of greenhouse snapdragon rust. *Contr. Boyce Thompson Inst. Pl. Res.*, **13**, 367–84.

— & WELLMAN, R. H. (1943). A greenhouse method of evaluating fungicides by means of tomato foliage diseases. *Contr. Boyce Thompson Inst. Pl. Res.*, **13**, 93–134.

McCUBBIN, W. A. (1918). Dispersal distance of urediniospores of *Cronartium ribicola* as indicated by their rate of fall in still air. *Phytopathology*, **8**, 35–6.

— (1944a). Relation of spore dimensions to their rate of fall. *Phytopathology*, **34**, 230–4.

— (1944b). Airborne spores and plant quarantines. *Sci. Mo, New York*, **59**, 149–52.

— (1954). *The Plant Quarantine Problem*. Munksgaard, Copenhagen, 255 pp.

McCULLY, C. R., FISHER, M., LANGER, G., ROSINSKI, J., GLAESS, H. & WERLE, D. (1956). Scavenging action of rain on airborne particulate matter. *Industr. Engng. Chem.*, **48**, 1512–6.

McDERMOTT, W. (Ed.) (1961). Conference on airborne infection, Miami Beach, Florida, December 1960. *Bact. Rev.*, **25**, 173–382.

McDONALD, J. E. (1962). Collection and washout of airborne pollen and spores by raindrops. *Science*, **135**, 435–7.

— (1964). Pollen wettability as a factor in washout by raindrops. *Science*, **143**, 1180–1.

McDONOUGH, E. S. & LEWIS, A. L. (1968). The ascigerous stage of *Blastomyces dermatitidis. Mycologia*, **60**, 76–83.

McGOVERN, J. P., HAYWOOD, T. J. & McELHENNEY. T. R. (1966). Airborne algae and their allergenicity. II Clinical and laboratory multiple correlation studies with four genera. *Ann. Allergy*, **24**, 145–9.

McKISSICK, G. E., WOLFE, L. G., FARRELL, R. L., GRIESEMER, R. A. & HELLMAN, A. (1970). Aerosol transmission of oncogenic viruses. *See* Silver (1970), pp. 233–7.

McLEAN, A. L. (1918). Bacteria of ice and snow in Antarctica. *Nature, Lond.*, **102**, 35–9.

McLEAN, R. C. (1935). Bacteriology of the atmosphere. *Nature, Lond.*, **136**, 880.

— (1943). Microbiology of the air. *Nature, Lond.*, **152**, 258–9.

MEHTA, K. C. (1933). Rusts of wheat and barley in India. A study of their annual recurrence, life-histories and physiologic forms. *Indian J. agric. Sci.*, **3**, 939–62.

— (1940). Further studies on cereal rusts in India. *Sci. Monogr. Coun. Agric. Res. India.*, No. 14, 1–224.

— (1952). Further studies on cereal rusts in India, Part II. *Sci. Monogr. Coun. Agric. Res. India.*, No. 18, 1–368.

MEIER, F. C. (1935a). Micro-organisms in the atmosphere of arctic regions. *Phytopathology*, **25**, 27.

— (1935b). Collecting micro-organisms from the Arctic atmosphere. *Sci. Mo. New York*, **40**, 5–20.

— (1936a). Collecting micro-organisms from winds above the Caribbean sea. *Phytopathology*, **26**, 102.

— (1936b). Effects of conditions in the stratosphere on spores of fungi. *Nat. Geog. Soc. Stratosphere Series*, No. 2, 152–3.

— & ARTSCHWAGER, E. (1938). Airplane collections of sugar-beet pollen. *Science, N.Y.*, **88**, 507–8.

— STEVENSON, J. A. & CHARLES, V. K. (1933). Spores in the upper air. *Phytopathology*, **23**, 23.

MENNA, M. E. DI (1955). A quantitative study of airborne fungus spores in Dunedin, New Zealand. *Trans. Br. mycol. Soc.*, **38**, 119–29.

MERCER, T. T. (1963). On the calibration of cascade impactors. *Ann. occup. Hyg.*, **6**, 1–14.

— (1965). The interpretation of cascade impactor data. *Am. Industr. Hyg. Ass. J.*, **26**, 236–41.

MEREDITH, D. S. (1962a). Report of Plant Pathologist. *Rep. Banana Bd. Res. Dep. Jamaica, 1961*, pp. 16–21.

— (1962b). Some components of the air spora in Jamaican banana plantations. *Ann. appl. Biol.*, **50**, 577–94.

— (1963). Violent spore release in some Fungi Imperfecti. *Ann. Bot.*, **27**, 39–47.

— (1965). Violent spore release in *Helminthosporium turcicum*. *Phytopathology*, **55**, 1099–1102.

— (1967). Conidium release and dispersal in *Cercospora beticola*. *Phytopathology*, **57**, 889–93.

MILLER, L. P., McCALLAN, S. E. A. & WEED, R. M. (1953). Quantitative studies on the role of hydrogen sulfide formation in the toxic action of sulfur to fungus spores. *Contr. Boyce Thompson Inst. Pl. Res.*, **17**, 151–71.

MINERVINI, R. (1900). Einige bakteriologische Untersuchungen über Luft und Wasser inmitten des Nord-Atlantischen Oceans. *Z. Hyg. InfektKrankh.*, **35**, 165–94.

MIQUEL, P. (1878–99). Annual reports in *Annu. Obs. Montsouris*.

— (1883). *Les organismes vivants de l'atmosphère*. Gauthier-Villars, Paris, 310 pp.

— & BENOIST, L. (1890). De l'enregistrement des pousières atmosphériques brute et organisées. *Annal. Micrographie*, **1**, 572–9.

MISCHUSTIN, E. (1926). Zur Untersuchung der Mikroflora der höheren Luftschichten. *Zentbl. Bakt. Parasitkde. Abt. II*, **67**, 347–51.

MOLISCH, H. (1920). Biologie des atmosphärischen Staubes (Aëroplankton). In *Populäre biologische Vorträge*, Fischer, Jena, pp. 209–26.

MONTEITH, J. L. (1960). Micrometeorology in relation to plant and animal life. *Proc. Linn. Soc. Lond.*, **171**, 71–82.

MORROW, M. B., MEYER, G. H. & PRINCE, H. E. (1964). A summary of airborne mold surveys. *Ann. Allergy*, **22**, 575–87.

MOULTON, S. (Ed.) (1942). *Aerobiology*. *Amer. Assoc. Adv. Sci.*, Pub. No. 17, Washington, 289 pp.

— PUCK, T. T. & LEMON, H. M. (1943). An apparatus for determination of the bacterial content of air. *Science*, **97**, 51–2.

NÄGELI, C. VON (1877). *Die niederen Pilze*. Oldenbourg, Munich, 285 pp.

NAUMOV, N. A. (1934). *Bolezni sadovykh i ovoshchnykh rastenii*. Sel'khozgiz, Leningrad/Moscow, 344 pp.

NEUBERGER, H., HOSLER, C. L. & KOCMOND, W. C. (1967). Vegetation as an aerosol filter. In Tromp and Weeks (Eds.), *Biometeorology*, Vol. 2, Part 2, pp. 693–701. Pergamon Press, Oxford.

NEVILLE, A. C. (1970). 'Airborne organism' identified. *Nature, Lond.*, **225**, 199.

NEWMAN, I. V. (1948). Aerobiology on commercial air routes. *Nature, Lond.,* **161**, 275–6.

NEWMARK, F. M. (1968). Pollen aerobiology – the need for research and compilation. *Ann. Allergy,* **26**, 358–73.

— (1970). Recent developments in pollen aerobiology. *Ann. Allergy,* **28**, 149–52.

NISIKADO, Y., INOUYE, T. & OKAMOTO, Y. (1955). Conditions of the spores of the scabbed wheat ear suspended in raindrops. *Ber. Ohara Inst.,* **10**, 125–34.

NOBLE, W. C. & CLAYTON, Y. M. (1963). Fungi in the air of hospital wards. *J. gen. Microbiol.,* **32**, 397–402.

NORRIS, K. P. & HARPER, G. J. (1970). Windborne dispersal of foot and mouth virus. *Nature, Lond.,* **225**, 98–9.

NUTMAN, F. J., ROBERTS, F. M. & BOCK, K. R. (1960). Method of uredospore dispersal of the coffee leaf rust fungus, *Hemileia vastatrix. Trans. Br. mycol. Soc.,* **43**, 509–15.

OGAWA, J. M. & ENGLISH, H. (1955). The efficiency of a quantitative spore collector using the cyclone method. *Phytopathology,* **45**, 239–40.

OGDEN, E. C. & RAYNOR, G. S. (1967). A new sampler for airborne pollen: the rotoslide. *J. Allergy,* **40**, 1–11.

— RAYNOR, G. S. & HAYES, J. V. (1971). Travels of airborne pollen. *Progress Report,* No. 11. Oct. 1 1969–Sept. 30 1970. New York State Museum & Science Service, Albany, New York.

OGILVIE, L. & THORPE, I. G. (1961). New light on epidemics of black stem rust of wheat. *Science Progress,* **49**, 209–27.

OLIVE, L. (1964). Spore discharge mechanism in basidiomycetes. *Science,* **146**, 542–3.

— & STOIANOVITCH, C. (1966). A simple new mycetozoan with ballistospores. *Am. J. Bot.,* **53**, 344–9.

OORT, A. J. P. (1952). Taksterfte bij Bramen: veroorzakt door *Septocyta ramealis* (Rab.) Pat. *Tijdschr. PlZiekt.,* **58**, 247–50.

OPARIN, A. I. (1957). *The Origin of Life on the Earth.* (Translated by Ann Synge.) Oliver & Boyd, Edinburgh, 495 pp.

OVEREEM, M. A. VAN (1936). A sampling-apparatus for aeroplankton. *Proc. Acad. Sci. Amst.,* **39**, 981–90.

— (1937). On green organisms occurring in the lower troposphere. *Rec. Trav. Bot. Néerl.,* **34**, 388–442.

PADY, S. M. (1951). Fungi isolated from Arctic air in 1947. *Can. J. Bot.,* **29**, 46–56.

— (1957). Quantitative studies of fungus spores in the air. *Mycologia,* **49**, 339–53.

— & KAPICA, L. (1953). Airborne fungi in the Arctic and other parts of Canada. *Can. J. Bot.,* **31**, 309–23.

— & KAPICA, L. (1955). Fungi in air over the Atlantic Ocean. *Mycologia,* **47**, 34–50.

— & KAPICA, L. (1956). Fungi in air masses over Montreal during 1950 and 1951. *Can. J. Bot.,* **34**, 1–15.

— & KELLY, C. D. (1953). Studies on micro-organisms in Arctic air during 1949 and 1950. *Can. J. Bot.,* **31**, 107–22.

— & KELLY, C. D. (1954). Aerobiological studies of fungi and bacteria over the Atlantic Oceon. *Can. J. Bot.,* **32**, 202–12.

— KELLY, C. D. & POLUNIN, N. (1948). Arctic Aerobiology II Preliminary report on fungi and bacteria isolated from the air in 1947. *Nature, Lond.*, 162, 379–81.

— & KRAMER, C. L. (1967). Diurnal periodicity in airborne bacteria. *Mycologia*, 59, 714–6.

— KRAMER, C. L. & CLARY, (1969). Aeciospore release in *Gymnosporangium*. *Can. J. Bot.*, 47, 1027–32.

— PETURSON, B. & GREEN, G. J. (1950). Arctic aerobiology, III The presence of spores of cereal pathogens on slides exposed from aeroplanes in 1947. *Phytopathology*, 40, 632–41.

PALMÉN, E. (1951). The role of atmospheric disturbances in the general circulation. *Q. Jl R. met. Soc.*, 77, 337–54.

PAPE, H. & RADEMACHER, B. (1934). Erfahrungen über Befall und Schaden durch den Getreidemehltau (*Erysiphe graminis* DC.) bei gleichzeitigen Anbau von Winterund Sommergerste. *Angew. Bot.*, 16, 225–50.

PAPPAGIANIS, D. (1969). Some characteristics of respiratory infection in man. *See* Dimmick & Akers (1969), pp. 390–406.

PARKER, B. C. (1968). Rain as a source of vitamin B_{12}. *Nature, Lond.*, 219, 617–8.

PARKER-RHODES, A. F. (1951). The basidiomycetes of Stokholm Island. V An elementary theory of anemophilous dissemination. *New Phytol.*, 50, 84–97.

PASQUILL, F. (1956). Meteorological research at Porton. *Nature, Lond.*, 177, 1148–50.

— (1961). The estimation of dispersion of windborne material. *Met. Mag. London*, 90, 33–49.

— (1962a). *Atmospheric Diffusion*. van Nostrand, London, 297 pp.

— (1962b). Some observed properties of medium-scale diffusion in the atmosphere. *Q. Jl R. Met. Soc.*, 88, 70–9.

PASTEUR, L. (1861). Mémoire sur les corpuscles organisés qui existent dans l'atmosphère. Examen de la doctrine des générations spontanées. *Ann. Sci. Nat. (Zool.)*, 4ᵉ sér., 16, 5–98.

PAWSEY, R. G. (1964). An investigation of the spore population of the air at Nottingham. II The results obtained with a Hirst spore trap, June–July, 1956. *Trans. Br. mycol. Soc.*, 47, 357–63.

PEIRSON, D. H. & CAMBRAY, R. S. (1965). Fission product fall-out from the nuclear explosions of 1961 and 1962. *Nature, Lond.*, 205, 433–40.

PEPYS, J. (1969). *Hypersensitivity Diseases of the Lungs Due to Fungi and Organic Dusts. Monographs in Allergy*, Vol. 6. S. Karger, Basel & New York, 147 pp.

— HARGREAVES, F. E., LONGBOTTOM, J. L. & FAUX, J. (1969). Allergic reactions of the lungs to enzymes of *Bacillus subtilis*. *Lancet*, 1969, ii, 1181–4.

— JENKINS, P. A., FESTENSTEIN, G. N., GREGORY, P. H., LACEY, M. E. & SKINNER, F. A. (1963). Farmer's lung: thermophilic Actinomycetes as a source of 'Farmer's lung hay' antigen. *Lancet*, 1963, ii, 607–11.

PERKINS, W. A. (1957). The rotorod sampler. *2nd Semiannual Rept. Aerosol. Lab.*, Dept. Chemistry & Chem. Engng., Stanford Univ. CML., 186, 66 pp.

PERSSON, H. (1944). On some species of *Aloina*, with special reference to their dispersal by the wind. *Svensk. Bot. Tidskr.*, 38, 260–8.

PETERSEN, L. J. (1959). Relations between inoculum density and infections of wheat by uredospores of *Puccinia graminis* var. *tritici*. *Phytopathology*, **49**, 607–14.

PETERSON, J. E. & COHRS, C. C. (1966). Preliminary clinical evidence for Myxomycete spores as human allergens. *Am. J. Bot.*, **53**, 628.

PETERSON, M. L., COOPER, J. P. & VOSE, P. B. (1958). Non-flowering strains of herbage grasses. *Nature, Lond.*, **181**, 591–4.

PETRI, R. J. (1888). Eine neue Methode Bacterien und Pilzsporen in der Luft nachzuweisen und zu zählen. *Z. Hyg. InfektKrankh.*, **3**, 1–145.

PETTERSSON, B. (1936). Experimentella iakttagelser över den anemochora diasporspridningen och dess beroende an de atmosfäriska förhållandena. *Nordiska (19. Skandinaviska) Naturforskarmötet i Helsingfors den 11–15 Aug. 1936*, pp. 467–9.

— (1940). Experimentelle Untersuchungen über die euanemochore Verbreitung der Sporenpflanzen. *Acta Bot. Fenn.*, **25**, 1–103.

PETURSON, B. (1931). Epidemiology of cereal rusts. *Dom. Canada Dept. Agric. Div. Bot., Rept. Dom. Botanist, 1930*, pp. 44–46.

PHELPS, E. B. & BUCHBINDER, L. (1941). Studies on micro-organisms in simulated room environments. I A study of the performance of the Wells air centrifuge and of the settling rates of bacteria through the air. *J. Bact.*, **42**, 321–44.

PIRIE, J. H. HARVEY (1912). Notes on antarctic bacteriology. In *Scottish National Antarctic Expedition. Report on the Scientific Results of the voyage of the S.Y. Scotia*, Vol. III, *Botany*, pp. 137–48.

PLANK, J. E. VAN DER (1946). A method for estimating the number of random groups of adjacent plants in a homogeneous field. *Trans. R. Soc. S. Africa*, **31**, 269–78.

— (1960). Analysis of epidemics. *See* Horsfall & Dimond (1960), pp. 229–89.

— (1963). *Plant Diseases: Epidemics and Control*. Academic Press, New York & London, 349 pp.

POHL, F. (1937). Die Pollenkorngewichte einiger windblütigen Pflanzen und ihre ökologische Bedeutung. *Beih. Bot. Zbl. Abt. A*. 57, 112–72.

POLLEN ET SPORES (Journal 1959 to present). Editions du Muséum, Paris.

POLUNIN, N. (1951a). Seeking airborne botanical particles about the North Pole. *Svensk. bot. Tidskr.*, **45**, 320–54.

— (1951b). Arctic aerobiology. Pollen grains and other spores observed on sticky slides exposed in 1947. *Nature, Lond.*, **168**, 718–21.

— (1954). Progress in arctic aero-palynology. *Huitième Congrès International de Botanique, Paris, 1954, Rapports et Communications aux Sections, 2, 4, 5, et 6*, pp. 279–81.

— (1955a). Arctic aeropalynology. Spora observed on sticky slides exposed in various regions in 1950. *Can. J. Bot.*, **33**, 401–15.

— (1955b). *Botanical studies on ice-island T-3*. Final Report under Contract No. AF19(604)–1144, Yale University and Air Force Cambridge Research Center, 54 pp.

— (1960). Appendix 2: Details of aeropalynological collection. *See* United States Air Force, Bedford, Massachusetts (1960), pp. 108–11.

— & KELLY, C. D. (1952). Arctic aerobiology: fungi and bacteria, etc., caught in the air during flights over the geographical North Pole. *Nature, Lond.*, **170**, 314–6.

— PADY, S. M. & KELLY, C. D. (1947). Arctic aerobiology. *Nature, Lond.*, **160**, 876–7.

— PADY, S. M. & KELLY, C. D. (1948). Aerobiological investigations in the Arctic and Subarctic. *Arctic*, **1**, 60–61.

— PRINCE, A. E. & BAKANAUSKAS, S. (1960). Appendix 3: Details of collecting and processing living micro-organisms. *See* United States Air Force, Bedford, Massachusetts (1960), pp. 112–14.

POTTER, L. D. & ROWLEY, J. (1960). Pollen rain and vegetation, San Augustin plains, New Mexico. *Bot. Gaz.*, **122**, 1–25.

PREECE, T. F. (1963). Micro-exploration and mapping of apple scab infections. *Trans. Br. mycol. Soc.*, **46**, 523–9.

PRINCE, A. E. & BAKANAUSKAS, S. (1960). Airborne viable micro-organisms spores collected near sea level. *See* United States Air Force, Bedford, Massachusetts (1960), pp. 92–4.

PROCTOR, B. E. (1934). The microbiology of the upper air I. *Proc. Am. Acad. Arts Sci.*, **69**, 315–40.

— (1935). The microbiology of the upper air II. *J. Bact.*, **30**, 363–75.

— & PARKER, B. W. (1938). The microbiology of the upper air III. *J. Bact.*, **36**, 175–86.

— & PARKER, B. W. (1942). Micro-organisms in the upper air. *See* Moulton (1942), pp. 48–54.

PROCTOR, D. F. (1966). Airborne disease and the upper respiratory tract. *Bact. Rev.*, **30**, 498–513.

PROETZ, A. W. (1953). *Essays on the applied physiology of the nose.* Annals Publishing Co., St Louis, 425 pp.

PUSCHKAREW, B. M. (1913). Über die Verbreitung der Süsswasserprotozöen durch die Luft. *Arch. Protistenk.*, **28**, 323–62.

QUIMBY, F. H. (Ed.) (1964). *Concepts for detection of extra terrestrial life.* NASA SP-56, Washington, D.C., 53 pp.

RACK, K. (1959). Untersuchungen über die elektrostatische Ladung der *Lophiodermium-sporen*. *Phytopath. Z.*, **35**, 439–44.

RAMSBOTTOM, J. (1934). L. G. Windt and heterothallism. *Trans. Br. mycol. Soc.*, **19**, 128–38.

RANZ, W. E. & JOHNSTONE, H. F. (1952). Some aspects of the physical behaviour of atmospheric aerosols. *Proc. 2nd. Nat. Air Pollution Symposium, Pasadena*, pp. 35–41.

RAPER, J. R., KRONGELB, G. S. & BAXTER, M. G. (1958). The number and distribution of incompatibility factors in *Schizophyllum*. *Am. Nat.*, **92**, 221–32.

RAYNOR, G. S. (1970). Variation in entrance efficiency of a filter sampler with air speed, flow rate, angle and particle size. *Amer. Ind. Hyg. Assoc. J.*, **31**, 294–304.

— & OGDEN, E. C. (1965). Twenty-four-hour dispersion of ragweed pollen from known source. Brookhaven National Laboratory, BNL 957 (T–398).

— & OGDEN, E. C. (1970). The swing-shield: an improved shielding device for the intermittent rotoslide sampler. *J. Allergy*, **45**, 329–32.

— OGDEN, E. C. & HAYES, J. V. (1968). Effect of a local source on ragweed pollen concentrations from background sources. *J. Allergy*, **41**, 217–25.

— OGDEN, E. C. & HAYES, J. V. (1970). Dispersion and deposition of ragweed pollen from experimental sources. *J. appl. Meteorol.*, **9**, 885–95.

REMPE, H. (1937). Untersuchungen über die Verbreitung des Blütenstaubes durch die Luftstromungen. *Planta*, **27**, 93–147.

RETTGER, L. F. (1910). A new and improved method of enumerating air bacteria. *J. med. Res.*, **22**, 461–8.

RICH, S. & WAGGONER, P. E. (1962). Atmospheric concentration of *Cladosporium*. *Science*, **137**, 962–5.

RICHARDS, M. (1954a). Seasonal periodicity in atmospheric mould spore concentrations. *Acta allerg.*, **7**, 357–66.

— (1954b). Atmospheric mould spores in and out of doors. *J. Allergy*, **25**, 429–39.

— (1955). A water-soluble filter for trapping airborne micro-organisms. *Nature, Lond.*, **176**, 559.

RIDLEY, H. N. (1930). *The Dispersal of Plants Throughout the World*. Reeve, Ashford, Kent, 744 pp.

RILEY, R. L., MILLS, C. C., O'GRADY, F., SULTAN, L. U., WITTSTADT, F. & SHIVPURI, D. N. (1962). Infectiousness of air from a tuberculosis ward. Ultra-violet irradiation of infected air: comparative infectiousness of different patients. *Am. Rev. resp. Dis.*, **85**, 511–25.

RIMINGTON, C., STILLWELL, D. E. & MAUNSELL, K. (1947). The allergen(s) of house dust. Purification and chemical nature of active constituents. *Brit. J. exp. Path.*, **28**, 309–25.

RISHBETH, J. (1958). Detection of viable airborne spores in air. *Nature, Lond.*, **181**, 1549.

— (1959). Dispersal of *Fomes annosus* (Fr.) and *Peniophora gigantea* (Fr.) Massee. *Trans. Br. mycol. Soc.*, **41**, 243–60.

RITCHIE, J. C. & LICHTI-FEDEROVICH, S. (1967). Pollen dispersal phenomena in arctic-subarctic Canada. *Rev. Palaeobot. Palynol.*, **3**, 255–6.

RITTENBERG, S. C. (1939). Investigations on the microbiology of marine air. *J. Mar. Res.*, **2**, 208–17.

ROELFS, A. P., ROWELL, J. B. & ROMIG, R. W. (1970). Sampler for monitoring cereal rust uredospores in rain. *Phytopathology*, **60**, 187–8.

ROEMER, T. (1932). Über die Reichweite des Pollens beim Roggen. *Z. Zücht.* A, **17**, 14–35.

ROGERS, L. A. & MEIER, F. C. (1936a). An apparatus for collecting bacteria in the stratosphere. *J. Bact.*, **31**, 27.

— & MEIER, F. C. (1936b). The collection of micro-organisms above 36 000 feet. *Nat. Geog. Soc. Stratosphere Series*, **2**, 146–51.

ROMBAKIS, S. (1947). Über die Verbreitung von Pflanzensamen und Sporen durch turbulente Luftströmungen. *Z. Met.*, **1**, 359–63.

ROOKS, R. (1954). Airborne *Histoplasma capsulatum* spores. *Science*, **119**, 385–6.

ROSEBURY, T. (1947). *Experimental airborne infection*. Williams & Wilkins, Baltimore, 222 pp.

— (1949). *Peace or pestilence: biological warfare and how to avoid it*. McGraw-Hill, New York, 218 pp.

— KABAT, E. A. & BOLDT, M. H. (1947). Bacterial warfare (A critical analysis of the available agents, their possible military applications, and the means for protecting against them). *J. Immun.*, **56**, 7–96.

ROWELL, J. B. & OLIEN, C. R. (1957). Controlled inoculation of wheat seedlings with urediospores of *Puccinia graminis* var. *tritici*. *Phytopathology*, **47**, 650–5.

— & Romig, R. W. (1966). Detection of uredospores of wheat rusts in spring rains. *Phytopathology*, **56**, 807–11.

Saito, K. (1904). Untersuchungen über die atmosphärischen Pilzkeime. I Mitt. *J. Coll. Sci. Tokyo*, **18** (Art. 5), 58 pp. (1903–4).

— (1908). Untersuchungen über die atmosphärischen Pilzkeime. II Mitt. *J. Coll. Sci. Tokyo*, **23**, 1–78.

— (1922). Untersuchungen über die atmosphärischen Pilzkeime. III Mitt. *Jap. J. Bot.*, **1**, 1–54.

Sakula, A. (1967). Mushroom-worker's lung. *Br. med. J.*, **111**, 708–10.

Salimovskaja-Rodina, A. G. (1936). Über die Mikroflora des farbigen Schnees. (In Russian, German summary.) *Arch. Sci. Biol. Ser.* 2–3, **43**, 229–38.

Salisbury, J. H. (1866). On the cause of intermittent and remittent fevers, with investigations which tend to prove that these affections are caused by certain species of *Palmella. Am. J. Med. Sci.*, **51**, 51–75.

Sartory, A. & Langlais, M. (1912). *Poussières et Microbes de l'Air*. Poinat, Paris. 237 pp.

Savile, D. B. O. (1954). Cellular mechanics, taxonomy and evolution in the uredinales and ustilaginales. *Mycologia*, **46**, 736–61.

— (1964). Geographic variation and gene flow in *Puccinia. Mycologia*, **56**, 240–8.

Savulescu, T. (1941). *Mana Vitei de Vie*. Imprimeria Nationala, Bucharest, 214 pp.

Sayer, W. J., Shean, D. B. & Ghosseiri, J. (1969). Estimation of airborne fungal flora by the Andersen sampler versus the gravity settling culture plate. *J. Allergy*, **44**, 214–27.

Schenck, N. C. (1964). A portable, inexpensive, and continuously sampling spore trap. *Phytopathology*, **54**, 613–4.

Scheppegrell, W. (1922). *Hayfever and Asthma*. Lea & Febiger, Philadelphia, 274 pp.

— (1924). Airplane tests of hayfever pollen density in the upper air. *Med. J. Rec.*, **119**, 185–9.

— (1925). Hayfever pollens in the upper air. *Med. J. Rec.*, **121**, 661–3.

Schisler, L. C., Sinden, J. W. & Sigel, E. M. (1963). Transmission of a virus disease of mushrooms by infected spores. *Phytopathology*, **53**, 888.

Schlichting, H. E. (1969). The importance of airborne algae and protozoa. *Air Pollut. Cont. J.*, **19**, 946–51.

Schmidt, F. H. (1967). Palynology and meteorology. *Rev. Palaeobot. Palyn.*, **3**, 27–45.

Schmidt, W. (1918). Die Verbreitung von Samen und Blütenstaud durch die Luftbewegung. *Öst. Bot. Z.*, **67**, 313–28.

— (1925). Der Massenaustausch in frier Luft und verwandte Erscheinungen. *Probl. Kosm. Phys.*, **7**, 1–118.

Schrödter, H. (1954). Die Bedeutung von Massenaustausch und Wind für die Verbreitung von Pflanzenkrankheiten. Ein Beitrag zur Epidemiologie. *NachrBl. Dtsch. PflSchDienst.*, N.S. 8, 166–72.

— (1960). Dispersal by air and water – the flight and landing. *See* Horsfall & Dimond (1960), Vol. III, pp. 169–227.

Scorer, R. S. (1954). The nature of convection as revealed by soaring birds and dragonflies. *Q. Jl R. met. Soc.*, **80**, 68–77.

SCOTT, R. F. (1913). *Scott's Last Expedition*. Smith Elder, London, Vol. I, p. 269.

SCRASE, F. J. (1930). Some characteristics of eddy motion in the atmosphere. *Gt. Brit. Met. Off. Geophys. Mem.*, **52**, 3–16.

SELL, W. (1931). Staubausscheidung an einfachen Körpern und in Luftfiltern. *Forschungs. Ver. Dtsch. Ing.*, **347**, 1–22.

SERNANDER, R. (1927). Zur Morphologie und Biologie der Diasporen. *Nova Acta Soc. Sci. Upsal. Ser. 4*, Extra vol. 1927, pp. 1–104.

SHADOMY, H. J. (1970). Clamp connections in two strains of *Cryptococcus neoformans*. In D. G. Ahearn (Ed.), *Recent Trends in Yeast Research*. Georgia State University, Atlanta, pp. 67–79.

SIANG, WAN-NIEN (1949). Are aquatic phycomycetes present in the air? *Nature, Lond.*, **164**, 1010–1.

SIDDIQI, M. A. (1961). Annual report of the Fusarium Scheme Pathologist. *Ann. Rept. Dep. Agric. Nyasaland, 1959/60*, Pt. II, pp. 150–1.

SILVER, I. H. (Ed.) (1970). *Aerobiology: Proceedings of the Third International Symposium*. Academic Press, London & New York, 278 pp.

SMITH, C. V. (1964). Some evidence for the windborne spread of fowl pest. *Met. Mag. London*, **93**, 257–63.

SMITH, D. S. (1970). 'Airborne organism' identified. *Nature, Lond.*, **225**, 199.

SMITH, K. M. (1937). An airborne plant virus. *Nature, Lond.*, **139**, 370.

SMITH, L. P. & HUGH-JONES, M. E. (1969). The weather factor in foot and mouth disease epidemics. *Nature, Lond.*, **223**, 712–5.

SNELL, W. H. (1941). Two pine plantings near cultivated red currents in New York. *J. Forestry*, **39**, 537–41.

SOPER, G. A. (1908). *The Air and Ventilation of Subways*. Wiley, New York, 244 pp.

SOULE, M. H. (1934). A micro-organism carried by the dust storm. *Science*, **80**, 14–15.

SPECTOR, W. S. (1956). *Handbook of Biological Data*. Saunders, Philadelphia, 584 pp.

SREERAMULU, T. (1958a). Effect of mowing grass on the concentrations of certain constituents of the air spora. *Curr. Sci.*, **27**, 61–3.

— (1958b). Spore content of air over the Mediterranean Sea. *J. Indian bot. Soc.*, **37**, 220–8.

— (1963). Observations on the periodicity in the airborne spores of *Ganoderma applanatum*. *Mycologia*, **55**, 371–9.

— (1964). Incidence of conidia of *Erysiphe graminis* in the air over a mildew-infected barley field. *Trans. Br. mycol. Soc.*, **47**, 31–8.

— & RAMALINGAM, A. (1961). Experiments on the dispersion of *Lycopodium* and *Podaxis* spores in the air. *Ann. appl. Biol.*, **49**, 659–70.

— & RAMALINGAM, A. (1966). A two-year study of the air spora of a paddy field near Visakhapatnam. *Indian J. agric. Sci.*, **36**, 111–32.

— & SESHAVATARAM, V. (1962). Spore content of air over paddy fields. I Changes in a field near Pentapadu from 21 September to 31 December, 1957. *Indian Phytopath.*, **15**, 61–74.

ŚRODOŃ, A. (1960). Pollen spectra from Spitsbergen. *Folia Quaternaria*, **3**, 1–15.

STAKMAN, E. C. & CHRISTENSEN, C. M. (1946). Aerobiology in relation to plant disease. *Bot. Rev.*, **12**, 205–53.

— & HAMILTON, L. M. (1939). Stem rust in 1938. *Pl. Dis. Reptr. Supp.*, **117**, 69–83.

— & HARRAR, J. G. (1957). *Principles of Plant Pathology*. Ronald, New York, 581 pp.

— HENRY, A. W., CURRAN, G. C. & CHRISTOPHER, W. N. (1923). Spores in the upper air. *J. agric. Res.*, **24**, 599–606.

STANIER, R. Y. (1960). Carotenoid pigments: problems of synthesis and function. *Harvey Lect.*, **54**, 219–55.

— & COHEN-BAZIRE, G. (1957). The role of light in the microbial world: some facts and speculations. In R. E. O. Williams and C. C. Spicer (Eds.), *Microbial Ecology. Symp. Soc. gen. Microbiol.*, **7**, 56–89.

STARR, J. R. (1967a). Deposition of particulate matter by hydrometeors. *Q. Jl R. met. Soc.*, **93**, 516–21.

— (1967b). Inertial impaction of particulates upon bodies of simple geometry. *Ann. occup. Hyg.*, **10**, 349–61.

— & MASON, B. J. (1966). The capture of airborne particles by water drops and simulated snow crystals. *Q. Jl R. met. Soc.*, **92**, 490–9.

STEPANOV, K. M. (1935). Dissemination of infective diseases of plants by air currents. (In Russian, English title.) *Bull. Pl. Prot. Leningr.*, Ser. 2, *Phytopathology*, No. 8, 1–68.

— (1962). *Gribnȳe épifitotii*. Sel'khozizdat, Moscow. 471 pp.

STEPHENS, D. E. & WOOLMAN, H. M. (1922). The wheat bunt problem in Oregon. *Oregon Agr. Exp. Sta. Bull.*, **188**, 42 pp.

STEWART, N. G. & CROOKS, R. N. (1958). Long-range travel of the radio-active cloud from the accident at Windscale. *Nature, Lond.*, **182**, 627–8.

STOVER, R. H. (1962). Intercontinental spread of banana leaf spot (*Mycosphaerella musicola* Leach). *Trop. Agric. Trin.*, **39**, 327–38.

STRAND, L. (1957). Pollen dispersal. *Silvae Genetica*, **6**, 129–36.

STRAND, R. D., NEUHAUSER, E. B. D. & SORNBERGER, C. F. (1967). Lycoperdonosis. *New Engl. J. Med.*, **277**, 89–91.

SUTTON, O. G. (1932). A theory of eddy diffusion in the atmosphere. *Proc. R. Soc. A.*, **135**, 143–65.

— (1947). The theoretical distribution of airborne pollution from factory chimneys. *Q. Jl R. met. Soc.*, **73**, 426–36.

— (1953). *Micrometeorology*. McGraw-Hill, London & New York, 333 pp.

SUZUKI, H. (1965a). An improved rotorod-sampler for the study of aerobiology in *Piricularia*. *Ann Phytopath. Soc. Japan*, **31**, 296–9. (In Japanese, English title.)

— (1965b). The falling of *Piricularia oryzae* spores by rain. *Ann. Rept. Soc. Plant Protect. North Japan*, No. 15, 135–7.

SWINBANK, P., TAGGART, J. & HUTCHINSON, S. A. (1964). The measurement of electrostatic charges on spores of *Merulius lacrymans* (Wulf.) Fr. *Ann. Bot.*, **28**, 239–49.

SYKES, G. (1970). The control of airborne contamination in sterile areas. *See* Silver (1970), pp. 146–56.

SYMPOSIUM ON AEROBIOLOGY (1963). Naval Biological Laboratory, Oakland, California.

SYMONS, G. J. (Ed.) (1888). The eruption of Krakatoa and subsequent phenomena. *Rept. Krakatoa Comm.*, Royal Society, London, 494 pp.

TAGGART, J., HUTCHINSON, S. A. & SWINBANK, P. (1964). Spore discharge in

Basidiomycetes. III Spore liberation from narrow hymenial tubes. *Ann. Bot.*, **28**, 607–18.

TANNER, F. W. & RYDER, E. (1923). Action of ultra-violet light on yeast-like fungi, II. *Bot. Gaz.*, **75**, 309–17.

TARGOW, A. M. (1966). Studies on atopic skin test reactions: I Survey of reactions to the fungi: evidence for existence of two different groups of allergies. II Fallacies that underlie the concept that fungi contain species-specific allergens. *Ann. Allergy*, **24**, 112–27.

TASUGI, H. & KUROSAWA, E. (1938). Investigations of fungi in the air. *Ann. Phytopath. Soc. Japan*, **8**, 35–42. (In Japanese.)

TAUBER, H. (1965). Differential pollen dispersion and the interpretation of pollen diagrams. *Danmarks Geologiske Undersøgelse*, II Raekke, Nr. 89, pp. 1–69.

— (1967). Investigations of the mode of pollen transfer in forested areas. *Rev. Palaeobot. Palynol.*, **3**, 277–86.

TAYLOR, G. I. (1915). Eddy motion in the atmosphere. *Phil. Trans. R. Soc. Ser. A.*, **215**, 1–26.

TERVET, I. A., RAWSON, A. J., CHERRY, E. & SAXON, R. B. (1951). A method for the collection of microscopic particles. *Phytopathology*, **41**, 282–5.

THOMPSON, W. R. (1924). La théorie mathématique de l'action des parasites entomophages et le facteur du hasard. *Ann. Fac. Sci. Marseille*, 2ᵉ Sér., **2**, 69–89.

THOROLD, C. A. (1955). Observations on black-pod disease *(Phytophthora palmivora)* of cacao in Nigeria. *Trans. Br. mycol. Soc.*, **38**, 435–52.

TILAK, S. T. & KULKARNI, R. L. (1970). A new air sampler. *Experientia*, **26**, 443–4.

TIMMONS, D. E., FULTON, J. D. & MITCHELL, R. B. (1966). Micro-organisms of the upper atmosphere. I Instrumentation for the isokinetic sampling at altitude. *Appl. Microbiol.*, **14**, 229–31.

TRANSEAU, E. N. (1949). Fruiting patterns of *Coprinus variagatus* Peck. *Am. J. Bot.*, **36**, 596–602.

TREUB, M. (1888). Notice sur la nouvelle flore de Krakatau. *Ann. Jard. bot. Buitenzorg*, **7**, 213–23.

TRILLAT. A. & FOUASSIER, M. (1914). Action du refroidissement sur les gout-telettes microbiennes. *C.R. Acad. Sci., Paris*, **158**, 1441–4.

TSUCHIYA, H. M. & BROWN, A. H. (Editors) (1964). *Proceedings of the Atmospheric Biology Conference.* April 1964, University of Minnesota, Minneapolis, 235 pp.

TURNER, D. M. (1956). Studies on cereal mildew in Britain. *Trans. Br. mycol. Soc.*, **39**, 495–506.

TYLDESLEY, J. B. (1967). Movement of particles in the lower atmosphere. *See* Gregory & Monteith (1967), pp. 18–30.

TYNDALL, J. (1881). *Essays on the Floating-Matter of the Air in Relation to Putrefaction and Infection.* Longmans, London, 338 pp.

TYRRELL, D. A. J. (1967). The spread of viruses of the respiratory tract by the airborne route. *See* Gregory & Monteith (1967), pp. 286–306.

UKKELBERG, H. G. (1933). The rate of fall of spores in relation to the epidemiology of black stem rust. *Bull. Torrey Bot. Club*, **60**, 211–28.

UNITED NATIONS (1969). *Chemical and bacteriological (biological) weapons and the effects of their possible use. Report of the Secretary-General,* New York. (United Nations Publication, Sales No. E.69.I.24.)

342 BIBLIOGRAPHY

UNITED STATES AIR FORCE, BEDFORD, MASSACHUSETTS (1960). *Scientific studies on Fletcher's Ice Island, T-3, 1952–1955*. Vol. III (Geophysics Research Papers No. 63, Geophysics Research Directorate, Air Force Cambridge Research Center, Air Research and Development Command).

UNITED STATES ATOMIC ENERGY COMMISSION (1968). D. H. Slade (Ed.), *Meteorology and Atomic Energy*. TID-24190, U.S. Dept. of Commerce, Springfield, Virginia.

VAN ARSDEL, E. P. *See* Arsdel, E. P. van.

VAN BRONSWIJK, J. E. M. H. *See* Bronswijk, J. E. M. H. van.

VAN DER PLANK, J. E. *See* Plank, J. E. van der.

VAN DER WERFF, P. J. *See* Werff, P. J. van der.

VAN DER ZWERT, T. *See* Zwert, T. van der.

VAN EIJNATTEN, C. L. M. *See* Eijnatten, C. C. M. van.

VAN LEEUWEN, W. M. DOCTERS. *See* Leeuwen, W. M. Docters van.

VAN OVEREEM, M. A. *See* Overeem, M. A. van.

VARESCHI, V. (1942). Die Pollenanalytische Untersuchung der Gletscherbewegung. *Veröff. geobot. Inst. Rubel.*, **19**, 1–144.

VOISEY, P. W. & BASSETT, J. J. (1961). A new continuous pollen sampler. *Can. J. Pl. Sci.*, **41**, 849–53.

VON NÄGELI, C. *See* Nägeli, C. von.

WACE, N. M. (1961). The vegetation of Gough Island. *Ecol. Monogr.*, **31**, 337–67.

— & DICKSON, J. H. (1965). The biological report of the Royal Society expedition to Tristan da Cunha 1962. Part II The terrestrial botany of the Tristan da Cunha Islands. *Phil. Trans. R. Soc. Ser. B.*, **249**, 273–360.

WADDY, B. B. (1957). African epidemic cerebro-spinal meningitis. *J. Trop. Med. Hyg.*, **60**, 218–23.

WADLEY, F. M. & WOLFENBARGER, D. O. (1944). Regression of insect density on distance from center of dispersion as shown by a study of the smaller European elm bark beetle. *J. agric. Res.*, **69**, 299–308.

WAGGONER, P. E. (1952). Distribution of potato late blight around inoculum sources. *Phytopathology*, **42**, 323–8.

— (1962). Weather, space, time, and chance of infection. *Phytopathology*, **52**, 1100–08.

— & TAYLOR, G. S. (1958). Dissemination by atmospheric turbulence: spores of *Peronospora tabacina*. *Phytopathology*, **48**, 46–51.

WALKER, G. (1935). Bacterial content of the air at high altitudes. *Science*, **82**, 442–3.

WALKER, J. C. & PATEL, P. N. (1964). Splash dispersal and wind as factors in epidemiology of halo blight of bean. *Phytopathology*, **54**, 140–1.

WARD, H. MARSHALL (1882). Researches on the life-history of *Hemileia vastatrix*, the fungus of 'coffee-leaf disease'. *J. Linn. Soc. (Bot.)*, **19**, 299–335.

— (1893). Further experiments on the action of light on *Bacillus anthracis*. *Proc. R. Soc.*, **53**, 23–44.

WASTIE, R. L. (1962). Mechanism of action of an infective dose of *Botrytis* spores on bean leaves. *Trans. Br. mycol. Soc.*, **45**, 465–73.

WATKINS, H. M. S. *et al.* (1970). Epidemiological investigations in Polaris submarines. *See* Silver (1970), pp. 9–53.

WATSON, H. H. (1936). The dust-free space surrounding hot bodies. *Trans. Faraday Soc.*, **32**, 1073–81.

WEBB, S. J. (1961). Factors affecting the viability of airborne bacteria. V The effect of desiccation on some metabolic systems of *Escherichia coli. Can. J. Microbiol.*, **7**, 621–32.

— (1965). *Bound Water in Biological Integrity.* Thomas Springfield, Illinois, 187 pp.

— (1967). The influence of oxygen and inositol on the survival of semidried micro-organisms. *Can. J. Microbiol.*, **13**, 733–42.

WEBSTER, J. (1952). Spore projection in the hyphomycete *Nigrospora sphaerica. New Phytol.*, **51**, 229–35.

WEDUM, A. G. & KRUSE, R. H. (1969). *Assessment of risk of human infection in the microbiological laboratory* (Second edition). Miscellaneous Publication 30, Dept. of the Army, Fort Detrick, Frederick, Maryland, 89 pp.

WEINHOLD, A. R. (1955). Rate of fall of urediospores of *Puccinia graminis tritici* Erikss. & Henn. as affected by temperature and humidity. *Tech. Rept. Office of Naval Research,* ONR Contract No. N90nr, 82400, 104 pp.

WEINZERL, J. & FOS, M. V. (1910). Bacteriological methods for air analysis. *Am. J. publ. Hlth.*. **20**, 633–8.

WELLOCK, C. E. (1960). Epidemiology of Q fever in the urban East Bay area. *Calif. Health,* **18**, 73–6.

WELLS, A. C. & CHAMBERLAIN, A. C. (1967). Transport of small particles to vertical surfaces. *Br. J. appl. Phys.*, **18**, 1793–9.

WELLS, W. F. (1933). Apparatus for study of bacterial behaviour of air. *Am. J. publ. Hlth.*, **23**, 58–99.

— (1955). *Airborne Contagion and Air Hygiene: an Ecological Study of Droplet Infections.* Harvard University Press, Cambridge, Massachusetts, 423 pp.

— & WELLS, M. W. (1936). Airborne infection. *J. Am. med. Ass.*, **107**, 1698–1703, 1805–9.

WELSH, H. H., LENNETTE, E. M., ABINANTI, F. R. & WINN, J. F. (1958). Airborne transmission of Q fever: the role of parturition in the generation of infective aerosols. *Ann. N.Y. Acad. Sci.*, **70**, 528–40.

WENZEL, F. J. & EMANUEL, D. A. (1967). The epidemiology of maple bark disease. *Arch. Environmental Hlth.*, **14**, 385–9.

WERFF, P. J. VAN DER (1958). *Mould Fungi and Bronchial Asthma.* Stenfert Kroese, Leiden, 174 pp.

WESTON, W. A. R. DILLON (1929). Observations on the bacterial and fungal flora of the upper air. *Trans. Br. mycol. Soc.*, **14**, 111–7.

— (1931). The effect of ultra-violet radiation on the urediniospores of some physiological forms of *Puccinia graminis. Scient. Agric.*, **12**, 81–87.

— (1932). The reaction of disease organisms to certain wave-lengths in the visible and invisible spectrum. II *Phytopath. Z.*, **4**, 229–46.

— & HALNAN, E. T. (1930). The fungicidal action of ultra-violet radiation. *Phytopathology,* **20**, 959–65.

— & TAYLOR, R. E. (1948). *The Plant in Health and Disease,* pp. 35–6. Crosby Lockwood, London, 173 pp.

WESTWOOD, J. C. N. (1970). Résumé on rapid diagnosis. *See* Silver (1970), pp. 222–5.

WHEELER, S. M., FOLEY, G. E. & JONES, T. D. (1941). A bubbler pump method for quantitative estimations of bacteria in air. *Science*, **94**, 445–6.

WHINFIELD, B. (1947). Studies in the physiology and morphology of *Penicillium notatum*. I Production of penicillin by germinating conidia. *Ann. Bot.*, **11**, 35–9.

WHISLER, B. A. (1940). The efficiency of ultra-violet light sources in killing bacteria suspended in air. *Iowa St. Coll. J. Sci.*, **14**, 215–31.

WHITEHOUSE, H. L. K. (1949). Heterothallism and sex in fungi. *Biol. Rev.*, **24**, 411–47.

WHYTLAW-GRAY, R. & PATTERSON, H. S. (1932). *Smoke: A Study of Aerial Disperse Systems*. Arnold, London, 192 pp.

WIFFEN, R. D. & HEARD, M. J. (1969). Unidentified airborne organism. *Nature, Lond.*, **224**, 715–6.

WILCOX, J. D. (1953). Design of a new five-stage cascade impactor. *Archs ind. Hyg.*, **7**, 376–82.

WILDING, N. (1970). *Entomophthora* conidia in the air spora. *J. gen. Microbiol.*, **62**, 149–57.

WILLIAMS, C. B. (1947). The logarithmic series and its application to biological problems. *J. Ecol.*, **34**, 253–72.

— (1960). The range and pattern of insect abundance. *Am. Nat.*, **94**, 137–51.

WILLIAMS, R. E. O. (1967). Spread of airborne bacteria pathogenic for man. In Gregory & Monteith (1967), pp. 268–85.

WILLIS, J. C. (1940). *The Course of Evolution*. Cambridge University Press, Cambridge, England, 207 pp.

WILSON, E. E. & BAKER, G. A. (1946a). Some features of the spread of plant diseases by airborne and insect-borne inoculum. *Phytopathology*, **36**, 418–32.

— & BAKER, G. A. (1946b). Some aspects of the aerial dissemination of spores, with special reference to conidia of *Sclerotinia laxa*. *J. agric. Res.*, **72**, 301–27.

WILSON, G. S. & MILES, A. A. (1964). *Topley and Wilson's: Principles of Bacteriology and Immunity*, 5th edition. Arnold, London, p. 1609.

WINDT, L. G. (1806). *Der Berberitzenstrauch, ein Feind des Wintergetreides*. Bückeburg and Hanover (cited from Ramsbottom, 1934).

WODEHOUSE, R. P. (1945). *Hayfever Plants*. Chronica Botanica, Waltham, Massachusetts, 245 pp.

— (1965). *Pollen Grains*. Hafner, New York, 574 pp.

WOLF, F. T. (1943). The microbiology of the upper air. *Bull. Torrey bot. Cl.*, **70**, 1–14.

— (1969). Observations on an outbreak of pulmonary aspergillosis. *Mycopath. Mycol. appl.*, **38**, 359–61.

WOLF, H. W., SKALIY, P., HALL, L. B., HARRIS, M. M., DECKER, H. M., BUCHANAN, L. M. & DAHLGREN, C. M. (1959). *Sampling microbiological aerosols*. *United States Publ. Hlth. Monogr.* No. 60, 53 pp.

WOLFENBARGER, D. O. (1946). Dispersion of small organisms. Distance dispersion rates of bacteria, spores, seeds, pollen and insects, incidence rates of diseases and injuries. *Am. Midl. Nat.*, **35**, 1–152.

— (1959). Dispersion of small organisms. Incidence of viruses and pollen; dispersion of fungus spores and insects. *Lloydia*, **22**, 1–106.

WOOD, R. K. S. (1967). *Physiological Plant Pathology*, pp. 104–14. Blackwells, Oxford, 570 pp.

WOODS, R. D. (1955). Spores and pollen – a new stratigraphic tool for the oil industry. *Micropaleontology*, **1**, 368–75.

WOODHAM-SMITH. C. (1962). *The Great Hunger: Ireland 1845–9*. Hamish Hamilton, London, 510 pp.

WORLD HEALTH ORGANIZATION (1970). *Health aspects of chemical and biological weapons*. Rept. of a WHO Group of consultants, 132 pp.

WORTHINGTON, A. M. (1908). *A Study of Splashes*. Longmans, London.

— & COLE, R. S. (1897). Impact with liquid surface, studied by the aid of instantaneous photography. *Phil. Trans. R. Soc. A.*, **189**, 137–48.

WRIGHT, J. W. (1952). Pollen dispersion of some forest trees. *N.E. Forest Expt. Sta., Paper*, **46**, 1–42.

WRIGHT, S. (1943). Isolation by distance. *Genetics, Princeton*, **28**, 114–38.

— (1946). Isolation by distance under diverse systems of mating. *Genetics, Princeton*, **31**, 39–59.

YARWOOD, C. E. (1952). Some water relations of *Erysiphe polygoni* conidia. *Mycologia*, **44**, 506–22.

— & HAZEN, W. E. (1942). Vertical orientation of powdery mildew conidia during fall. *Science*, **96**, 316–7.

— & SYLVESTER, E. S. (1959). The half-life concept of longevity of plant pathogens. *Pl.Dis. Reptr.*, **42**, 125–8.

YATES, A. H. (1953). Atmospheric convection: the structure of thermals below cloud base. *Q. Jl R. met. Soc.*, **79**, 420–4.

YOKOTA, S. (1963). Studies on shoot-blight disease of larch trees III *Bull. Govt. Forest Exp. Sta.* No. 151, 44 pp. (Tokyo: Japanese, English summary.)

ZADOKS, J. C. (1967). International dispersal of fungi. *Neth. J. Pl. Path.*, **73** (Suppl. 1), pp. 61–80.

ZELENY, J. & MCKEEHAN, L. W. (1910). Die Endgeschwindigkeit des Falles kleiner Kugeln in Luft. *Phys. Z.*, **11**, 78–93.

ZENTMYER, G. A., WALLACE, P. P. & HORSFALL, J. G. (1944). Distance as a dosage factor in the spread of Dutch elm disease. *Phytopathology*, **34**, 1025–33.

ZHUKOVA, A. I. (1963). Methods of microbiological investigation of air. *Microbiology*, **31**, 605–12. (Translation from *Mikrobiologia*, **31**, 745–57.)

— & KONDRATIEV, I. I. (1964). (Species composition of microflora of land atmosphere layers in Moscow City.) *Mikrobiologia*, **33**, 1022–6. (Russian, English summary.)

ZOBELL, C. E. (1942). Micro-organisms in marine air. *See* Moulton (1942), pp. 55–68.

— (1946). *Marine Microbiology*. Chronica Botanica, Waltham, Massachusetts, 240 pp.

— & MATHEWS, H. M. (1936). A qualitative study of the bacterial flora of sea and land breezes. *Proc. natn. Acad. Sci. U.S.A.*, **22**, 567–72.

ZOBERI, M. H. (1961). Take-off of mould spores in relation to wind speed and humidity. *Ann. Bot.*, **25**, 53–64.

— (1964). Effect of temperature and humidity on ballistospore discharge. *Trans. Br. mycol. Soc.*, **47**, 109–14.

— (1965). Spore discharge in *Lepiota konradii*. *Nature, Lond.*, **208**, 405.

ZOGG, H. (1949). Untersuchungen über die Epidemiologie des Maisrostes *Puccinia sorghi* Schw. *Phytopath. Z.*, **15**, 143–92.

ZWERT, T. VAN DER & LEWIS, W. A. (1963). Simple technique for studying dispersal pattern of ascospores of *Mycosphaerella aleuritidis*. *Phytopathology*, **53**, 734–5.

Subject Index

Author Index